Process and Method in Canadian Geography

Weather and Climate

Process and Method in Canadian Geography

Weather and Climate

Selected Readings

edited by
J. G. Nelson
M. J. Chambers
R. E. Chambers

Methuen

Toronto London Sydney Wellington

Library of Congress Catalog Card Number 77-94431

ISBN 0-458-90460-0 (Hardbound)

ISBN 0-458-90470-8 (Paperback)

Design by Hiller Rinaldo

Printed and Bound in Canada

74 73 72 71 70 1 2 3 4 5 6

Preface

This volume is part of a series on Canadian geography. Companion works include *Geomorphology, Vegetation, Soils and Wildlife,* and *Water*. Volumes on other topics are being planned.

Each volume comprises a collection of recent articles gathered, for the most part, from learned journals. The series is Canadian in the sense that the papers deal with Canadian topics or have been prepared by geographers working in Canada; it is geographic in the sense that they are concerned with the landscape, and the physical and cultural processes at work on it. For purposes of general definition the landscape can be thought of as *that zone at or near the surface of the earth that is perceived, used or affected by man.* This zone obviously includes surface phenomena such as vegetation, animals, and soils. The zone changes, however, with changes in knowledge and technology, so that more and more contemporary geographers are becoming concerned about human penetration into the oceans and into space. A process can be thought of as *a change that varies in space and time.* A process can be considered as cultural when it is *totally or primarily a result of the activities of man.* Examples are: urbanization with its cities and towns; agriculture with its fields and crops. A process can be considered as physical when it is *totally or primarily a result of forces independent of man.* Examples are ocean currents and hurricanes. As the articles in these volumes will show, however, the distinction between cultural and physical processes is, in many cases, increasingly difficult to draw. Floods are influenced by cultural activities, such as lumbering, as well as by physical processes, such as thunderstorms. Climate is affected by cities as well as by the general circulation of the atmosphere.

In recent years geographers have placed increasing emphasis on the study of processes. While this is partly a result of improved instrumentation, research methods, and technique, it also reflects a general trend in geography toward a more analytical approach. In this and companion volumes, stress has therefore been placed on articles illustrating research techniques and the study of processes. An attempt, however, has also been made to add to the usefulness of the volumes by including studies that are concerned with many parts of Canada in order to provide knowledge and appreciation of the Canadian landscape.

In our initial thinking about the volumes, we recognized that it would be necessary and desirable to include relevant research undertaken by biologists, geologists, and members of other landscape-oriented disciplines. Geography, as with other disciplines, interrelates with many fields of study, and a considerable amount of cross-fertilization of thought and technique occurs. Now that a number of volumes have been formulated and organized, we find that some of them contain more work by practitioners in related disciplines than we had originally anticipated. Our concern, however, is to deal with those processes, techniques, problems, and topics that we believe to be of interest and importance to geographers in the wide sense. Moreover, the broader content of some of the volumes may make them extremely useful to members of related disciplines, as well as to those many scholars and students whose work and interest carry them into the so-called interdisciplinary areas where so many important contemporary problems lie.

Aside from their possible interest and value to some members of the public, the various volumes in this series should be useful at a variety of academic levels. Many of the articles in each volume will be of interest to the beginning student. They would also seem to be suitable for more advanced undergraduate courses, for example, the first courses exclusively on biogeography, geomorphology or climatology, or for broader courses in resources or conservation. The volumes could also be valuable in certain graduate courses as they describe important recent developments in research fields. They are a guide to the state of the art—or the science—in Canada. For those wishing to investigate a topic further, all the original bibliographies have been retained. These references are considered to be of greater value to the reader than other material that might be introduced, such as comments and reviews interjected between articles by the editors. Most of the maps, diagrams, and photographs have been retained in each article.

Contents

Contributors

Ashwell, I. Y.
Department of Geography, University of Calgary.

Baier, W.
Agrometeorology Section, Plant Research Institute,
Central Experimental Farm, Ottawa.

Barry, R. G.
Institute of Arctic and Alpine Research, University of Boulder, Colorado.

Boyd, D. W.
Division of Building Research, National Research Council, Ottawa.

Brinkmann, W.
Department of Geography, University of Calgary.

Brown, R. J. E.
Northern Research Group, Division of Building Research, National Research Council, Ottawa.

Carder, A. C.
Canada Department of Agriculture, Beaverlodge Research Station, Alberta.

Chapil, R. C.
Meteorological Branch, Department of Transport, Toronto.

Cork, H. F.
Meteorological Branch, Department of Transport, Toronto.

Gold, L. W.
Division of Building Research, National Research Council, Ottawa.

Hennig, A. M. F.
Canada Department of Agriculture, Beaverlodge Research Station, Alberta.

Holmes, R. M.
Inland Waters Branch, Department of Energy, Mines and Resources, Calgary.

Hopkins, J. W.
Division of Biosciences, National Research Council, Ottawa.

Kates, R. W.
Department of Geography, University of Victoria.

Laycock, A. H.
Department of Geography, University of Alberta.

Lee, R.
Meteorological Branch, Department of Transport, Toronto.

Longley, R. W.
Department of Geography, University of Alberta.

Maunder, W. J.
Department of Geography, University of Victoria.

Paul, A. H.
Department of Geography, University of Alberta.

Phillips, L. E.
Department of Geography, University of Victoria.

Richards, T. L.
Hydrometeorology Section, Meteorological Branch, Department of Transport, Toronto.

Robertson, G. W.
Agrometeorology Section, Plant Research Institute, Central Experimental Farm, Ottawa.

Sewell, W. R. D.
Department of Geography, University of Victoria.

Storr, D.
Meteorological Branch, Department of Transport, Calgary.

Summers, P. W.
Alberta Hail Studies, Research Council of Alberta.

Terasmae, J.
Department of Geological Sciences, Brock University, St. Catharines.

Thomas, M. K.
Meteorological Branch, Department of Transport, Toronto.

Thompson, C. E.
Meteorological Office, Department of Transport, Edmonton.

Villimow, J. R.
Division of Geography, Department of Earth Sciences, Northern Illinois University, Illinois.

Williams, G. P.
Snow and Ice Section, Division of Building Research, National Research Council, Ottawa.

Introduction

The compilation of a series of readings concerned with such a broad topic as weather and climate into a coherent volume of practical size has created serious problems. Even with the limitations of an expressly Canadian content, and an emphasis on processes and the research techniques being employed to examine them, there is a wealth of excellent material which could fill several volumes such as this. Selection of articles has, therefore, been somewhat arbitrary and no attempt has been made to include either a complete regional coverage of Canada or a comprehensive discussion of any single topic. The aim has been to emphasize the great breadth of the subject and the implications of the various climatic phenomena for the inhabitants of Canada. At the same time, articles have been chosen that best explain research techniques, and that are useful for teaching purposes.

There are some topics remarkable by their absence from this volume—urban climate is perhaps the most important of these. The most relevant Canadian research in this area was not available for reprinting, so it is with regret that this important item has been omitted. There are also some exceptions to the general principles that have guided the compilation of this collection, including some review articles of a very comprehensive nature. In fact the first paper, by Lee, falls into this category in that he presents a broad summary of knowledge of Arctic circulation up to 1960. Of particular value to this volume is his discussion of the techniques used by the various researchers whose work he reviews. The use of physical and numerical models in the elucidation of atmospheric processes is clearly demonstrated.

Barry's work focusses attention on the position of the Arctic front over North America and its ecological implications. Techniques of frontal contour analysis are described and the concept of a single Arctic front is questioned. Another aspect of circulation is taken up by Villmow, who discusses upper-air features over the Canadian Dry Belt—originally described as Palliser's Triangle—and then emphasizes the expression of these features in surface moisture. He examines a number of criteria in order to define the position of this region, and uses them as drought indicators to correlate with various atmospheric conditions.

The relationship between climate and agriculture is so obvious as to be scarcely worth noting, yet few people have actually attempted to quantify the part that weather plays in food production—as compared with human factors such as fertilization, tillage, and irrigation. Maunder reviews this basic relationship, showing that it is not quite as simple as might be imagined. Although his prime example is not specifically Canadian in content, the principles outlined are relevant to this country, where no similar work has been carried out. One particular aspect of the agroclimatological relationship is examined by Holmes and Robertson, who discuss the influence of temperature on plant growth. Their objective is that optimum use of the growing season may be made by recognition of the heat unit concept. In this way the accurate prediction of planting and harvesting times can maximise land-use efficiency.

Water supply is of equal importance to heat in the prosperity of Canada's agriculture and all aspects of the general moisture budget have been the subject of widespread research. Shortage of water is a limiting factor throughout the Prairie Provinces and Laycock presents a series of maps representing various depths of moisture deficit on a regional scale. These can be used to indicate drought susceptibility in the light of local crop and soil conditions. Another useful tool in planning the irrigation of Canada's water-deficient areas is the concept of latent evaporation, but the determination of this function is rendered very difficult where weather records are limited. Baier and Robertson review the deficiencies of the more common methods used for this calculation and present an improved technique that requires only a short run of standard weather observations. The value of this article is increased by the detailed discussion of the derivation and testing of the multiple regression equations. The work by Carder and Hennig deals with soil moisture and the moisture requirements of wheat, grass and fallow in the Peace River region. The calibration and use of Coleman blocks in a series of replicated plots is described in detail, providing a useful study of instrumentation and experiment design.

Instrumentation also plays an important part in Storr's work in the Marmot Creek Basin, where precipitation has been integrated into a study of the total water budget. This is but one facet of a major interdisciplinary project designed to stimulate scientific watershed management in Canada. Another project of nationwide significance is the subject of the papers by Summers and Paul, and Longley and Thompson. The first discusses distribution variability of hail determined from data gathered from 25,000 farmers in central Alberta. The work stresses the importance of a large number of observations before significant distribution characteristics can be recognized. The second paper approaches the problem from a meteorological point of view, examining numerous elements of the atmospheric structure in which hail occurs.

It is not coincidental that Alberta plays host to students of both hail and the Chinook, for these phenomena may be directly connected. The article by Brinkmann and Ashwell is the result of data gathered from a string of meteorological stations between Calgary and the Rockies, revealing patterns of intensity and movement of the Chinook. These observations are related to various theories of formation of the Chinook and its influence upon snow cover in the foothills of the mountains and on the western Great Plains. Air temperature on the Great Plains is the subject of the next paper, by Hopkins, who isolates the influence of elevation, latitude and longitude into the long-term air temperature normals for a large number of stations in this region. While he presents a precise explanation of statistical procedures, he is careful to relate numerical results to actual landscape features such as lakes and soils.

The importance of the Great Lakes on the climate and economy of Eastern Canada has promoted increasing interest and research in these vast water bodies. Richards describes the development of an airborne technique for the rapid measurement of water surface temperatures over large areas. This parameter is of vital importance in weather forecasting and the assessment of water level fluctuations and ice conditions in the region. The problem of forecasting ice conditions is illustrated by Williams, who uses a small lake near Ottawa to test various predictions. His work points out the need for a combination of accurate local instrumentation and observation with long term forecasting. Ice formation has many serious implications for water transport and hydro power production in Eastern Canada during the winter, and Cork and Chapil's instrumental study examines the conditions under which destructive ice-jams form, with a view to aiding forecasts and efficient dispersion. The difficulties involved in accurate weather observations under winter conditions are also illustrated by Williams in his study of evaporation from snow cover. His own work is compared with other experimental and statistical studies with the conclusion that an empirical approach is the most practical in view of present techniques of field measurement. The following paper by Gold and Boyd provides valuable source material in the use of Fourier analysis, as they examine the various factors affecting the heat and mass budget of an Ottawa site. Brown is also concerned with the thermal regime, but his work is in a region of permafrost where the cover-type is far more critical. He compares measurements of evapotranspiration, radiation, ground temperature, and depth of thaw and determines their relative importance in a permafrost environment under various types of vegetation.

There have been several broad topics within the general realm of geographic interest that have been difficult to categorize into any one of the volumes of this series. Thus the papers by Storr and Richards could well have been included in the volume dealing with water in all of its aspects and Brown's article could have found place in the geomorphology volume. Similarly the subject of climatic change, included here, transgresses the artificial classification of the series and could equally well have been located elsewhere. Terasmae outlines the late- and post-glacial history of Canada's climate and presents a comprehensive review of the techniques by which such information is gathered. Thomas' paper illustrates how modern climatic records can be used to try and ascertain the minor climatic fluctuations that have occurred in more recent decades, in this case in the Canadian prairies. He concludes that although patterns can be discerned from past records, techniques are not available for extrapolating these trends as reliable predictions.

It is appropriate that this volume should not end with man's efforts merely to study and predict the weather, but with his attempts to modify it to suit his own purposes. The review by Sewell, Kates and Phillips soon reveals from many sources of geographical writing that whilst man is busy modifying weather in many areas, at the same time weather has a subtle and pervasive influence on many areas of man's life and activities. The great majority of the research behind the articles in this collection of readings has been motivated by a desire to further this dual process of adaptation and modification.

J. G. Nelson
M. J. Chambers
R. E. Chambers

1

The Circulation of the Arctic

R. Lee

Before trying to present a review of recent research on the general circulation of the Arctic, it is necessary to discuss in some detail the meaning of this expression. In any attempt to give a systematic and consistent picture of a general subject, it is important to keep the basic questions in mind and proceed to discuss the observed facts in the light of physical principles. The basic question to be considered here is the description and explanation of the large-scale (time and space) behaviour of the Arctic atmosphere, i.e., atmospheric flow patterns having physical dimensions of the continents and of the earth itself. The approach taken is to describe the observed mean large-scale behaviour of the Arctic atmosphere in the light of four fundamental physical factors: namely, differential solar heating, the rotation of the earth, the location of mountain barriers and the geostrophic thermal wind equation.

The physical features of the Arctic regions may be briefly described as follows. The polar cap consists mainly of the Arctic Ocean which is almost completely landlocked except for one main access to the warmer waters of the Atlantic between Greenland and Norway. The central Arctic basin is covered with a permanent ice pack which extends to the continents during the winter. The land masses in the Arctic are Siberia, Alaska, northern Canada and Greenland. These land masses are far from uniform in elevation, rather, there are large mountain barriers along the east coast of Asia and in western North America. Finally, the warm waters of the Pacific and Atlantic Oceans extend to high latitudes.

Reprinted from *Canadian Geographer,* no. 16 (July, 1960), pp. 1-13.

It is desirable to examine the mean large-scale air motions in the Arctic against this background of land, sea and mountains. A model of the circulation of the atmosphere in meridional planes is shown in Figure 1.

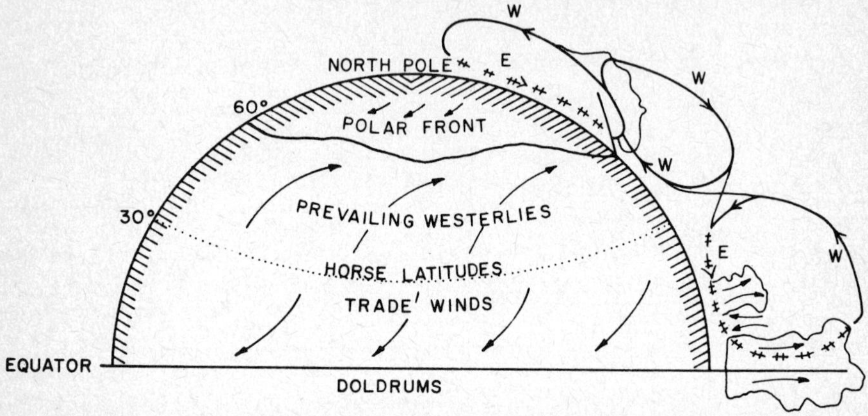

Figure 1. The final cellular meridional circulation on a rotating earth: Convection near the equator, a clear zone of descending air motion north of it (about latitude 30° N.), and heavy slanting cloud masses with accompanying precipitation in the polar-front zone (55°–60° N). (After Rossby.)

This is a schematic diagram presented by Rossby in 1941[10] showing the classical three-cell structure of the atmospheric circulation. (It should be noted that this model has not gained universal acceptance among meteorologists.) The area of interest here is the northernmost of these three cells, in which there is considered to be general descending motion near the pole, a surface easterly circulation in the polar cap and a westerly circulation aloft. The purpose of this schematic picture was to describe in very general terms the planetary meridional circulation averaged with respect to latitude and time; hence, one must not infer that the daily wind patterns will necessarily conform to this picture. The magnitudes of the vertical movements are in general too small to be detectable from daily observations; nevertheless, on the planetary scale, this model would appear to have a general degree of validity.

The Mean Sea Level Pressure Distribution

One can examine the mean sea level pressure distribution to see how the actual mean surface wind field compares with this schematic picture described by Rossby. The normal pattern of the mean sea level circulation is well established by the work of many authors including Baur,[1] U.S. Weather Bureau[12] and more recently by Namias[7] who showed eight-year averages of mean sea level pressure and 700-mb height over the Arctic for January, April, July and October.

The January mean sea level chart is shown in Figure 2, from which one can observe immediately that there is not a mean surface easterly flow throughout the polar cap as one might be led to infer from Rossby's schematic diagram; rather, there is a distinctly cellular pattern in the wind field with large mean cyclones over the warm ocean and large anticyclones over the cold continents. Mean surface easterly winds occur only to the north of the oceanic low pressure cells. There is, moreover, a marked asymmetry in the wind field insofar as the belt of mean easterlies north of the Icelandic low extends beyond the pole. The centre of circulation appears to lie along the 180° meridian and has been called by La Seur the circulation pole. An average of the surface wind around latitude circles centred with respect to the circulation pole would permit one to infer the existence of the net polar easterly circulation which Rossby showed in his diagram.

One might be led to inquire if there is a rational basis for the existence of a cellular structure in the mean sea level circulation or for the precise location of the Icelandic and Aleutian lows and Siberian high. The discussion of the cellular structure will be deferred for a moment in order to consider the results of a theoretical investigation by Smagorinsky.[11] He investigated the relation between the location of these pressure features and the location of the heat sources and sinks in the northern hemisphere and showed on theoretical grounds that mean surface low pressure cells should occur about 25 degrees of longitude east of the heat sources and that mean surface high pressure cells should occur about 25 degrees of longitude east of the cold sources. Referring to the January mean sea level map, it is observed that the Icelandic low is found approximately this distance east of the region of maximum heating due to the warm ocean off the North American coast and the latent heat of condensation in the belt of maximum precipitation just inland. In the far western Pacific the position of maximum precipitation coincides with that of the maximum heating from the warm Kuroshio current. The Aleutian low appears to have the correct location with respect to these heat sources. The Siberian high pressure cell may be considered to be generated by the cold source some 25 to 30 degrees of longitude to the west of it, as inferred from mean surface isotherms and isobars.

Other features of the January mean sea level chart are a ridge of high pressure extending across northern Alaska connecting the continental high pressure cells. Finally, there are mean troughs of low pressure extending from the Aleutian low toward the Gulf of Alaska and from the Icelandic low along the coast of Norway and the U.S.S.R. One can infer from the mean January chart that anticyclones frequently develop in the Alaska area. These anticyclones are associated with masses of cold Arctic air which

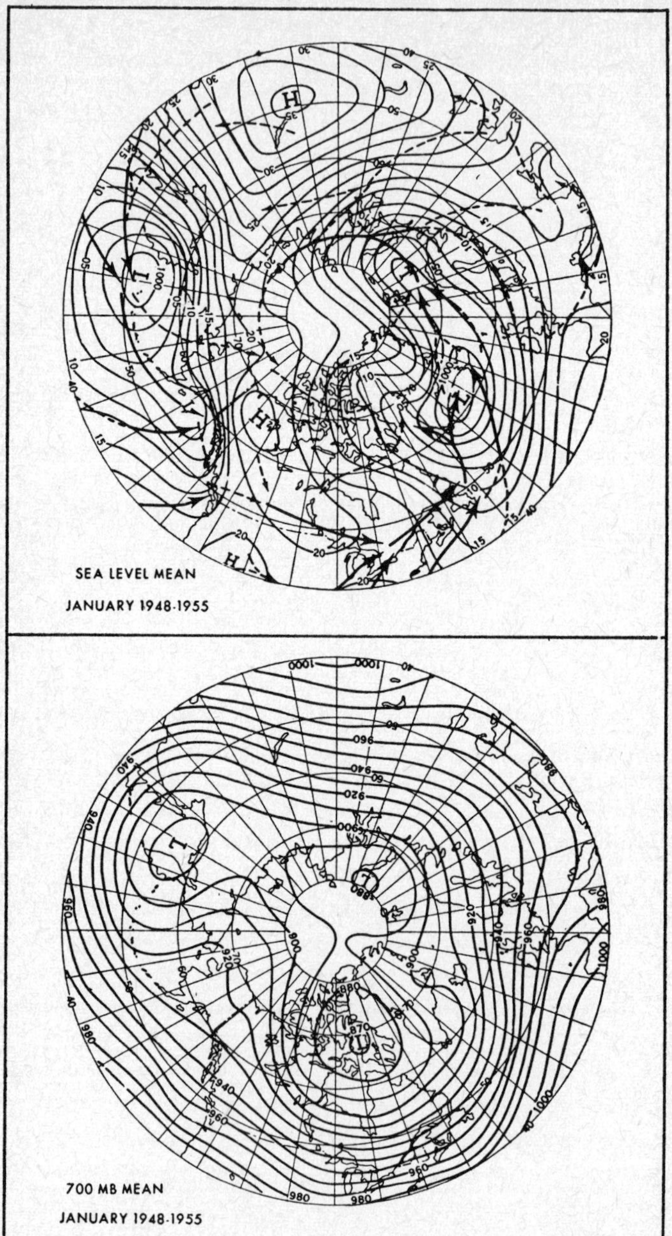

Figure 2. 'Normal' distribution of sea-level pressure and 700-mb height for January. (After Namais, Trans. Amer. Geophys. Union.)

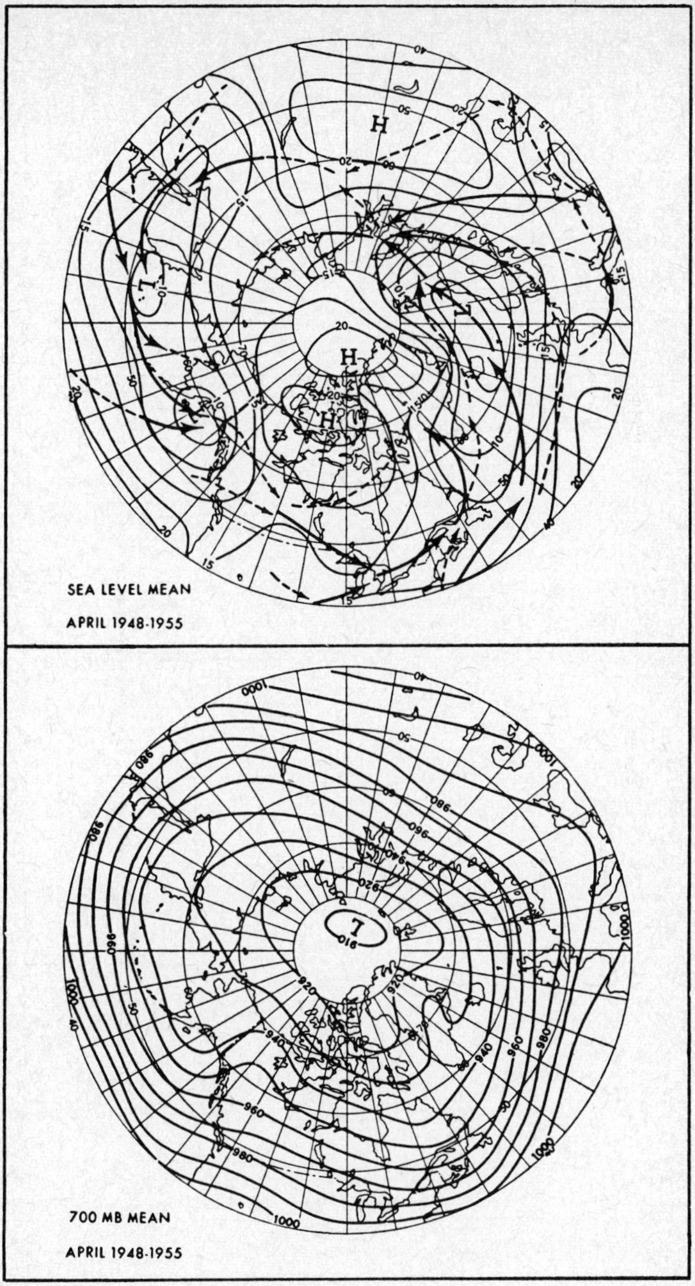

Figure 3. 'Normal' distribution of sea-level pressure and 700-mb height for April. (After Namias, Trans. Amer. Geophys. Union.)

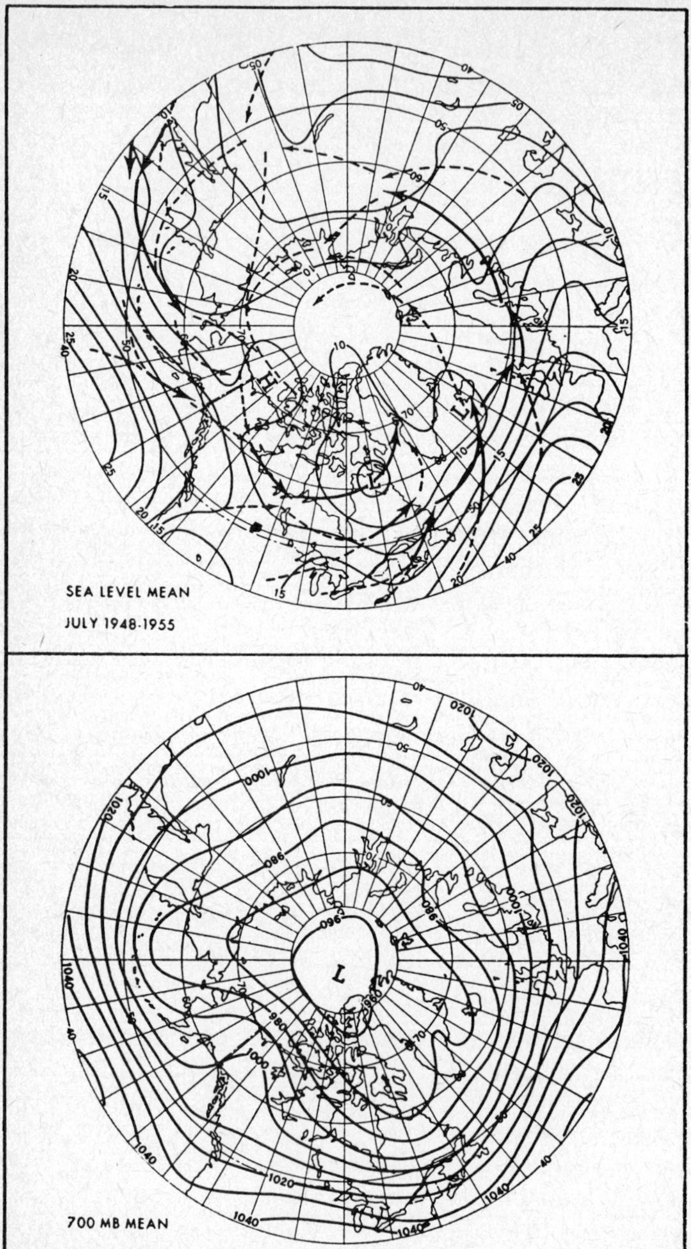

Figure 4. 'Normal' distribution of sea-level pressure and 700-mb height for July. (After Namias, Trans. Amer. Geophys. Union.)

sweep southward in the wake of depressions of the more southerly latitudes. The mean troughs of low pressure in the Gulf of Alaska and along the northern coast of the U.S.S.R. suggest belts of frequent cyclonic activity, verification for which can be found in the locations of the storm tracks in Figure 2. The principal storm track occurs along the oceanic fringe of Norway and the U.S.S.R. and spirals in toward the North Pole. The occurrence of cyclonic activity in the vicinity of the pole was confirmed by weather records from the Russian expedition "North Pole" in 1937–39. For example, on one occasion a 24-hour temperature change of 52°F accompanied by a pressure change of 1 inch of mercury was reported. Secondary storm tracks occur across Bering Strait, along the Arctic coast of Canada and northward from Labrador through Davis Strait to Baffin Bay.

Figure 3 illustrates the mean sea level pressure in April which shows marked changes in the general circulation from January, the most noticeable of which is the relaxation in pressure gradient. This relaxation can be attributed to the weakening of the heat sources along the coastlines as the temperature difference from land to ocean diminishes with the season. At the same time, the character of the Arctic basin sea-level circulation becomes more anticyclonic so that the mean pressure there is some 3–4 mb higher than in January.

The July mean sea level map in Figure 4 shows a relatively flat pressure field in the Arctic. This is largely due to the highly variable pressures during this season, when weak cyclones and anticyclones occur very frequently. Evidence for cyclonic activity in the basin itself in the summertime is shown by the secondary storm tracks which exhibit a tendency to spiral in toward the North Pole. The mean low pressure centre in the Gulf of Alaska vanishes from spring to summer probably owing to the lack of temperature difference between Alaska and the Pacific Ocean in July. The main belts of cyclonic activity shift south from winter to summer especially in the European Russian sector.

The Upper Tropospheric Circulation

Before discussing the upper tropospheric circulation, it is of interest to consider the temperature distribution at high latitudes and its relationship to the wind field. Clearly, the shape of the earth is one factor controlling the temperature and its daily and seasonal variations. Since the earth is nearly spherical, the net solar radiation per unit area of the surface varies with latitude which should theoretically result in isotherms coinciding with latitude lines. However, the presence of warm oceans at high latitudes introduces an asymmetry into the surface temperature field. Consequently, the pattern of mean isotherms is distorted to conform more or less to the shape of the coast-lines as shown in Figure 5. This mean sea level isotherm pattern persists to higher altitudes. The significance of this

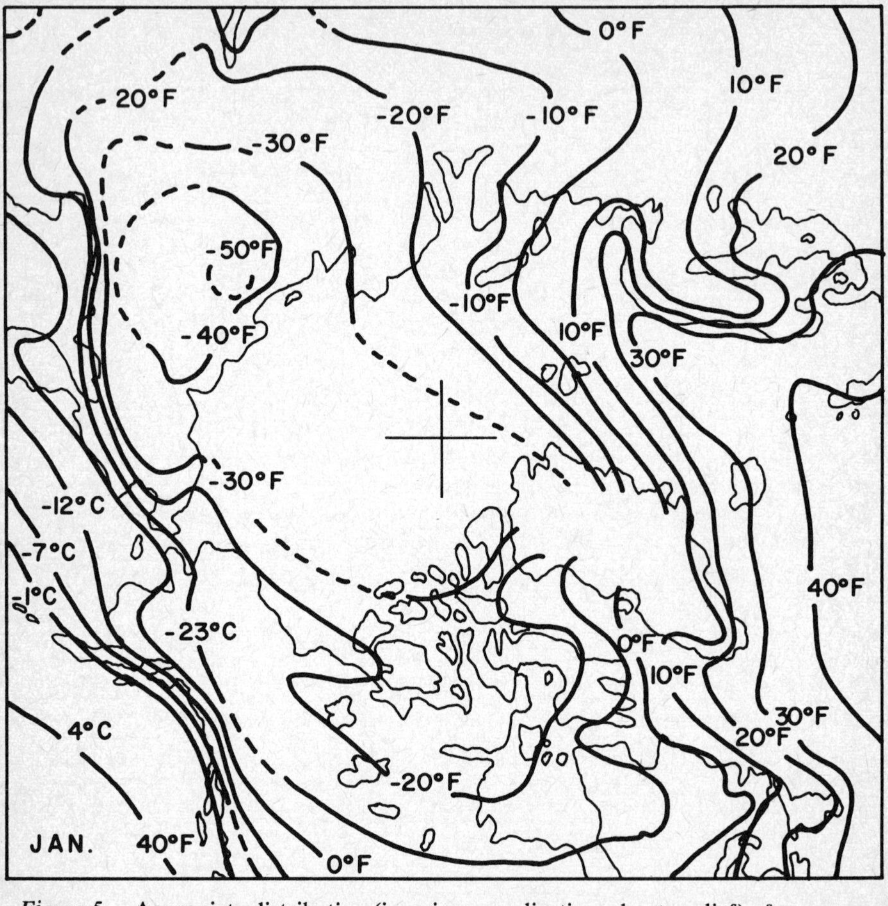

Figure 5. Approxiate distribution (ignoring complications due to relief) of mean daily temperature for January. (After Hare and Orvig, McGill Univ.)

thermal pattern is that it yields the variation of wind with height expressed quantitatively by the geostrophic thermal wind equation. The meaning of this equation can be described as follows. Suppose the horizontal wind at the base of an atmospheric layer is represented by a directed line segment or vector with length proportional to the wind speed (V_1). If the orientation and horizontal spacing of the mean isotherms in that layer are known, then one can visualize a thermal wind vector (V_T) parallel to the mean isotherms whose magnitude is inversely proportional to the distance between isotherms. The wind at the top of the layer (V_2) will then be found by the rule of vector addition shown in Figure 6.

Returning to the mean isotherm map for January, it is seen that colder air occurs toward the pole, consequently the thermal wind vector is gener-

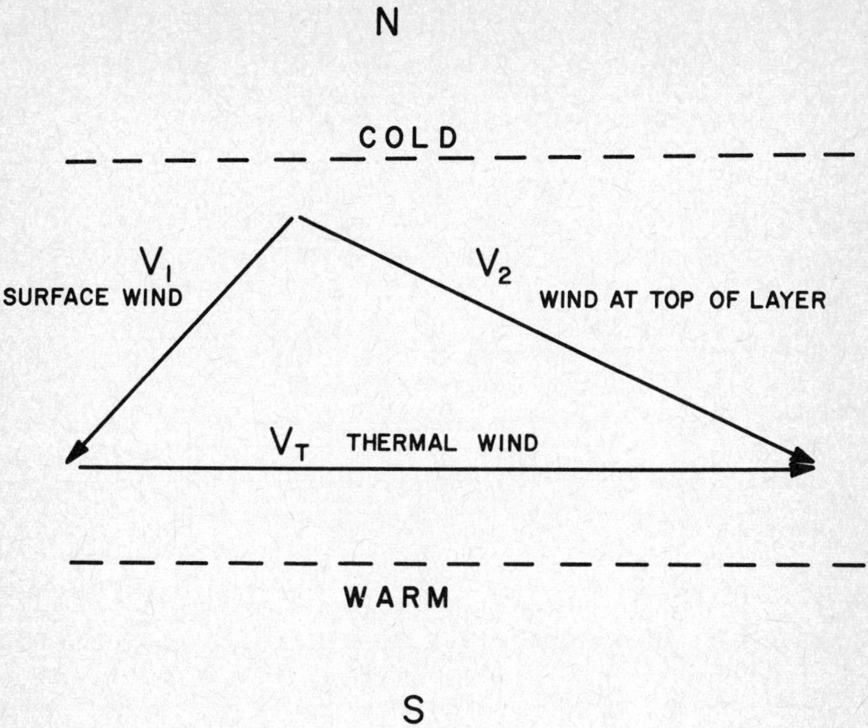

Figure 6. Relationship between surface wind (V_1), wind at top of layer (V_2) and the mean isotherms in the layer.

ally directed from west to east. The strength of the easterly circulation should then gradually diminish from the surface to higher levels and shift to westerly. A comparison of the flow at 700 mb (around 9,000 ft) with the mean sea level flow in Figure 2 will show this variation of wind with height, which is manifest in the belt of westerly winds around the hemisphere, often referred to as the circumpolar vortex. Another notable feature is that the mean westerly flow pattern is distinctly wave-shaped with three nearly symmetrical troughs and ridges around the hemisphere. In each case the troughs are located downstream from the main mountain barriers along the east coast of Asia and the Rockies.

A comparison of the January mean sea level map with the corresponding upper-level map will show that the main storm tracks are parallel to the upper flow with the main ones originating near the upper-level trough positions and ending near the upper-level ridge positions. This relationship is known as the principle of steering of surface cyclones by the upper flow.

The normal upper-air pattern in July shown in Figure 4 is somewhat similar to that in January with respect to the central Arctic vortex and the troughs near Kamchatka and eastern Canada. However, the flow is much weaker and one can detect evidence for the presence of four or possibly five waves. The number of waves is very closely related to the strength of the winds, in general, longer wavelengths are associated with stronger winds.

At this point, one is led to inquire: in what sense can we understand the existence of the circumpolar vortex, the large-scale train of waves which occurs in the westerly flow, and the cellular pattern of the surface wind field? Now, explanations in a physical science are characterized by one thing in common. An explanation of a phenomenon is said to be satisfactory if the phenomenon can be viewed as a logical consequence of established physical principles. For example, the motions of planets or the behaviour of moving objects are understood in so far as they can be deduced from the laws of classical mechanics expressed as Newton's laws of motion. In a similar manner one can speculate on the behaviour of a fluid on the rotating earth and ask: "What pattern of motion will a fluid exhibit under circumstances in which the physical factors can be controlled?" This problem has been treated theoretically but its complexities are such that a discussion of them is beyond the scope of this review. However, the same question may be asked from an experimental point of view. Such an approach has been followed at the University of Chicago by Riehl and Fultz.[9] Basically their experiment is arranged as shown schematically in Figure 7.

An ordinary circular vessel of about 16 cm radius, affectionately referred to as a dishpan, is filled with water to a depth of 4 cm and is supplied with a heat source with a nominal intensity of 150 watts at the equatorial rim. A cylindrical cold source with a radius of 6.5 cm is centred at the pole and provides an analogy with the cold source in the Arctic. When heat is applied at the rim, the resulting circulation of the water occurs largely in meridional planes. However, when the dishpan is rotated in the same sense as the earth, the flow pattern becomes zonal. A steady state can be attained depending on the intensity of the heat source and the rate of rotation, factors which can be varied at will. Now by an ingenious photographic arrangement, a movie camera can record the movement of the water by photographing floating aluminum powder. Thus the fluid motion as it would be seen by an observer standing on the dishpan can be shown. The object of this is to simulate the observation of the atmosphere relative to the earth.

To begin with, rotation of the dishpan immediately destroys the convective meridional circulation and establishes a strong westerly stream at the top surface in middle latitudes. Secondly, a Rossby wave pattern emerges and by altering the rate of rotation, the waves can be made sta-

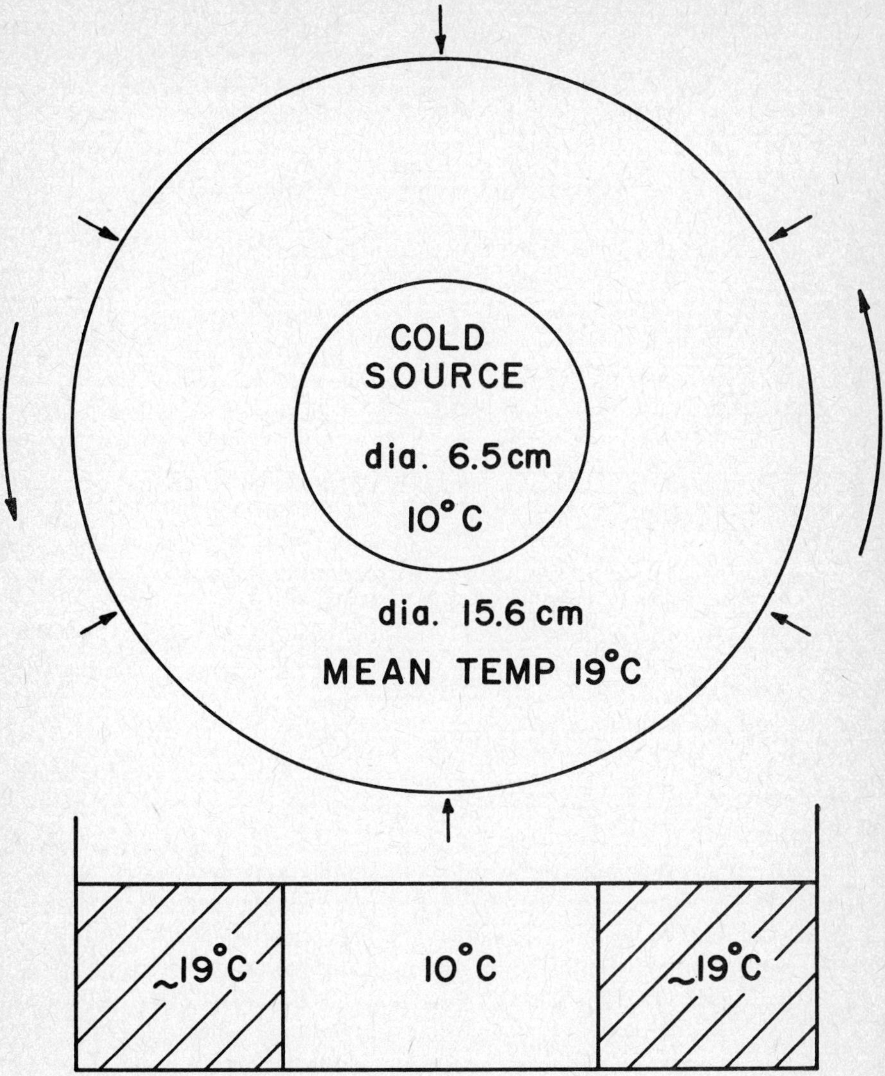

Figure 7. Schematic diagram of dishpan experiment apparatus used by Riehl and Fultz, University of Chicago.

tionary relative to the dishpan or can be made to propagate downstream, upstream or to change in number. To the north and south of the westerly current, the flow is sluggish but does exhibit a cellular pattern with cyclonic and anticyclonic eddies. Moreover, the original uniform horizontal temperature gradient is concentrated into a narrow zone in middle latitudes.

Figure 8 shows the jet stream clearly as a broad band with a speed of 20 per cent of the equatorial speed, comparable to 200 miles per hour, extending almost uniformly along the jet axis. Cyclonic eddies occur to the north of the jet stream while anticyclonic eddies occur around latitude 30 degrees. In this particular situation, three waves occur in the pattern.

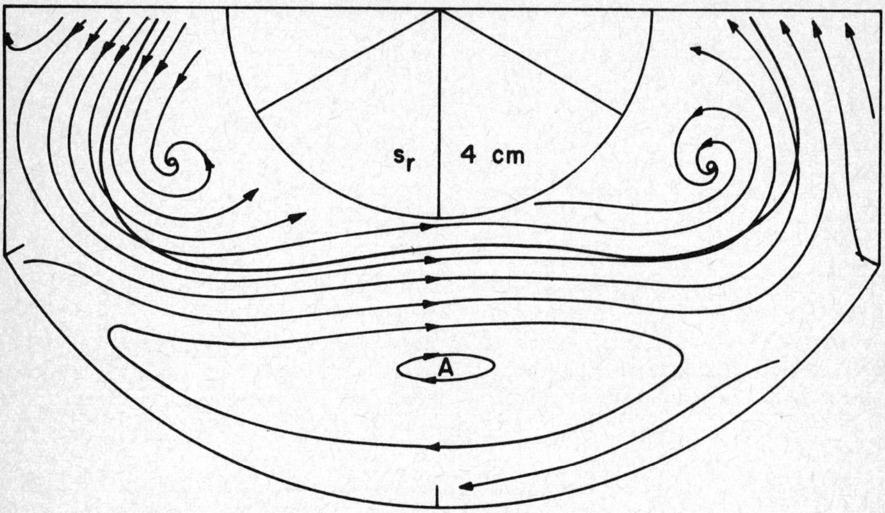

Figure 8. Relative streamlines on top surface. Heavy solid line is jet-stream axis. (After Riehl and Fultz, Quart. J. Roy. Met. Soc.)

The mean radial cross-section of temperature, given as departures from 16°C. is shown in Figure 9. It is obtained by averaging over all meridians and shows a strong north-south temperature gradient in middle latitudes which is strongest at the bottom and diminishes upward. This temperature field is again consistent with the wind field, as one can see by examining the calculated flow patterns at various levels in the dishpan shown in Figure 10. The jet stream decays downward and goes into a cellular pattern below 2 cm. A cyclone and an anticyclone are observed directly under the jet, with centres near the southwest and northwest inflection points of the upper current. This would be the analogue of the atmospheric sea level chart. The model also shows an easterly-westerly-easterly distribution of winds, comparable to that observed in the atmosphere when averaged zonally. On the bottom of the dishpan we find a sinusoidal easterly current, which would likely be obtained in the atmosphere in middle latitudes after downward extrapolation to about 5,000 ft below mean sea level.

Owing to the influence of the frictional boundary layer there were some winds at the bottom in the region just under and just ahead of the

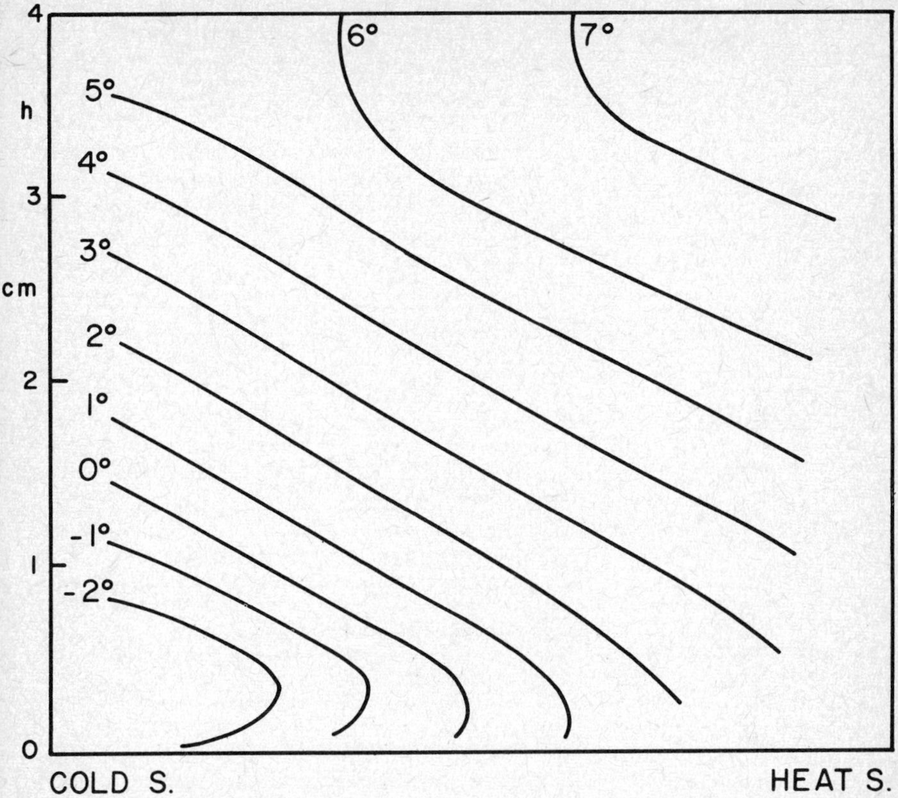

Figure 9. Zonally-averaged distribution of temperature (departure from 16°C) with radius and height above bottom (cm). (After Riehl and Fultz, Quart. J. Roy. Met. Soc.)

upper trough not indicated by this analysis. These were mainly northwesterly winds that occurred in a narrow sector behind a shallow cold front which appeared to lie on the average about ten degrees of longitude ahead of the upper trough. This front extended only slightly above 0.5 cm in height, evidence for its existence was given by much sharper breaks in the temperature traces when this feature moved past the temperature sensing element.

The significance of these experiments is that a replica of the atmospheric sea-level chart can be produced in the laboratory in association with an upper flow pattern which is related to it through the thermal wind equation.

A slightly different approach to the study of the general circulation was undertaken by Phillips[8] when he was at the Institute for Advanced Study in Princeton. This approach was made possible by the development of the modern electronic computer. Whereas the properties of a rotating

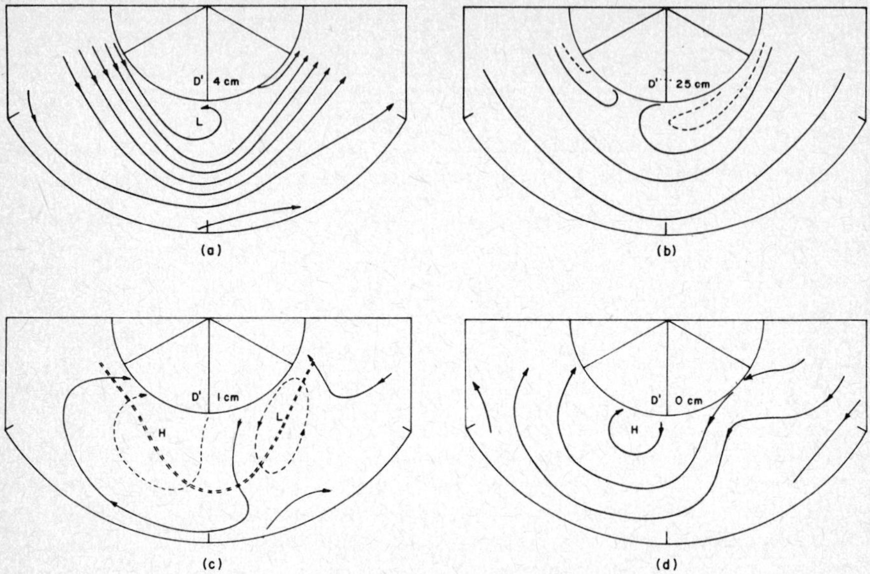

Figure 10. Contours on four selected isobaric surfaces (D′) shown as stream function of geostrophic velocity. Double-dashed line at 1 cm is jet-stream axis at top surface (heights are indicated on diagrams). (After Riehl and Fultz, Quart. J. Roy. Met. Soc.)

fluid subjected to heating and cooling were studied experimentally by Fultz and others, Phillips undertook an investigation of the same problem by means of an experiment based on recent experience in numerical forecasting. The problem was formulated thus: starting with an atmosphere initially at rest relative to a rotating earth which is subjected to differential heating, make a numerical forecast based on a simplified form of Newton's laws of motion. What type of flow pattern will develop in this atmosphere? At first the latitudinal heating and cooling should begin to create a mean latitudinal temperature gradient. After this has reached a certain critical value, large-scale eddies will presumably develop and it was hoped that some mean state of the atmosphere would finally evolve.

The results of the experiment were very striking. Figures 11 and 12 show the forecast 1,000-mb contours and 500-mb isotherms at 11 days and 26 days respectively. First of all, a cellular pattern develops at mean sea level which exhibits the easterly-westerly-easterly surface wind distribution. It is surmounted by an open wave pattern in the upper flow as revealed by the 500-mb isotherms. Again, a comparison of the 500-mb isotherms at 11 days and 26 days will reveal the intensification of the temperature gradient with time, or in effect, the development of a very strong jet stream aloft.

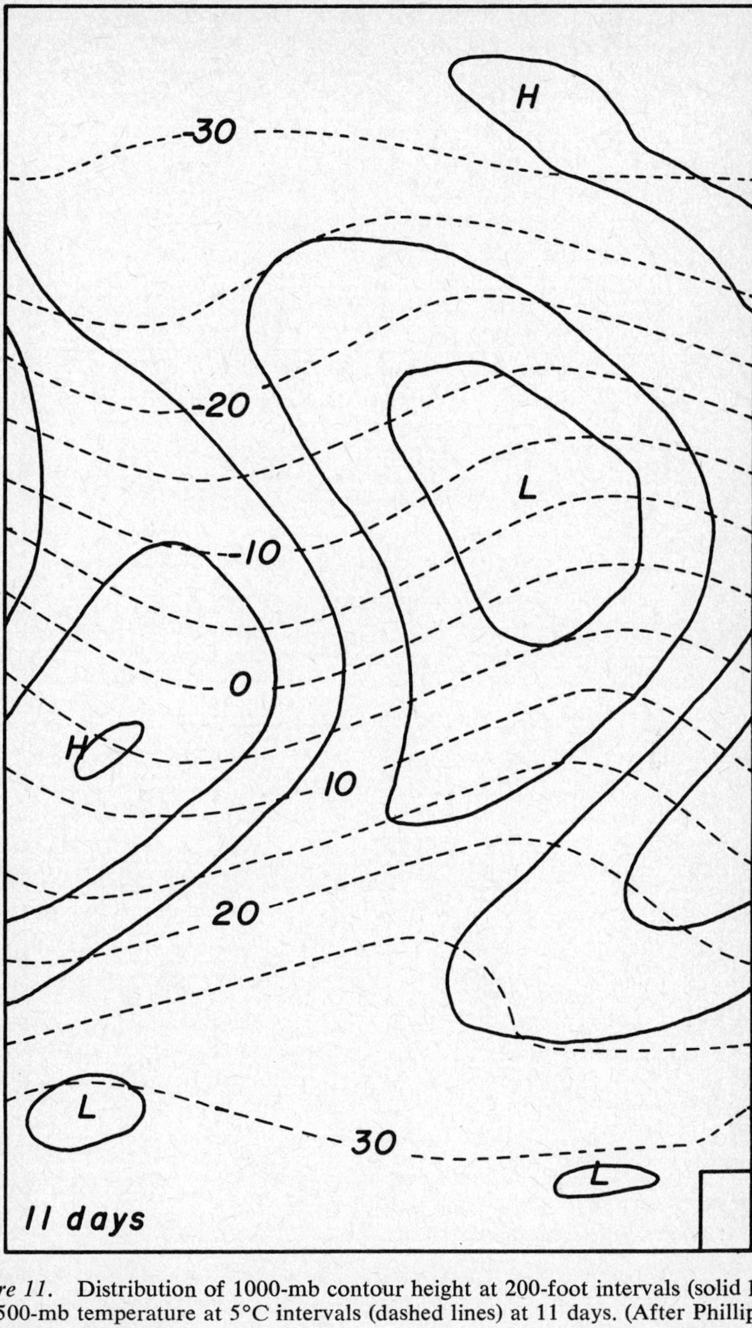

Figure 11. Distribution of 1000-mb contour height at 200-foot intervals (solid lines) and 500-mb temperature at 5°C intervals (dashed lines) at 11 days. (After Phillips, Quart. J. Roy. Met. Soc.)

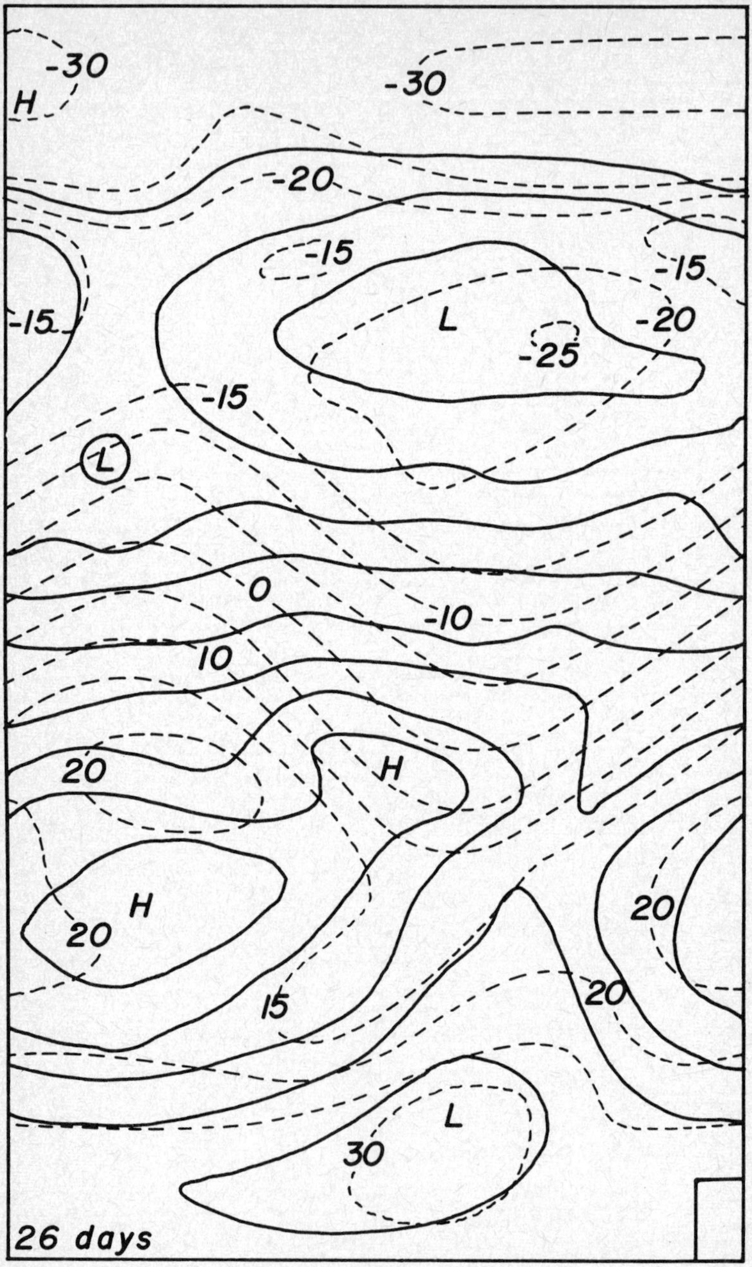

Figure 12. Distribution of 1000-mb contour height at 200-foot intervals (solid lines) and 500-mb temperature at 5°C intervals (dashed lines) at 26 days. (After Phillips, Quart. J. Roy. Met. Soc.)

During the period from 5 days to about 23 to 27 days, a very regular wave pattern developed having many of the features of the large-scale eddies seen on actual weather charts. Individual waves were observed to move eastward at about 1,800 km per day. These waves began as warm lows but a tongue of cold air gradually caught up with each deepening surface low centre and the tongue of warm air moved over the surface high centre so that in the final stages, the pattern looked very much like that of an occluded cyclone. The main surface troughs and ridges and also at upper levels were oriented so as to lean back toward the northwest in the northern half of the region and toward the southwest in the southern half of the region. Definite indications of something similar to cold and warm fronts are to be seen in the 1,000-mb contours (mean sea level) with the main temperature gradient occurring on the cold side of the frontal trough.

To recapitulate briefly then, these two investigations of fluid motion by H. Riehl, D. Fultz and N. A. Phillips in situations where the physical factors can be controlled, permit one to infer the following. A fluid on a rotating earth which is subjected to differential heating will develop an upper westerly flow around the pole in middle latitudes. In this westerly flow, an open wave pattern will appear and the waves will be distributed more or less symmetrically around the hemisphere. The individual troughs and ridges can occur anywhere on the earth, only their relative positions remain fixed. The effect of rotation is to intensify the existing temperature gradient into a very narrow zone, leading to the development of a jet stream at upper levels. At the surface, a cellular pattern of winds develops with a belt of high pressure around 30° latitude, south of which the winds are easterly, corresponding to the trades. Westerlies occur in middle latitudes with polar easterlies to the north. Features similar to low level fronts and cyclones also show up which are completely consistent with real atmospheric phenomena.

At this point one might ask why the individual upper troughs and ridges in the atmosphere take up such definite positions. Recalling that the major troughs occur just east of the Tibetan plateau and the Rockies which are separated by 120° longitude, it appears as though the orographic barriers determine the trough positions. It happens that these two barriers are separated by a distance corresponding to a wave number of 3. Thus the Kamchatka trough and Hudson Bay trough may be regarded as forced perturbations, a view that is supported by theoretical investigations. Indeed, one also observes on mean charts a lee trough east of the Andes in the southern hemisphere. The third trough in the Arctic occurs some 90 degrees downstream from Greenland, however, the strong westerlies are observed to pass south of Greenland so it appears that the icecap plays a relatively unimportant role in determining the positions of the large-scale planetary waves. The third trough near Novaya Zemlya can be considered to be a resonance wave induced by the other two perturbations.

The Stratospheric Circulation

The Arctic stratospheric circulation may now be examined in the light of these principles. Figure 13 illustrates the January temperature field along the 80°W meridian as is shown by Kochanski.[5] In this the typical features of the Arctic tropospheric temperature field which is largely characterized by near homogeneity at high latitudes can be observed. However, above the tropopause, there is a very strong temperature contrast from south to north. The reason for its existence lies in the fact that the ozone layer is found around this level. Although it occurs in minute amounts, ozone is extremely sensitive to solar radiation—if it encounters sunlight, it warms up very quickly. Now, the geographical location of this zone of maximum temperature gradient suggests that it is likely due to differential heating of the ozone layer along the earth's shadow cone. The fact that the Arctic stratosphere is appreciably warmer at the pole than at lower latitudes in July when the polar regions receive continuous sunlight lends support to this hypothesis (Figure 14). The existence of this

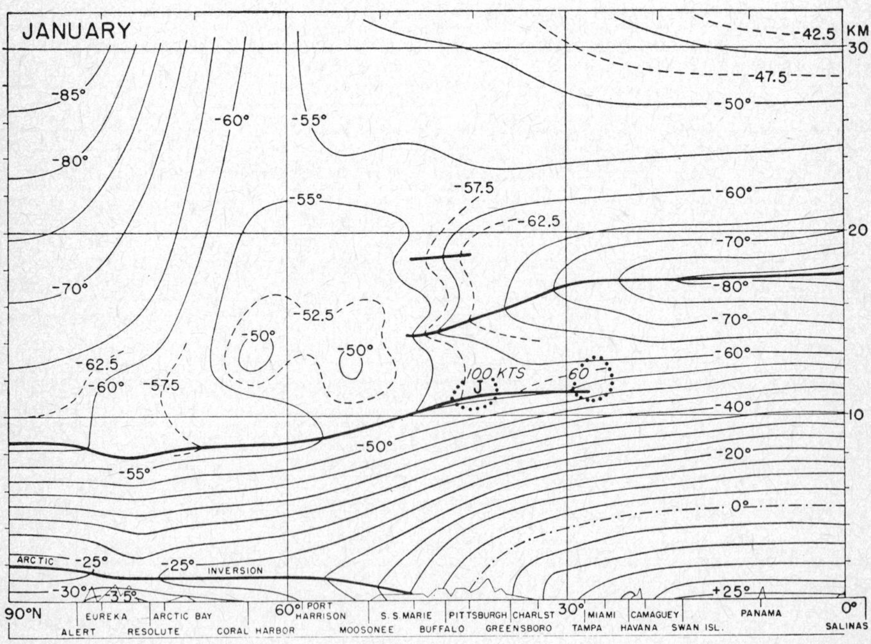

Figure 13. Profile of mean temperature along 80°W in January. Isotherms in deg.C. (After Kochanski, J. Met.)

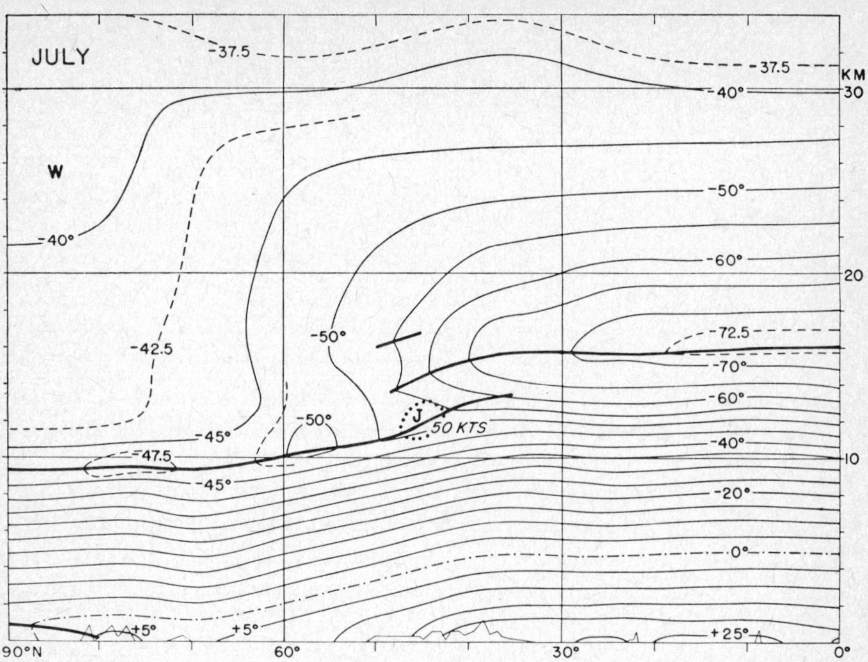

Figure 14. Profile of mean temperature along 80°W in July. Isotherms in deg.C. (After Kochanski, J. Met.)

horizontal temperature gradient through this deep layer during polar darkness should result in increasing westerly winds with height. The symmetry of the earth's shadow cone should lead us to expect a circumpolar vortex in the Arctic stratosphere with predominantly westerly winds. The effect of the earth's rotation is to intensify the existing horizontal temperature gradient. Furthermore, there will be a tendency to establish a Rossby wave pattern.

Figure 15 shows the mean January 25-mb contours (approximately 80,000 ft) over the North American sector. A mean trough will be observed in the vicinity of Hudson Bay which is similar to the tropospheric pattern. Figure 16 shows a vertical cross-section of the jet stream at 1500 GCT 20 November 1955. Stations selected for the cross-section are Eureka (917), Isachsen (074), Mould Bay (072), Sachs Harbour (051) and Norman Wells (043). The orientation of the cross-section is NE–SW so that the northwesterly winds at and above the 100-mb level (around 53,000 ft) are nearly at right angles to the plane of the cross-section. During November, the jet stream is in its initial stages of development and is therefore relatively weak. Nevertheless, the observations do define a wind speed maximum clearly in this cross-section. The temperature field

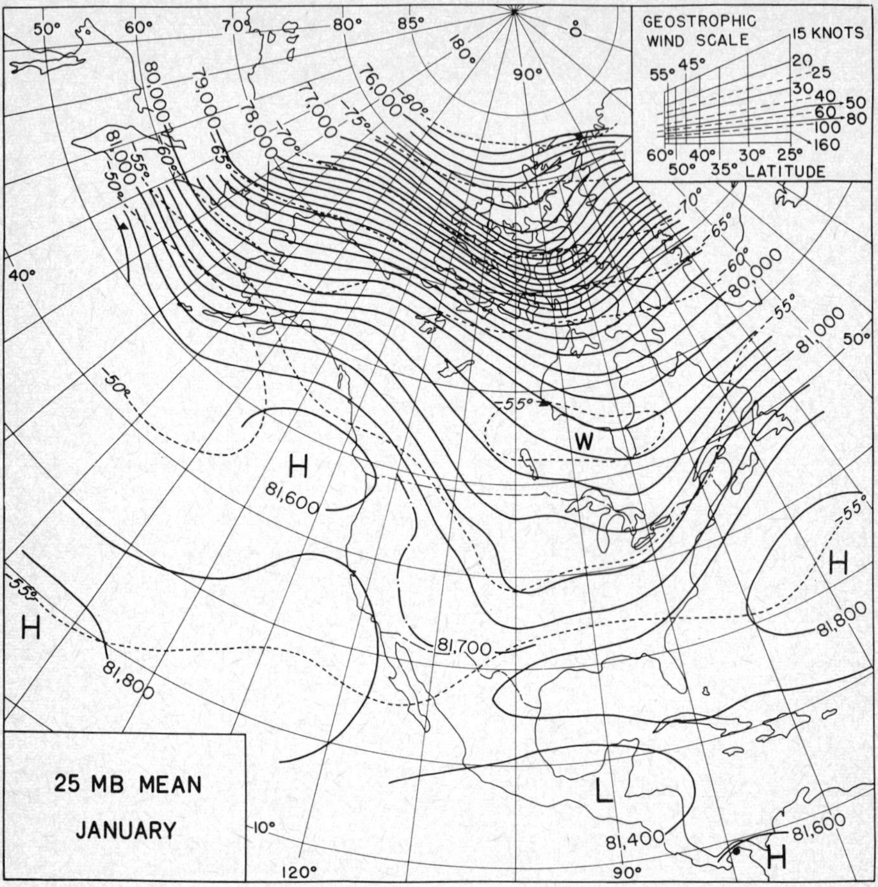

Figure 15. **25-mb mean flow in January. Contours at 200-foot intervals (solid lines) isotherms in 5-deg.C. intervals (broken lines). (After Kochanski and Wasko, Bull. Amer. Met. Soc.)**

around 65,000 ft is of particular interest. One can observe the greatest concentration of isotherms between Sachs Harbour (051) and Isachsen (074); farther south, the horizontal temperature gradient is quite small.

A vertical cross-section through the jet stream at 1500 GCT 26 February 1956 is shown in Figure 17. It was around this time of the 1955–56 winter that the jet stream was most intense. The orientation of this cross-section is also NE–SW and the strong northwesterly winds in the stratosphere are again nearly at right angles to the plane of the cross-section. The highest reported wind on this cross-section is 160 kt around the 80,000-ft level at Eureka (917). The strong horizontal temperature gradient below the jet stream is clearly evident; the temperature

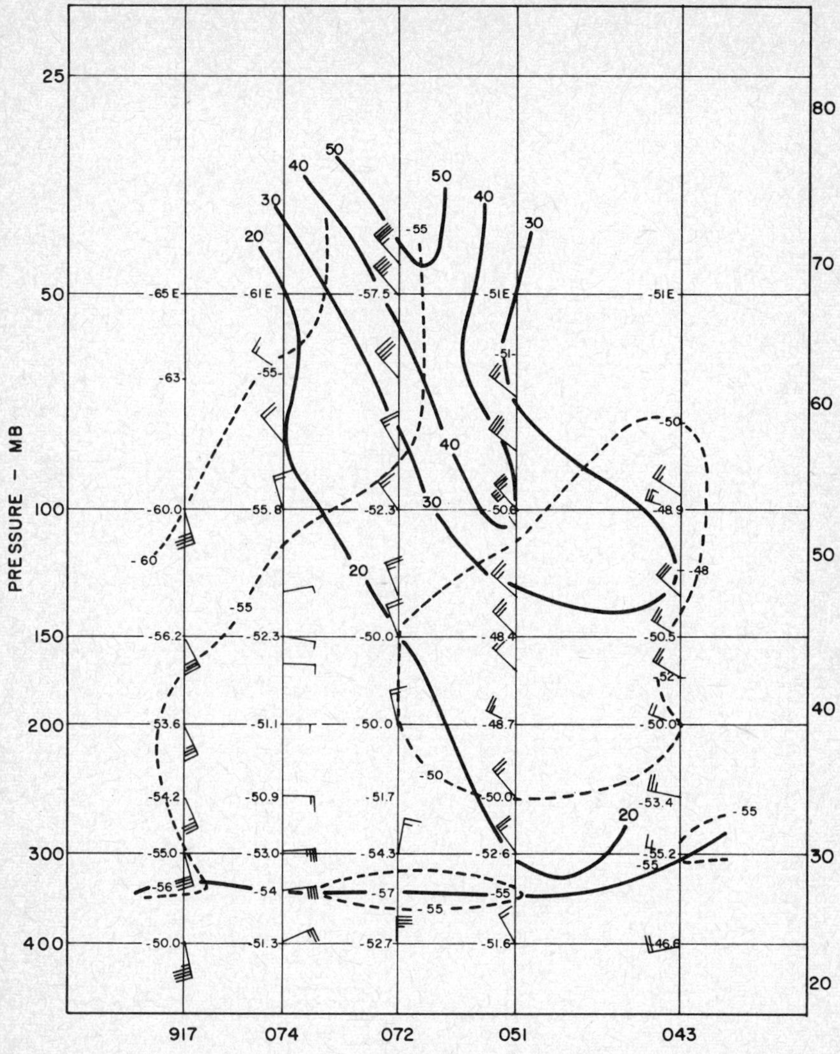

Figure 16. Vertical cross section of Arctic Stratospheric Jet Stream 1500 GCT 20 November 1955. Orientation is approximately NE–SW. Tropopause is indicated by thick solid line, isotherms (C) by broken lines, and isotachs of observed northwesterly winds in stratosphere by thin solid line (kn). (After Lee and Godson, J. Met.)

decreases from −70°C to −50°C in a distance of 800 miles. Owing to the orientation and irregularity of the cross-section, there is some doubt as to the reality of the double maximum.

A vertical cross-section for 0300 GCT 26 March 1956 is presented in Figure 18. The maximum wind occurs at or above the 80,000-ft level

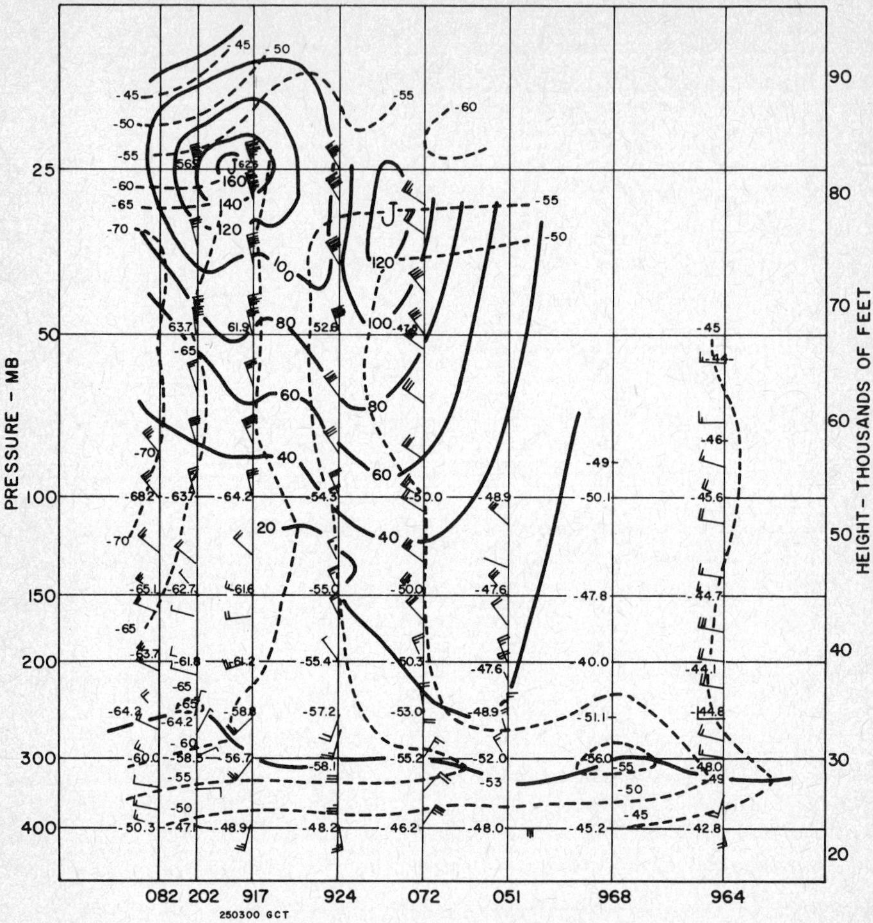

Figure 17. Vertical cross section of Arctic Stratospheric Jet Stream 1500 GCT 26 February 1956. Tropopause is indicated by thick solid line, isotherms (C) by broken lines, and isotachs of observed northwesterly winds in stratosphere by thin solid lines (kn). (After Lee and Godson, J. Met.)

over Isachsen (074) with a speed of around 120 kt. It is also noted that a maximum wind was reported in the Alert sounding at 75,000 ft indicating that the actual height of the jet axis is very nearly there on this occasion. The horizontal temperature gradient at 50,000 ft is 10°C over a distance of 860 miles, certainly much less than in the February cross-section where the maximum horizontal temperature gradient was estimated to be 20°C in 800 miles. The highest reported wind on this cross-section was slightly greater than 100 kt at 80,000 ft.

Consider now another planetary scale phenomenon, namely the ocean circulation, and ask if the general principles discussed above are not mani-

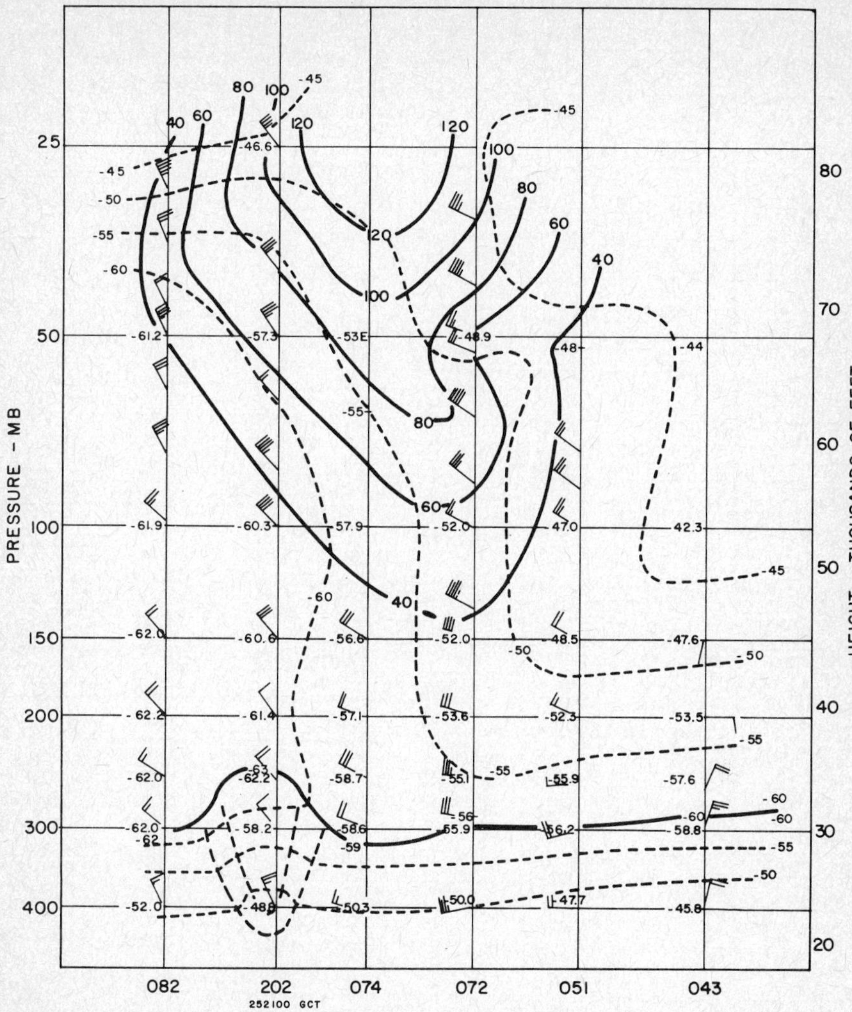

Figure 18. Vertical cross section of Arctic Stratospheric Jet Stream 0300 GCT 26 March 1956. Tropopause and frontal surfaces are indicated by thick solid and broken lines, respectively, isotherms (C) are indicated by broken lines, and isotachs of northwesterly flow in stratosphere by thin solid lines (kn). (After Lee and Godson, J. Met.)

fest in the ocean as well. The phenomenon in question is the Gulf Stream. It is well known that the Gulf Stream is a narrow, relatively fast-moving current off the east coast of North America. Figure 19 shows a vertical cross-section of the Gulf Stream off Chesapeake Bay. The lines of equal speed of the current are indicated by solid lines in cm per second. The dashed lines are isotherms in °C.

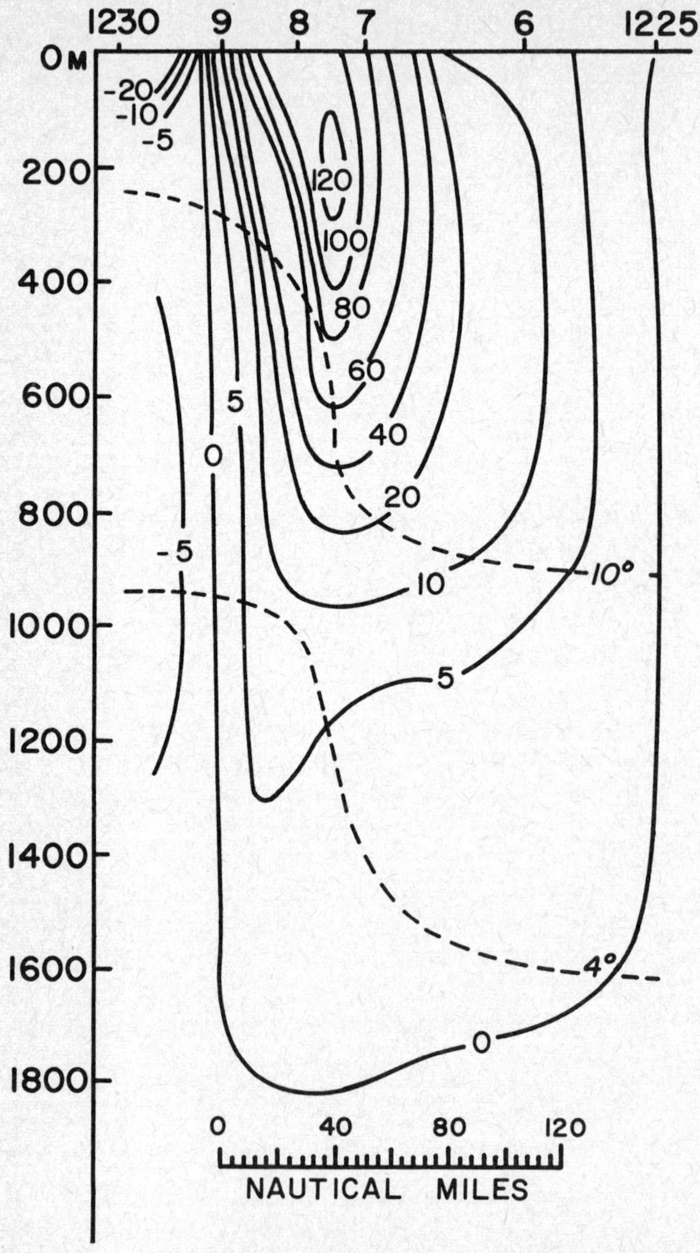

Figure 19. Relation between temperature distribution and current distribution in Gulf Stream off Chesapeake Bay. Isotherms are given by broken lines, lines of equal current velocity (in centimetres per second) by full lines. (After Staff Members, Dept. of Meteor., U. of Chicago; Bull. Amer. Met. Soc.)

It will be noted that there is a definite core of maximum speeds similar to the jet streams in the atmosphere. The greatest vertical increase in speed with height occurs in association with the thermocline, or the region where the isothermal surfaces are steepest. Finally, the current is extremely narrow, again possibly due to the intensification of the horizontal temperature gradient by the earth's rotation. Comparison of the slope of isotherms with those in the dishpan experiment will reveal a striking similarity.

Summary

To sum up, it is apparent that the January surface wind circulation in the Arctic is in no sense characterized by a general easterly current over the polar cap, but rather that the sea-level circulation exhibits a cellular structure. The cells of high and low are consistently located with respect to the heat sources and sinks. The surface wind field is surmounted by a circumpolar vortex within which there are on the average three long waves or Rossby waves in the winter, with two of the troughs located east of the Tibetan Plateau and the Rockies and the third near Novaya Zemlya. On the basis of the dishpan experiments, it may be inferred that differential heating of a fluid on a rotating earth is sufficient to demonstrate the development of a cellular structure in the surface circulation and the intensification of the existing temperature gradient into an upper narrow jet stream in middle latitudes. A Rossby wave pattern arises as a natural consequence of the motion. Similar features occur in the stratosphere. The temperature gradient which arises from differential solar heating of the ozone layer along the edge of the earth's shadow cone is intensified by the earth's rotation resulting in a strong westerly current in which a basic 3-wave Rossby pattern also develops. The ridges and troughs in this wave pattern are consistently located with respect to the main mountain barriers. Finally, the Gulf Stream structure manifests a narrow jet core within which the maximum vertical rate of increase in water speed is associated with the thermocline. The experimental and observational phenomena considered here appear to be quite dissimilar, nevertheless, viewed in the light of four physical factors, namely, differential solar heating, the rotation of the earth, the presence of mountain barriers and the geostrophic thermal wind equation, one observes a sense of physical unity.

References

[1] F. Baur, "Das Klima der bisher erforschten Teile der Arktis," *Arktis*, Vol. 2 (1929), pp. 77-89.

[2] B. Bolin, "On the influence of the earth's orography on the general character of the westerlies," *Tellus*, Vol. 2 (1950), pp. 184-95.

[3] B. Bolin, "Studies of the general circulation of the atmosphere," Advances in Geophysics, Vol. 1 (New York, Academic Press Inc., 1952), pp. 87-118.

[4] F. K. Hare and S. Orvig, "The Arctic Circulation," Arctic Meteorology Research Group Publication in Meteorology No. 12 (Montreal, McGill University, 1958).

[5] A. Kochanski, "Cross Sections of the mean zonal flow and temperature along 80° W," *J. Met.*, Vol. 12 (1955), pp. 95-106.
[6] R. Lee and W. L. Godson, "The Arctic Stratospheric jet stream during the winter of 1955-56," *J. Met.*, Vol. 14 (1957), pp. 126-35.
[7] J. Namias, "Synoptic and climatological problems associated with the general circulation at the Arctic," *Trans. Amer. Geophys. Union*, Vol. 39, no. 1 (1958), pp. 40-51.
[8] N. A. Phillips, "The general circulation of the atmosphere: a numerical experiment," *Q.J.R. Met. Soc.*, Vol. 82 (1956), pp. 123-63.
[9] H. Riehl and D. Fultz, "Jet streams and long waves in a steady rotating-dishpan experiment," *Q.J.R. Met. Soc.*, Vol. 83 (1957), pp. 216-31.
[10] C.-G. Rossby, *The scientific basis of meteorology*. Climate and Man. Yearbook of Agriculture, United States Department of Agriculture (1941), pp. 599-655.
[11] J. Smagorinsky, "The dynamical influence of large scale heat sources and sinks on the quasi-stationary mean motions of the atmosphere," *Q.J.R. Met. Soc.*, Vol. 79 (1953), pp. 342-66.
[12] U.S. Weather Bureau, "Normal weather charts for the Northern Hemisphere," Tech. Pap. 21 (1952).

2

Seasonal Location of the Arctic Front Over North America

R. G. Barry

Introduction

For over a decade the baroclinic structure of the atmosphere over North America has been considered by Canadian meteorologists (Anderson, Boville and McClellan, 1955; Penner, 1955), in terms of a three-front model. The fronts and the air masses they separate are:

the Polar front

*(mT—mP)**

the maritime Arctic front

*(mP—mA or wmP—cmP)**

the continental Arctic front

*(mA—cA)**

The typical seasonal locations of the fronts were noted by Anderson *et al.* (1955) but recent investigations by Bryson (1966) provide the first

*The air mass symbols used herein have the following meanings:
mT—maritime Tropical
mP—maritime Polar
mA—maritime Arctic
cA—continental Arctic
wmP—"warm" Polar maritime air (i.e., moving over a colder surface).
cmP—"cold" Polar maritime air (i.e., moving over a warmer surface).

Reprinted from *Geographical Bulletin*, Vol. 9, no. 2 (1967), pp. 79-95. By permission of The Queen's Printer.

detailed examination of the mean positions of the Polar and Arctic fronts over North America. Bryson employed three independent techniques to establish the frontal locations; surface air mass trajectories for July 1945-51 and 1954-56, frequency analysis of daily maximum temperatures for July 1948-57, and resultant streamline charts of surface wind for each month using data for 1930-45 (personal communication, R. A. Bryson, 1967). The considerable implications of the results for ecology and palaeoclimatology (Bryson and Wendland, 1966; Nichols, 1967) stimulated the present author to re-examine the occurrence of the Arctic front using frontal contour charts as a further source of information.

Frontal Contour Charts

The Meteorological Branch, Department of Transport, prepares frontal contour charts showing the location of the three fronts at the 850 mb,

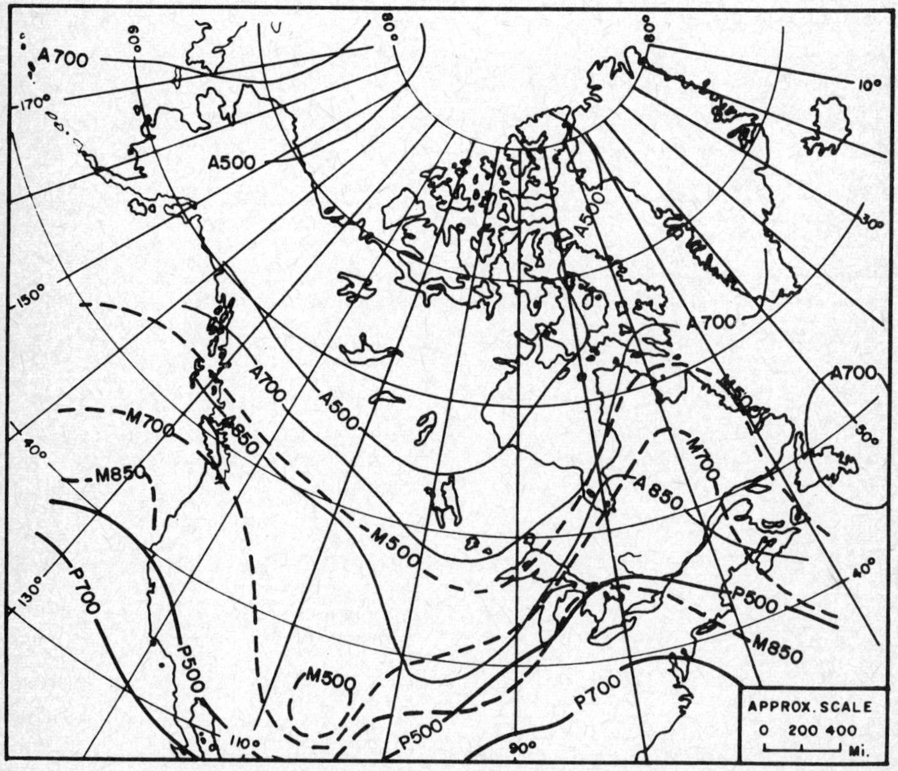

Figure 1. Specimen frontal contour chart for 00 GMT, 9 January 1965. The frontal contours are shown by heavy continuous lines (Polar front), broken lines (Maritime front) and fine continuous lines (Arctic front). A850, M850, P850 indicate the 850 mb locations of the Arctic, Maritime, and Polar fronts respectively; similarly for their locations at 700 mb and 500 mb.

700 mb and 500 mb pressure levels on a routine synoptic basis. It is pertinent at the outset to examine the nature of these analyses and the concepts on which they are based. Air mass characteristics and frontal boundaries are identified by analysis of aerological data in the form of tephigrams, hodographs and vertical cross-sections in order to determine the lines of intersection of frontal surfaces with pressure levels (Crocker, Godson and Penner, 1947). Contours of each front are then drawn at the three selected levels. Figures 1 and 2 illustrate two representative charts for January 1965 and Figures 3 and 4 south-north cross-sections along 70°W showing the frontal slopes. The technique ensures a large measure of horizontal and vertical congruity and indeed McIntyre (1959, p. 311) asserts that "this system, in the hands of a skilled analyst, yields a unique analysis which is consistent in space and time."

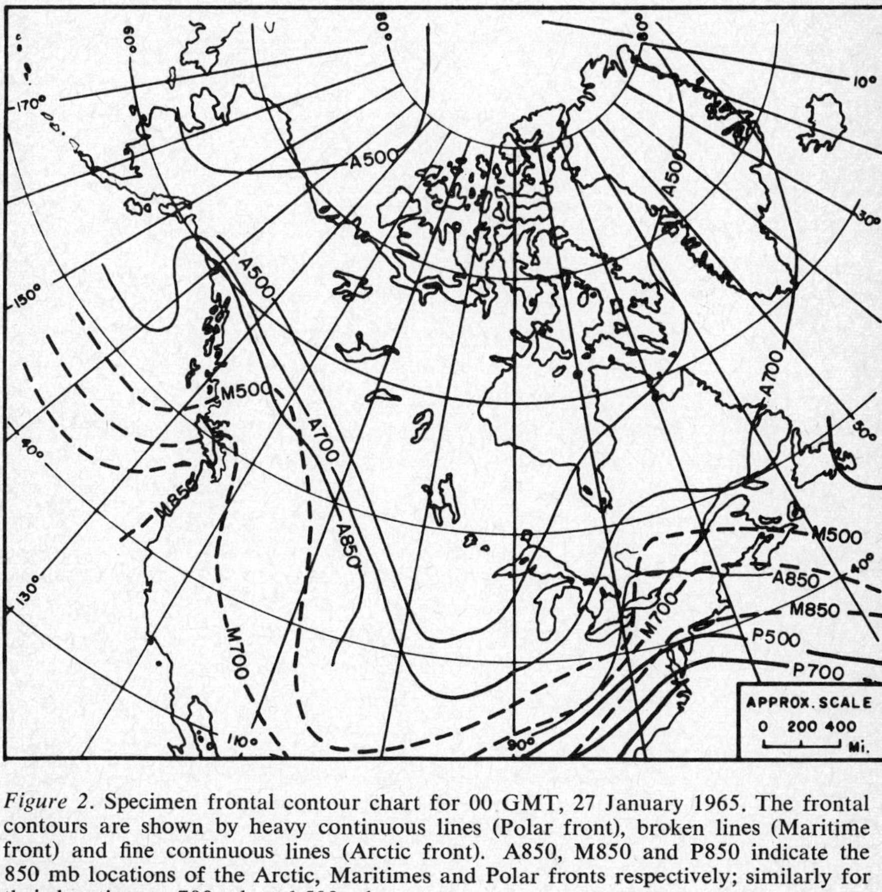

Figure 2. Specimen frontal contour chart for 00 GMT, 27 January 1965. The frontal contours are shown by heavy continuous lines (Polar front), broken lines (Maritime front) and fine continuous lines (Arctic front). A850, M850 and P850 indicate the 850 mb locations of the Arctic, Maritimes and Polar fronts respectively; similarly for their locations at 700 mb and 500 mb.

It is important to note that there are differences of opinion among meteorologists as to the validity of the three-front model (Galloway, 1958; Longley, 1959), but its continued use by the Meteorological Branch suggests that it is regarded of value for practical purposes of synoptic analysis. Since the analysis technique specifically sets out to achieve a consistent and coherent interpretation of the baroclinic structure, by using a wide variety of meteorological data—which includes wet-bulb potential temperature, the vertical wind shear and tropopause height—it is the writer's view that the frontal contour charts ought to provide a suitable information source for climatological analysis. Scrutiny of superimposed plots of the 00 GMT 850 mb contours for the Arctic front for each of the months examined (not shown) indicates that there is a generally well-maintained continuity of analysis even in July. For present purposes, therefore, these maps are preferable to surface weather maps because the major baroclinic zones are delimited while transient secondary frontal features are excluded. Although aerological stations are not numerous in Canada, only 24 in the area bounded by latitudes 55° and 72°N and longitudes 60° and 140°W, they have a quite even distribution. In fact, the total number of first class surface observing stations, at least in the sub-Arctic, is relatively little greater.

Method

The 850 mb position of the Arctic front has been determined along each 10 degrees meridian from 50° to 160°W on the daily 00 GMT maps for January, April, July and October 1961-65. The data extraction was from copies prepared from microfilm records of the Public Archives. These copies have approximate scales at 60°N of 1:21 million for the period 1961 to May 1963 and 1:35 million thereafter. Distortion of the maps due to their photographic enlargement and the reduction of the final analysis to a common scale cause an uncertainty of about 1 degree latitude in the calculated average locations.

The 850 mb level corresponds to a height of 1,457 metres (4,781 feet) in the Standard Atmosphere. Day-to-day and seasonal departures from this value are considered to be negligible on the scale of analysis used in this study. The mean surface front will be located at a distance of approximately 2 to 3 degrees of latitude (200-250 km) on the equatorial side of the mean 850 mb front. Figure 3 shows a greater difference than this between the 850 mb and surface positions, but on other occasions the two may be virtually coincident as Figure 4 demonstrates. In summer, when the southward penetration of shallow cold-air wedges is less pronounced and less frequent, the mean surface location is likely to be near that of the 850 mb frontal contour.

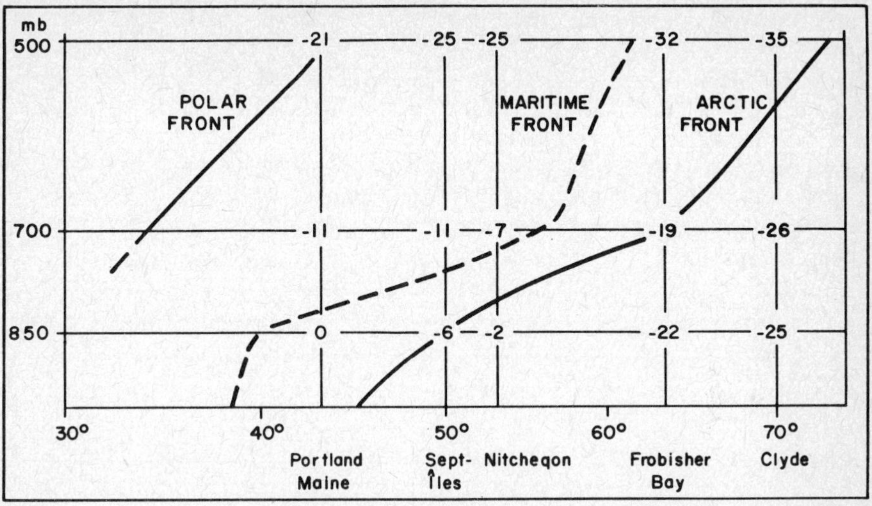

Figure 3. A cross-section along 70°W at 00 GMT, 9 January 1965 (see *Figure 1*). Air temperatures are in degrees Celsius. (Based on charts of the Meteorological Branch, Department of Transport.)

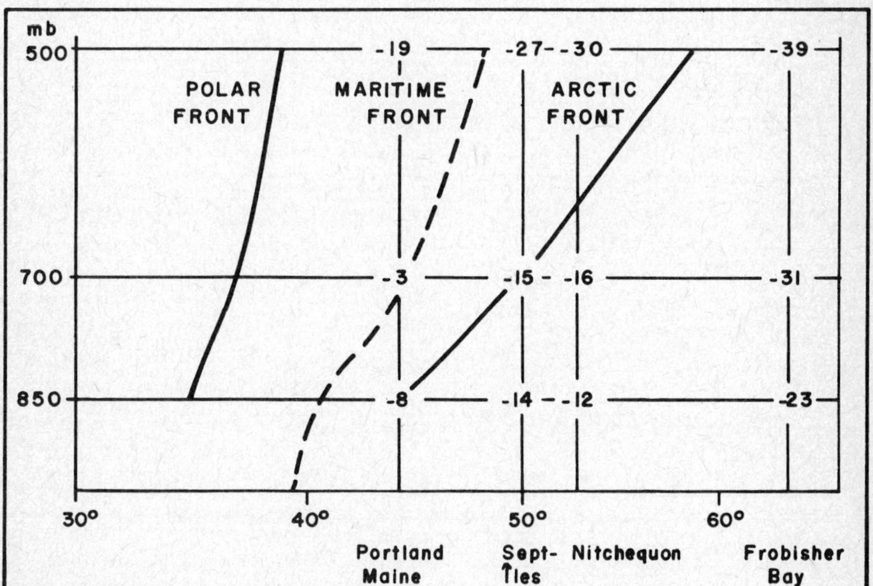

Figure 4. A cross-section along 70°W at 00 GMT, 27 January 1965 (see *Figure 2*). Air temperatures are in degrees Celsius. (Based on charts of the Meteorological Branch, Department of Transport.)

In most of the months under consideration, between 20 and 30 cases are available to determine the median location of the front. July 1962 was exceptional, however, in that no 850 mb Arctic front occurred in the area shown on the map between 16 and 26 July. At the western and eastern limits of the area the number of cases is only 10 to 18 in many months. In part this may reflect shortage of data for the original analyses. Table I summarizes the frequencies over the five-year period.

TABLE 1. THE FREQUENCY OF 850 MB ARCTIC FRONT CONTOURS, 1961-65

Longitude W	160°	150°	140°	130°	120°	110°	100°	90°	80°	70°	60°	50°
January	74	87	116	151	168	169	159	166	162	157	169	137
April	85	109	146	153	155	141	141	130	119	116	94	82
July	89	102	113	118	115	118	128	119	123	100	70	47
October	42	71	120	129	131	150	128	126	133	132	136	113

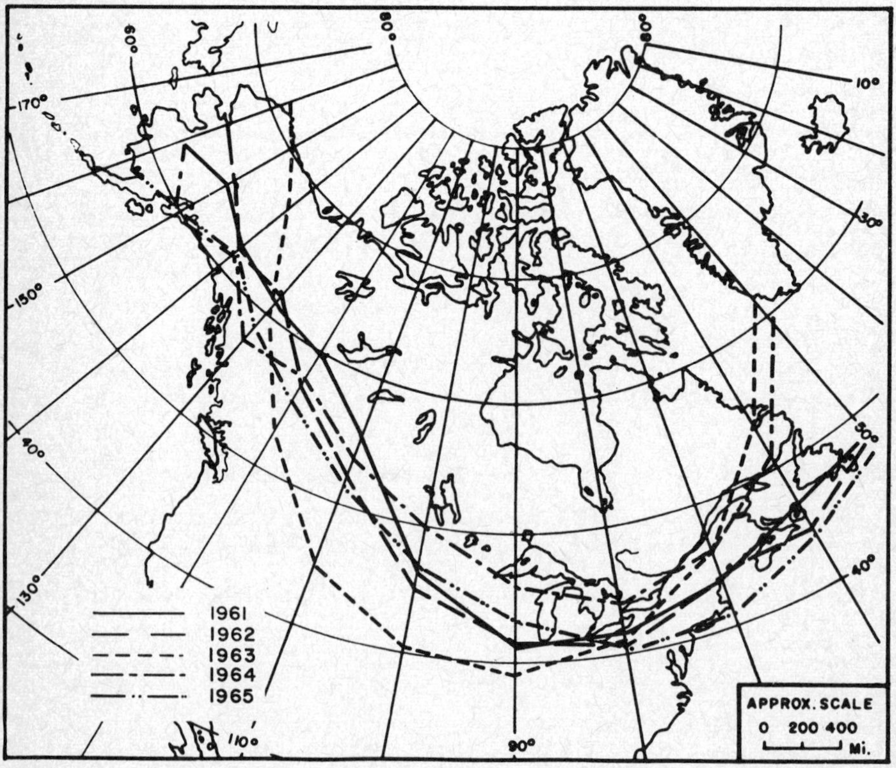

Figure 5. Median locations of the 850 mb Arctic front – January 1961-65.

Results

Figures 3 to 8 show the median positions of the 850 mb Arctic front in January, April, July and October 1961-65 and the variability of its location in January and July. The lines are dotted where the median is based on less than ten cases.

The *January* map (Figure 5) indicates that in most years a pronounced lobe of continental Arctic air covers the area north of a line from about southern Alaska to just south of the Great Lakes and then northeastward roughly across the Maritime Provinces. Figure 6 shows the quartile and decile lines for the daily positions of the front over the five years. These lines delimit, respectively, the central 50 per cent and 80 per cent of the frequency distribution. Great variability, both year-to-year and day-to-day, affects the region east of 70°W. The front is markedly farther north in this section, for example, in 1963 and 1964 (Figure 5). By contrast the least variability in the frontal location occurs between 120° and 140°W.

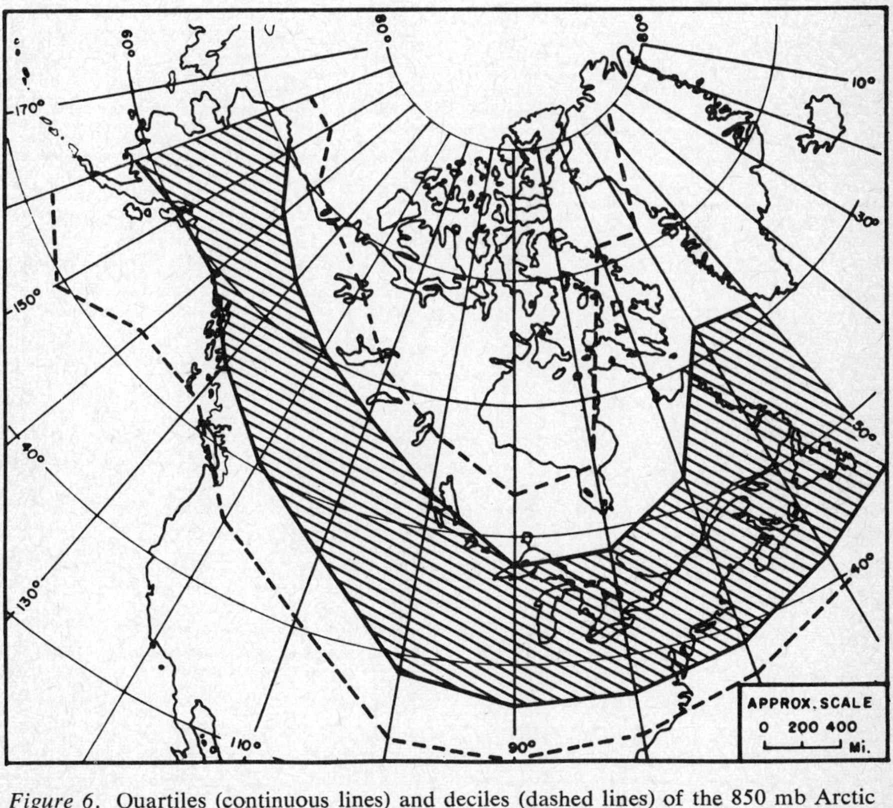

Figure 6. Quartiles (continuous lines) and deciles (dashed lines) of the 850 mb Arctic front location for the period January 1961-65. The central 50 per cent is hatched.

By *April* (Figure 7) the frontal zone has retreated to approximately James Bay in the interior of the continent, but along the Atlantic coast the movement has been negligible while over Alaska the median location is south of its January position in three of the years.

The *October* locations (Figure 8) are rather similar to those of April, although over the Maritimes the frontal zone is generally farther south in April. Examination of the daily location of the front indicates less day-to-day variability about the median location over eastern North America in both April and October than in January.

In *July* (Figure 9) the 850 mb Arctic front is situated more or less along the mainland Arctic coast from 160° to 120°W. Eastward of here, however, it may lie anywhere between approximately 60° and 70°N; a generalized line would pass along Hudson Strait-southern Baffin Island. Figure 8 showing the quartiles and deciles for July 1961-65 demonstrates the very evident contrast between eastern and western Canada. It is interesting to note that whereas over western North America the quartiles show

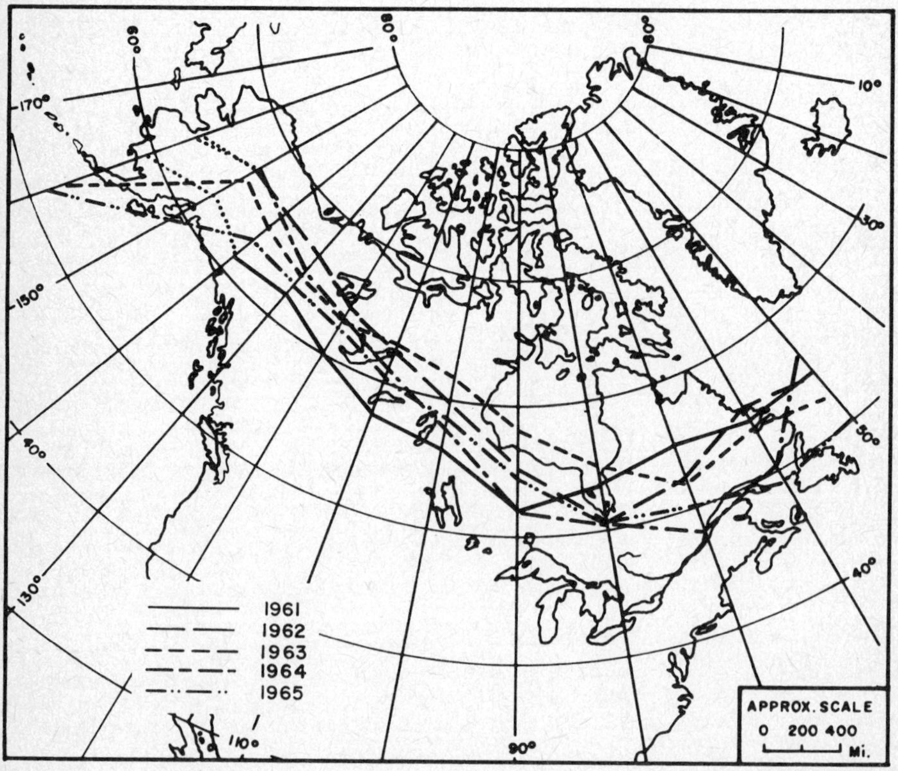

Figure 7. Median locations of the 850 mb Arctic front – April 1961-65.

greater spread in January (Figure 6) than in July (Figure 10), between 75° and 90°W the opposite is true.

Discussion

The status of the Arctic front presents special problems. Reed (1958, p. 55) asserts that "although there are fronts in the Arctic, there is no such thing as *the* 'Arctic front' ", a viewpoint which rejects the formalism of the three-front model. This question cannot be pursued here but the empirical results of Bryson (1966) certainly justify the recognition of an Arctic frontal zone in a climatological sense although its positioning is subject to some dispute.

The January map (Figure 5) is in reasonable agreement with Figure 16 except that the present results, other than 1964, tend to place the 850 mb front south of lakes Michigan and Erie rather than across lakes Huron and Superior. Figure 5 also indicates that in some years the front is farther north over Newfoundland and the Labrador Sea than Bryson's map suggests. The April map (Figure 7) is not readily comparable with the

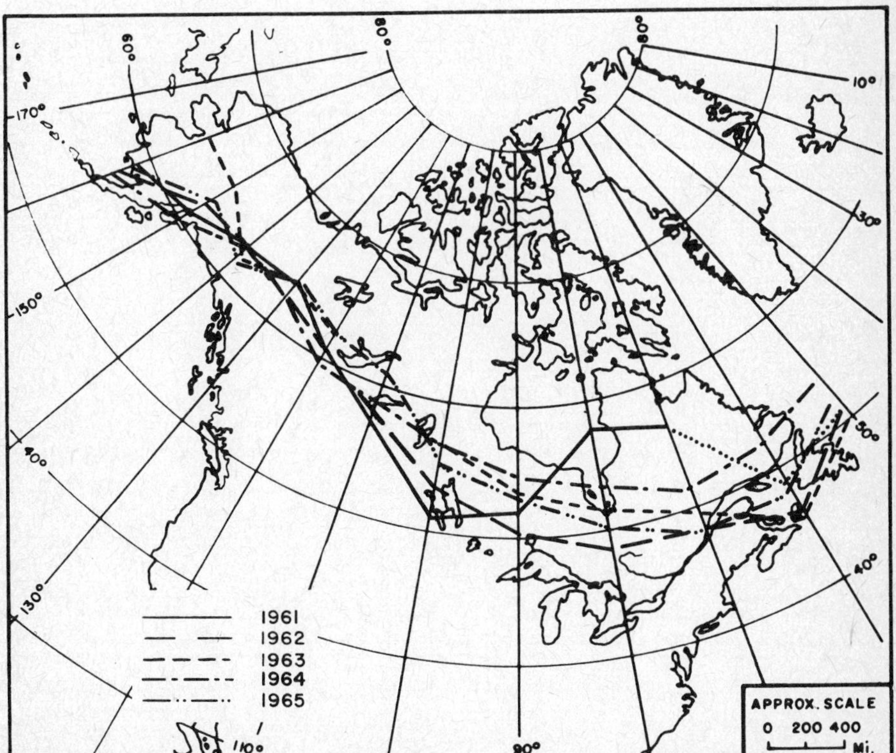

Figure 8. Median locations of the 850 mb Arctic front – October 1961-65.

streamline map of Figure 13 but there is no indication on that map of a confluence zone associated with the Arctic front. On the other hand, the October streamline map (Figure 15) is very similar to Figure 8.

The July location of the front is more problematic. Over the Northwest Territories the present results (Figure 9) are in accord with those of Bryson (Figures 12, 16) but there is a marked dissimilarity over Hudson Bay and Labrador-Ungava where Bryson shows the front at about 55°N. Part of this discrepancy is explained by the fact that Bryson includes Atlantic air with air of Arctic origin to obtain a composite 50 per cent line. This is based on the frequency analysis of maximum temperatures, although Bryson (1966, p. 234) states that the modal position of the front between Arctic and Pacific air is indefinite east of Hudson Bay. It must also be noted that the confluence zone evidenced by the surface resultant wind field in July (Figure 14) terminates in fact over western Labrador-Ungava.

Two difficulties arise in attempting to compare the present results with those of Bryson. The first is that there is probably no definite relationship

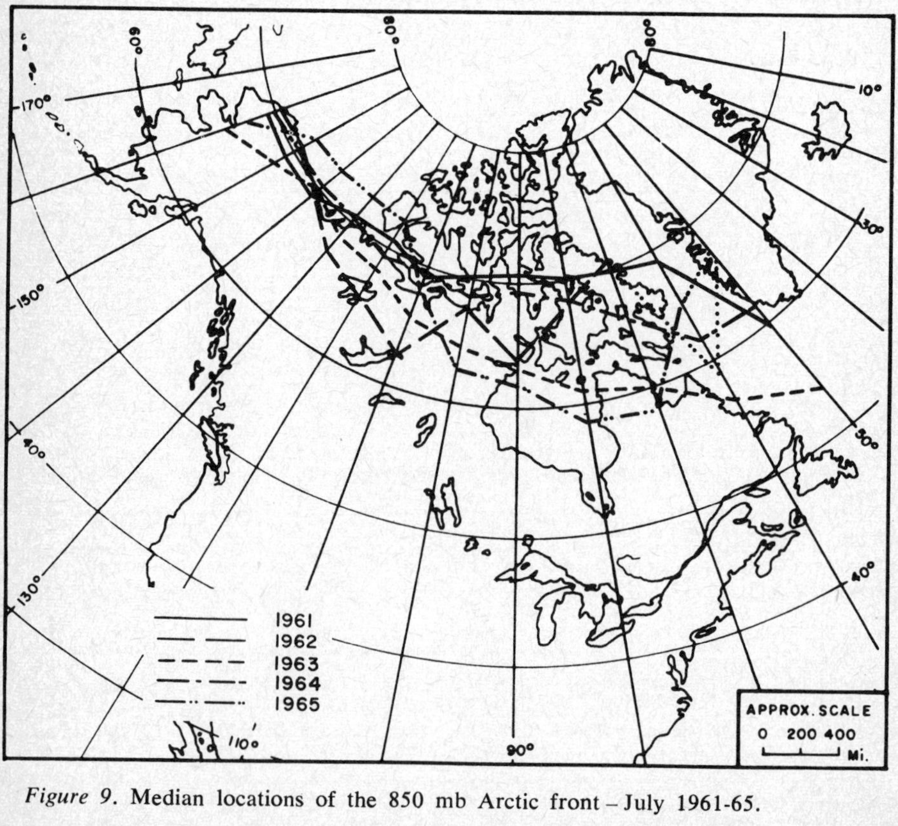

Figure 9. Median locations of the 850 mb Arctic front – July 1961-65.

between the four air-mass categories of the three-front model and the ones determined empirically by Bryson. The second is that the periods of analysis are different. These points are discussed in turn.

By the use of air-mass trajectories Bryson obtains for Canada, five groups of air masses (*Arctic, Pacific, Atlantic, Hudson Bay* and *"United States"* air) while, on the basis of temperature collectives, he delimits eight categories of air mass affecting Canada (*maritime Tropical, Canadian Rockies Pacific, Yukon Pacific, Alaskan, Eastern Arctic, Western Arctic, Hudson Bay* and *Atlantic*). It seems unlikely that, in terms of average conditions, any combination of these categories can be equated with *cA* and *mA* air. Consequently the mean Arctic front determined here may not be the *same* frontal zone along its length with respect to Bryson's categories. Nevertheless, the front should, by virtue of the methods used in the original analyses, indicate the primary northern zone of baroclinicity.

One significant point concerning the three-front model in summer is the lack of a source for *cA* air in the strict sense. Penner (1955, p. 90)

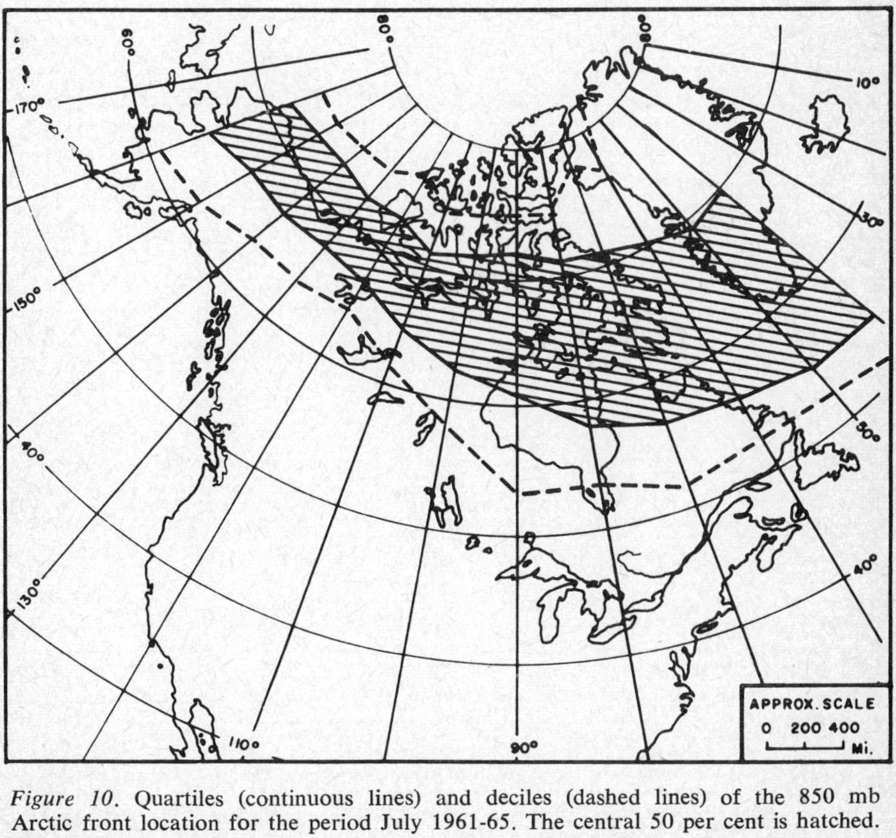

Figure 10. Quartiles (continuous lines) and deciles (dashed lines) of the 850 mb Arctic front location for the period July 1961-65. The central 50 per cent is hatched.

states that the Arctic front disappears or "retreats to the ice cap." All three frontal zones are present on most of the July maps examined here. Nevertheless, the possibility that the maritime Arctic front is the equivalent of the frontal zone on Bryson's summer maps must be examined.

Figure 11 shows the median locations of the 850 mb Maritime front for July 1961-65. It is evident that west of 115°W its occurrence bears no relation to Bryson's Arctic frontal zone. Indeed, the indicated front may involve different air masses in the sectors east and west of the Rocky Mountains in view of the marked kinks in the lines for 1963 and 1964. In the central sector from 115° to 90°W the 850 mb locations for 1961, 1963 and 1964 coincide with the surface position shown by Bryson, while over southern Labrador-Ungava the location in 1963 and 1964 is about 5 degrees south of the required one, if the difference between the 850 mb and surface front is taken into account.

The second consideration to be made in comparing the results is the possibility of different circulation regimes in the period 1961-65 and the years 1945-57 used by Bryson. For example, Lamb (1966) draws atten-

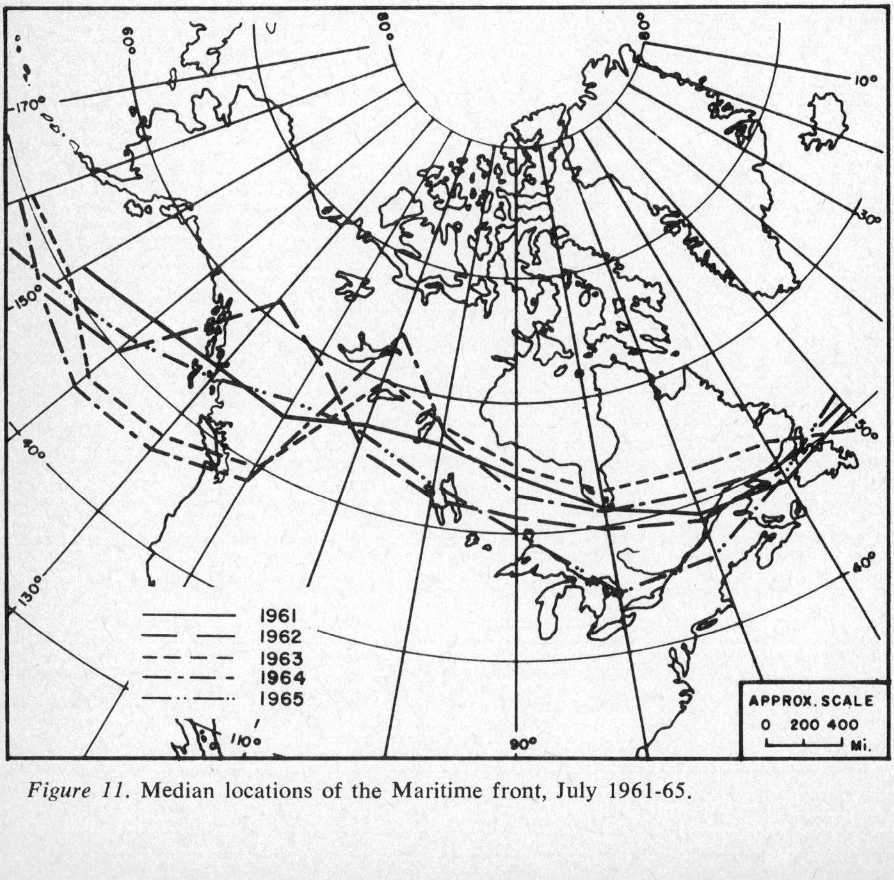

Figure 11. Median locations of the Maritime front, July 1961-65.

tion to the post-1960 changes of the general circulation, noting the tendency for greater meridional circulation in extra-tropical latitudes.

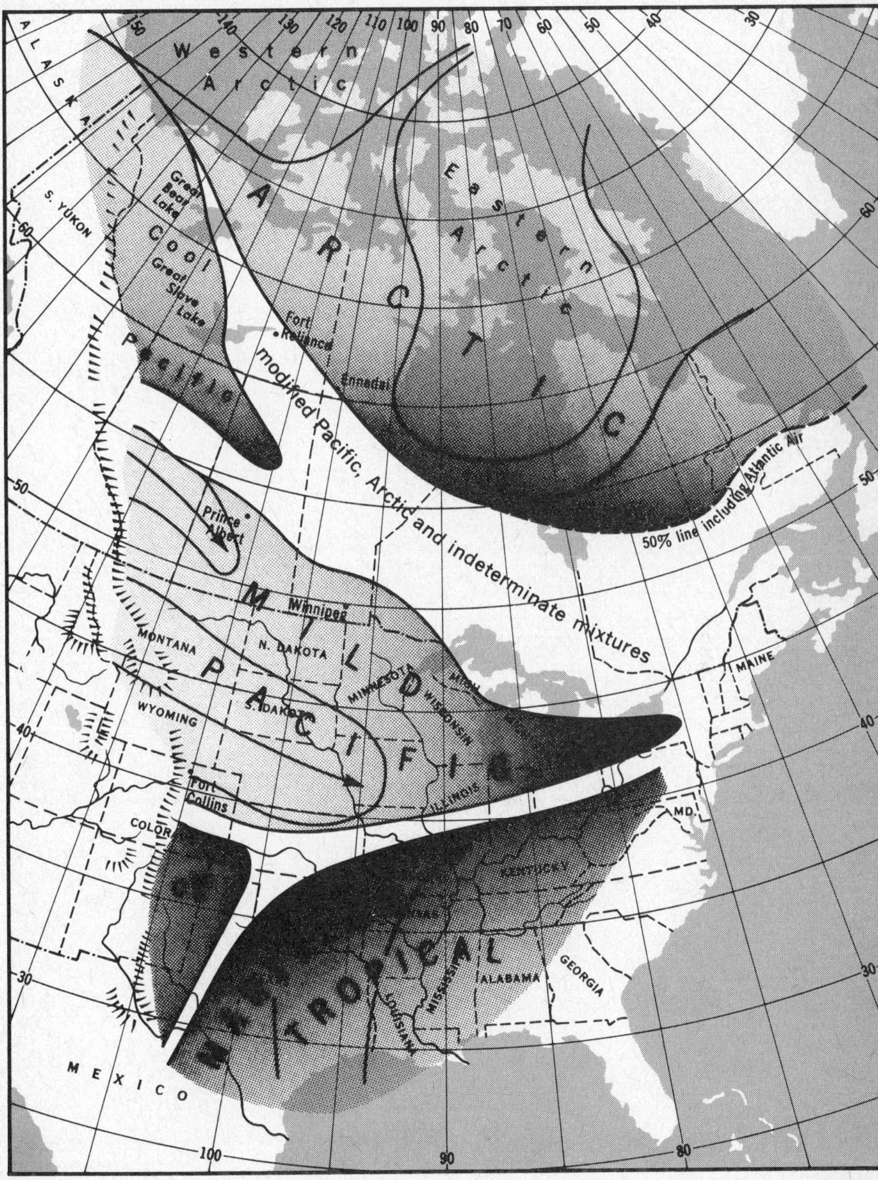

Figure 12. Composite chart of regions dominated by the various air mass types. The shaded regions are occupied more than 50 per cent of the time by the indicated air masses.

One readily available measure of the circulation is the zonal index for 35° to 55°N, between 5° and 185°W as calculated by the United States Weather Bureau—data not included in this paper. Examination of the highest and lowest monthly indices at the surface for January, April, July

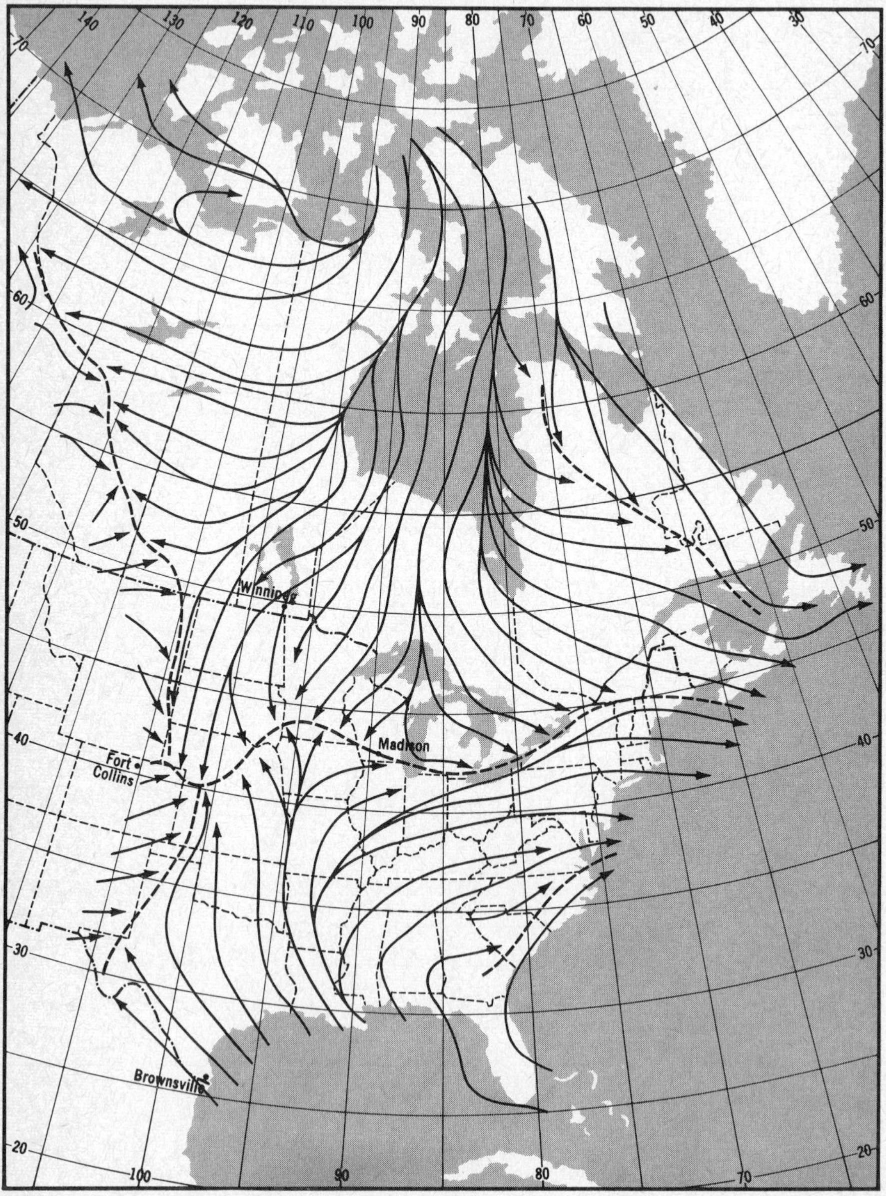

Figure 13. Streamlines of the surface resultant wind in April.

and October 1961-65 and the position of the Arctic front shows that the farthest north location occurs at 160°W with the lowest index values in all four months and at 60°W in all except July, whereas at 90°W it occurs with the highest zonal indices in all except July. This relationship may

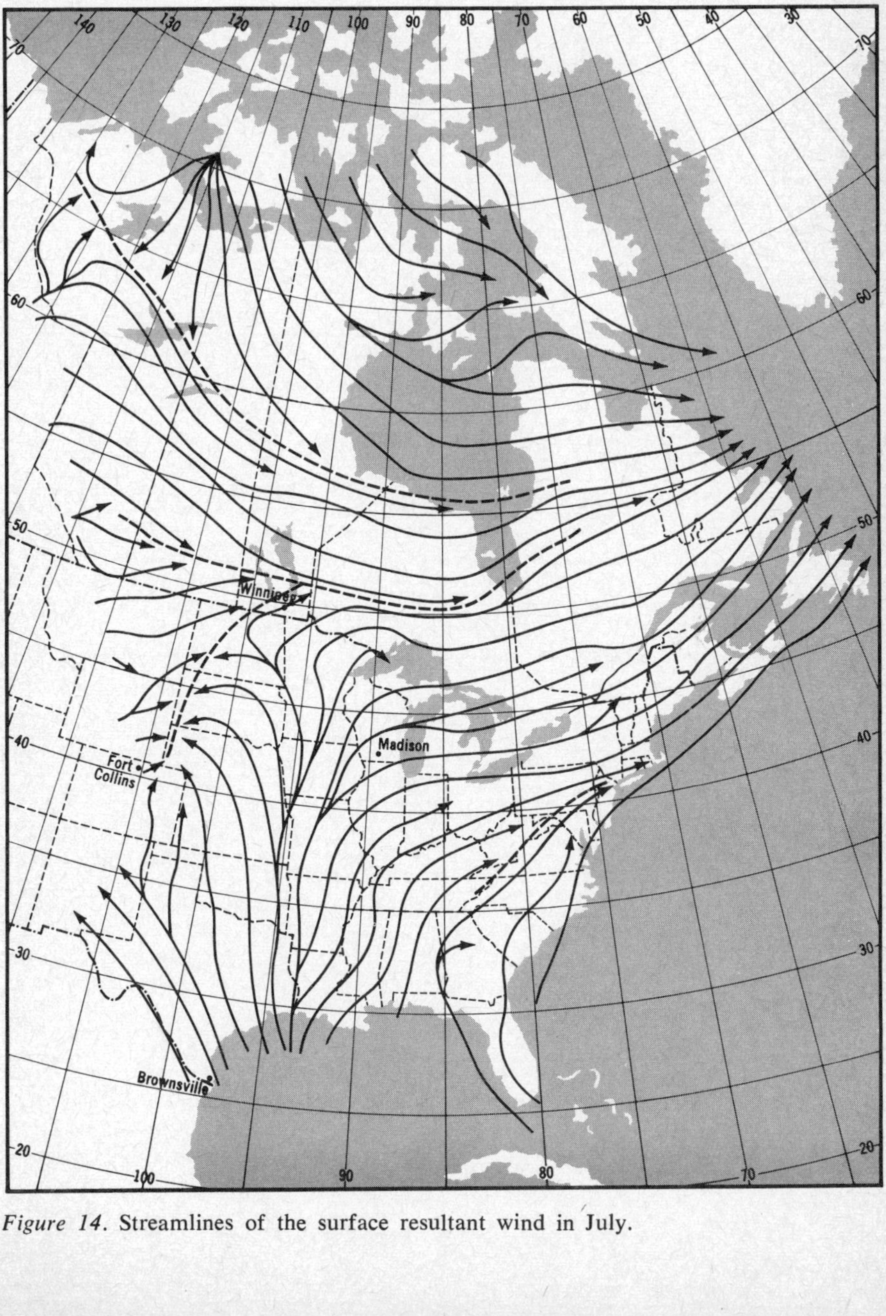

Figure 14. Streamlines of the surface resultant wind in July.

prove to be fortuitous but at least the pattern is consistent. Comparing the surface indices for 1961-65 with 1945-57, we may note that the mean January values for the two periods are virtually identical although the

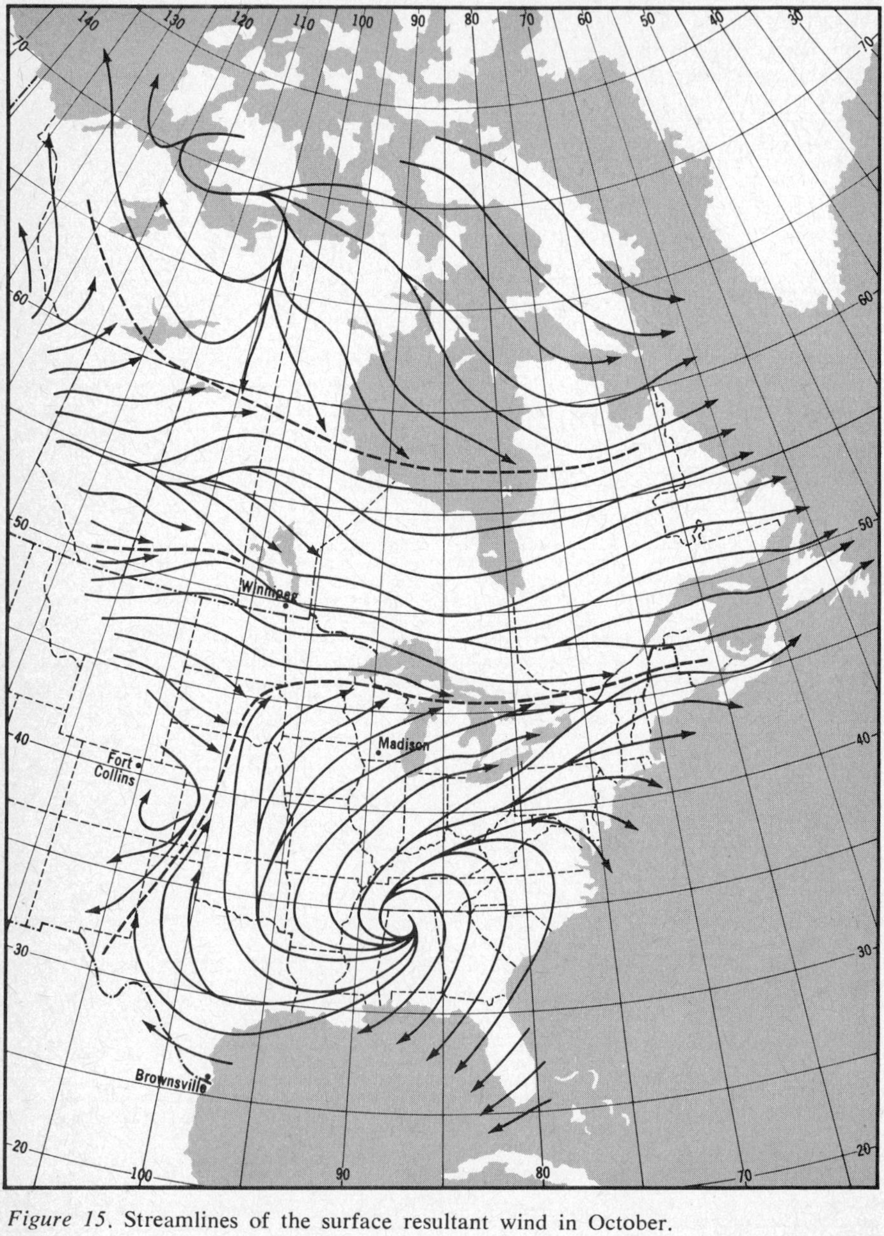

Figure 15. Streamlines of the surface resultant wind in October.

January 1963 index is the lowest on record. The July indices for 1961 and 1962 are also the two lowest on record and the mean for 1961-65 is only 2.12 m sec^{-1} compared with 2.55 m sec^{-1} for 1945-57, but in view

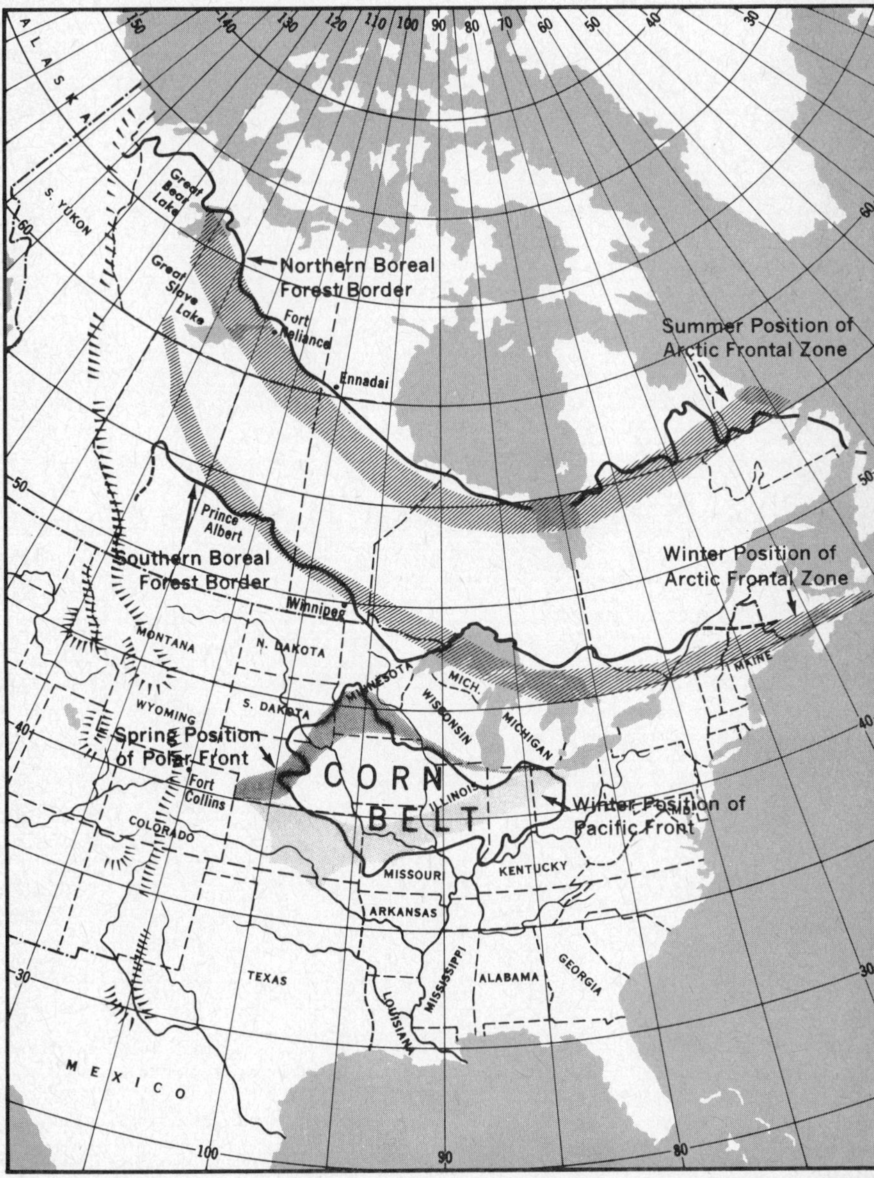

Figure 16. The coincidence of the "corn belt" and "boreal forest" biotic regions with meteorologically defined climatic (air mass) regions.

of the apparent absence of any relationship between the frontal location and zonal index in July this difference may be of little or no importance.

In the light of these results, three tentative conclusions are advanced:

1. The January location of the Arctic frontal zone shown by Bryson appears to represent the continental Arctic front, although the Newfoundland sector requires further checking.

2. The July or summer position shown by Figure 16 (Bryson, 1966) is equivalent to the continental Arctic front over the Northwest Territories and the maritime Arctic front from about 115° to 90°W. East of there it is apparently a median line relating to a broad baroclinic zone which is represented as a dual frontal zone by Canadian synopticians. It is beyond the scope of this study to consider which solution is more realistic. The problem could perhaps be resolved by an examination of the jet stream structure over the area.

3. The zonal westerlies were abnormally weak in January 1963, but there is no indication that the period 1961-65 as a whole is unrepresentative of the longer-term mean location of the frontal zones.

Ecological Implications

It is clear from the foregoing discussion that the mean location of the Arctic frontal zone over eastern Canada presents some problems, especially for the summer months. This is partly a question of definition, partly of the considerable day-to-day and year-to-year shifts in the position of the baroclinic zone. The latter aspect is by far the more important in its implications. Studies by the University of Wisconsin Meteorology Department have used the apparent correlations of the southern limit of Arctic air dominance in summer and winter with the respective northern and southern margins of the boreal forest zone in central Canada to reconstruct the locations of the frontal zone during late- and post-glacial periods from palynological evidence (Bryson and Wendland, 1966; Nichols, 1967). The findings of this study lend support to Bryson's results in northwest Canada where the forest-tundra boundary is extremely sharp, but suggest that the correlation may not be applicable in Labrador-Ungava. It is noteworthy that in this region where the location of the climatological front is, according to the present results, uncertain, the forest boundary is replaced by an extensive forest-tundra ecotone and the tree-line for individual species extends almost to Ungava Bay (Hare, 1950 and 1951). The forest boundary in Labrador-Ungava is undoubtedly greatly affected by topographic and edaphic factors. Marr (1948) shows that at Richmond Gulf on the east coast of Hudson Bay climate is apparently non-limiting where there is no dearth of soil, a fact which in turn reflects the short postglacial time available for soil to develop.

In conclusion, it is considered that although the median July position of the Arctic front in the northwest is in close agreement with Bryson's findings, the Arctic front—boreal forest/tundra boundary relationship proposed by Bryson warrants further investigation before it can be extended to eastern Canada. Moreover, if palaeoclimatological reconstructions are attempted it must be recognized that a variability of the frontal zone similar to that presently occurring is to be expected during postglacial time.

References

Anderson, R., Boville, B. W. and McClellan, D. E. 1955, "An operational frontal contour-analysis model," *Roy. Met. Soc. Q.J.* Vol. 81 (1955), pp. 588-99.

Bryson, R. A. 1966, "Air masses, streamlines and the boreal forest," *Geog. Bull.*, Vol. 8, no. 3 (1966), pp. 228-69.

Bryson, R. A. and Wendland, W. M., "Tentative climatic patterns for some late-glacial and postglacial episodes in central North America." Preprint of paper presented at the Conference on Environmental Studies of the Glacial Lake Agassiz Region, Winnipeg, Manitoba (4-6 November 1966), 33 pp.

Crocker, A. M., Godson, W. L. and Penner, C. M., "Frontal contour charts," *J. Meteorol.*, Vol. 4 (1947), pp. 95-99.

Galloway, J. L., "The three-front model: its philosophy, nature, construction and use," *Weather*, Vol. 13 (1958), pp. 3-10.

Hare, F. K., "Climate and zonal divisions of the boreal forest formations in eastern Canada," *Geog. Rev.*, Vol. 40 (1950), pp. 615-35; "Some climatological problems of the Arctic and sub-Arctic". Malone T. E. ed., *Compendium of Meteorology* (Boston, American Meteorology Society, 1951), pp. 952-64.

Lamb, H. H., "Climate in the 1960's. Changes in the world's wind circulation reflected in prevailing temperatures, rainfall patterns and the level of African lakes," *Geog. J.*, Vol. 132 (1966), pp. 183-212.

Longley, R. W., "The three-front model – a critical analysis," *Met. Br., Dept. Trans.*, Toronto, Circ. 3245, Tec. 309 (1959), 14 pp.

Marr, J. W., "Ecology of the forest-tundra ecotone on the east coast of Hudson Bay," *Ecol. Monogr.*, Vol. 18 (1948), pp. 117-44.

McIntyre, D. P., "The Canadian three-front, three-jet stream model," *Geophysica*, Vol. 6 (1959), pp. 309-24.

Nichols, H., "Pollen diagrams from sub-Arctic central Canada," *Science*, Vol. 155, no. 3770 (1967), pp. 1665-68.

Penner, C. M., "A three-front model for synoptic analyses," *Roy. Met. Soc. Q.J.*, Vol. 81 (1955), pp. 89-91.

Reed, R. J., "Arctic synoptic analysis" in *Contributions to the study of the Arctic circulation*, Wilson, C. V. ed., Arctic Met. Res. Group, pub. in Meteorology no. 9 (Montreal, McGill University, 1958), pp. 48-59.

3

The Nature and Origin of the Canadian Dry Belt

J. R. Villimow

Palliser's Triangle, a large dry area in southern Alberta, Saskatchewan, and Manitoba, extending northward from the United States border approximately to latitude 52°N. and westward from longitude 100°W. to 114°W., has been a regional geographic reference since Captain John Palliser[1] explored the vast unsettled interior of southern Canada nearly a century ago. Recognition of the distinctively dry character of this part of Canada is found in Palliser's reports made as a result of field work in 1857, 1858, 1859, and 1860.[2] Although Palliser's contemporaries varied in their opinions of his conclusions regarding the usability of the land for agricultural

1 John Palliser, *Papers Relative to the Exploration by Captain Palliser of that Portion of British North America which lies Between the Northern Branch of the River Saskatchewan and the Frontier of the United States; and Between the Red River and the Rocky Mountains* (London: Eyre and Spottiswoode, 1859), and John Palliser, *Exploration—British North America. Further Papers Relative to the Exploration by the Expedition under Captain Palliser of that Portion of British North America which lies between the Northern Branch of the River Saskatchewan and the Frontier of the United States; and Between the Red River and the Rocky Mountains, and thence to the Pacific Ocean* (London: Eyre and Spottiswoode, 1860).

2 John Palliser, *Exploration—British North America. The Journals, Detailed Reports, and Observations Relative to the Exploration, by Captain Palliser, of that Portion of British North America, which, in Latitude, lies Between the British Boundary Line and the Height of Land or Watershed of the Northern or Frozen Ocean Respectively, and in Longitude, Between the Western Shore of Lake Superior and the Pacific Ocean during the Years 1857, 1858, 1859, and 1860* (London: Eyre and Spottiswoode, 1863).

Reprinted from *Assoc. of Amer. Geographers Annals,* Vol. 46 (1956), pp. 211-32.

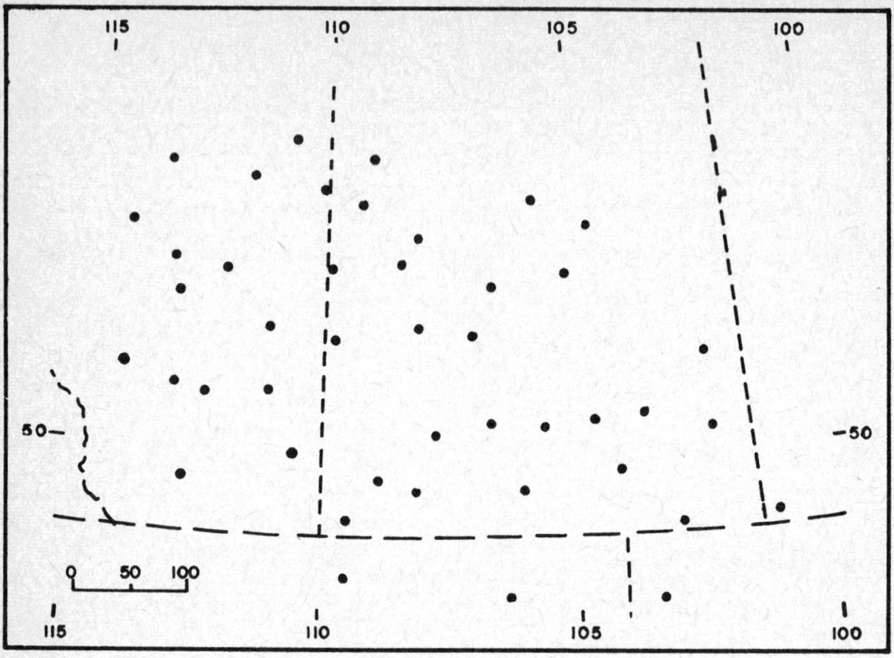

Figure 1. Locations of stations.

settlement,[3] opinion based upon climatological observations has been unanimous in its indictment of the area. Fifty years ago Stupart noted significant regionalisms in terms of temperature and precipitation characteristics in western Canada,[4] and later Connor, in the most painstaking research on Canadian climate, pointed to the problem nature of the "Dry Belt," as he called it, refining somewhat Palliser's conclusions concerning the extent of abnormally dry conditions.[5]

Since Connor's study of Canadian climate was made, a great body of surface meteorological data has accumulated on the Dry Belt; and to this information has been added the beginning of significant upper-air information particularly in the form of "normal" charts of pressure distribution at various levels above the surface of the earth.[6] It is the purpose of this

---

[3] Henry Youle Hind, "On the Commercial Progress and Resources of Central British America. The Lake Winnipeg and Saskatchewan Districts," *Journal of the Statistical Society of London,* Vol. 27 (1864), pp. 82–105; and John Macoun, *Manitoba and the Great Northwest* (London: Thomas C. Jack, 1883), pp. 142–154.

[4] R. F. Stupart, "The Canadian Climate," *Report of the 8th International Geographic Congress, 1904* (U.S. Government Printing Office, 1905), pp. 296–300.

[5] A. J. Connor, "The Climates of North America. Canada," in W. Köppen and R. Geiger, *Handbuch der Klimatologie*, Bd. II (Berlin, 1938), Teil J, pp. 359–364.

[6] See list of surface and upper air data sources at end of report.

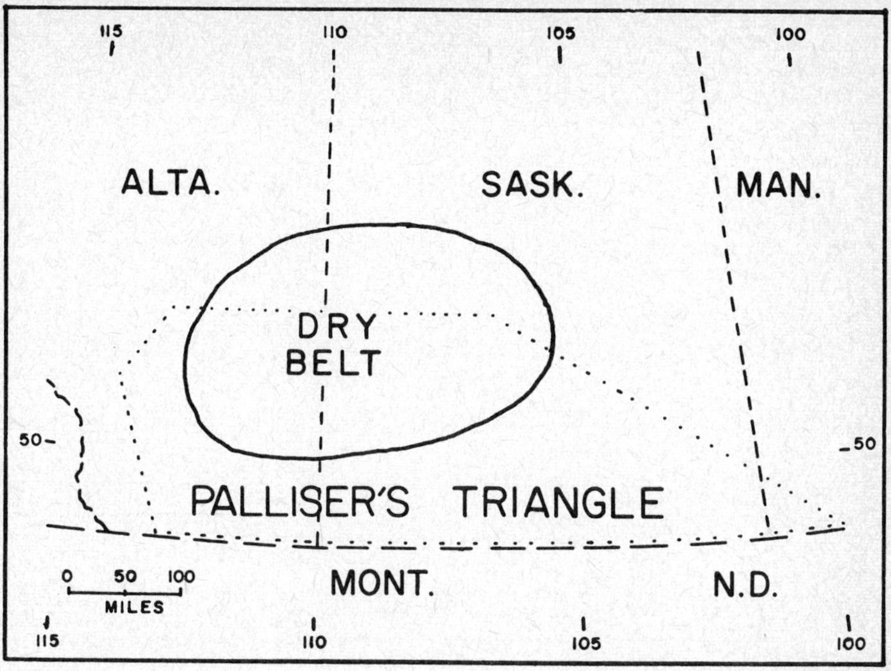

Figure 2. Location of Dry Belt (continuous line) and Palliser's Triangle (dotted line) in southern Canada.

paper to examine the recent data to define more exactly the extent of the Dry Belt in terms of its major uniquenesses and to determine, as far as data will allow, the causes for the distinctive climatic regionalism of the Dry Belt.

The distribution of stations whose climatological records are used in this study is shown in Figure 1. A serious attempt was made to abide by Conrad's general conclusion that "from all points of view, the length of normal period should be about 25 to 35 years."[7] Whenever this rule was violated it was because it was deemed wiser to include a station with a somewhat shorter record to make the climatological network more complete rather than to leave a significant part of the network void.

Regionalizing Features of Precipitation

Moisture characteristics form the primary basis for the contention that the Dry Belt is a unique and distinctive climatic region. Figure 2 shows the areal extent of the Dry Belt as defined by this study in comparison with Palliser's Triangle; this comparison indicates that the Dry Belt is approximately one-half the size of Palliser's Triangle, overlapping most of the

[7] Victor Conrad, *Methods in Climatology* (Harvard University Press, 1944), p. 148.

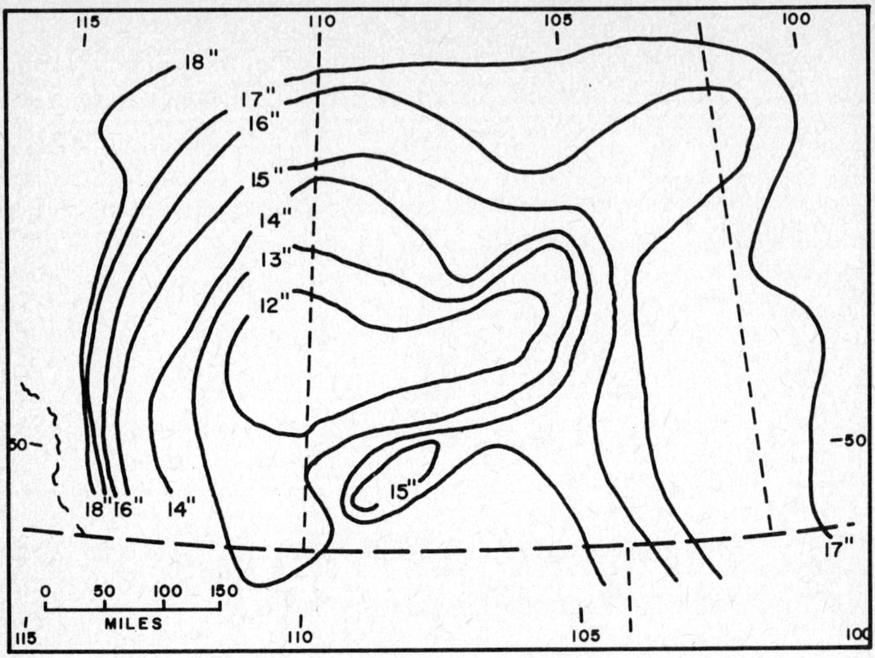

Figure 3. Average annual precipitation.

northern part of the region marked off by Captain Palliser as unfit for human settlement, and extending at least 100 miles beyond the northern limit of the Triangle. In total area the Dry Belt measures approximately 50,000 square miles, nearly equal to the size of Alabama or Greece.

The Canadian Dry Belt is the largest of several dry regions in the Great Plains of North America. Others, located in Montana, South Dakota, and Wyoming,[8] are not so marked by the frequency of stations reporting less than twelve inches of precipitation per year. Figure 3 on average annual precipitation clearly marks the Dry Belt as a region of little precipitation, having an elliptical shape, and trending in a west-east direction 150 miles north of the Canadian border. A broader view of the Dry Belt in North America (Figures 4 and 5) shows that it occupies low points along two precipitation profiles, one of which traverses North America from west to east along latitude 51°N., and the other parallels the east face of the Rocky Mountains from latitude 25°N. to latitude 60°N., but is located at a uniform distance of 250 miles east of the front of the Rockies (corresponding to the distance from the core of the Dry Belt to the Rockies). The slight increase in maximum precipitation immediately south of the Dry Belt

---

8 *Climate and Man,* Yearbook of Agriculture (U.S. Dept. of Agriculture, 1941), pp. 965, 1117, 1209.

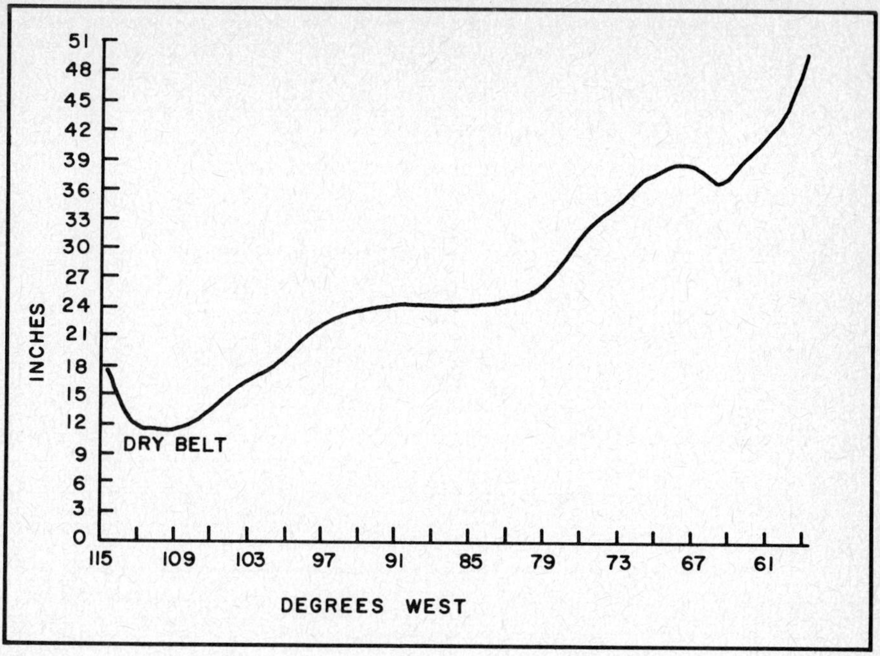

Figure 4. Precipitation profile along latitude 51°N from the Rocky Mountains to the Atlantic Ocean.

(Figure 5) reflects the localized effect of the Cypress Hills. Measurement of the rate of precipitation change per 100 miles of horizontal distance (Figure 6) reveals that the climatic boundary of the Dry Belt is most marked to the west, approximately equally marked to the north and east, and least marked to the south. The southern limit of the Dry Belt is better marked when certain other climatic features are examined.

When annual precipitation curves representative of the Dry Belt and its surroundings are studied, significant differentiating factors are added to the factor of total annual precipitation. Anglia, Saskatchewan, Edmonton, Alberta, Williston, North Dakota, and Winnipeg, Manitoba (Figures 7-10) represent conditions typical, respectively, of the Dry Belt and areas to the north, south, and east. Such comparison shows: (1) The Dry Belt's wettest months are distinctly drier than its surroundings—adjacent precipitation curves record two months each with more than 3 inches, but the Dry Belt's wettest month, June, has only 2.4 inches. (2) The wettest months in the Dry Belt are the warmest months, lessening the total effectiveness of the "wet season," but to the south—Williston, North Dakota—the "wet season" comes slightly earlier, May-July, at a time when the precipitation may be more effectively utilized. (3) Although the Dry Belt has a distinct secondary mid-winter maximum of precipitation, a regionalizing feature lack-

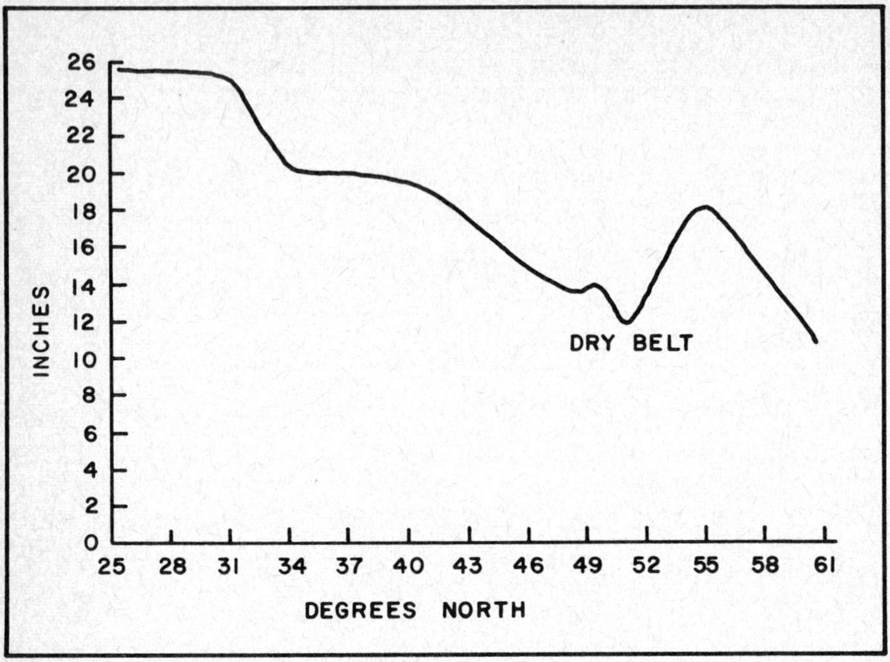

Figure 5. Precipitation profile 250 miles east of the Rocky Mountains.

ing to the south and east (Figure 7), it is not sufficiently large to be significant; moreover, as will be shown, chinooks here tend to nullify the fall of precipitation during the winter.

Variability of precipitation is an unfailing companion to dry regions and is marked in the Dry Belt in a number of ways. Both Currie[9] and Longley[10] have given statistical proof of the high variability of precipitation from year to year in the Dry Belt, and Connor's[11] study has taken note of the unusually dry conditions here during the drier cycles which characterize the Dry Belt. Studies by the writer have shown that the Dry Belt is specifically marked by fewer days with precipitation during both unusually dry years and unusually wet years, as well as a lower precipitation intensity during both abnormally dry and moist years.[12]

---

9 B. W. Currie, *Prairie Provinces and Northwest Territories: Precipitation* (Saskatoon: University of Saskatchewan, 1953), pp. 10–12 and Table I.

10 Richard W. Longley, "Variability of Annual Precipitation in Canada," *Monthly Weather Review*, Vol. 81 (1953), pp. 131–134.

11 A. J. Connor, "Droughts in Western Canada," *The Canada Year Book* (Dominion Bureau of Statistics, 1933), pp. 47–59.

12 The abnormally dry years included 1936, 1945, and 1949; the abnormally wet years were 1942, 1946, and 1947. During each of these six years the period of most noticeable regional pattern in the Dry Belt was April through August.

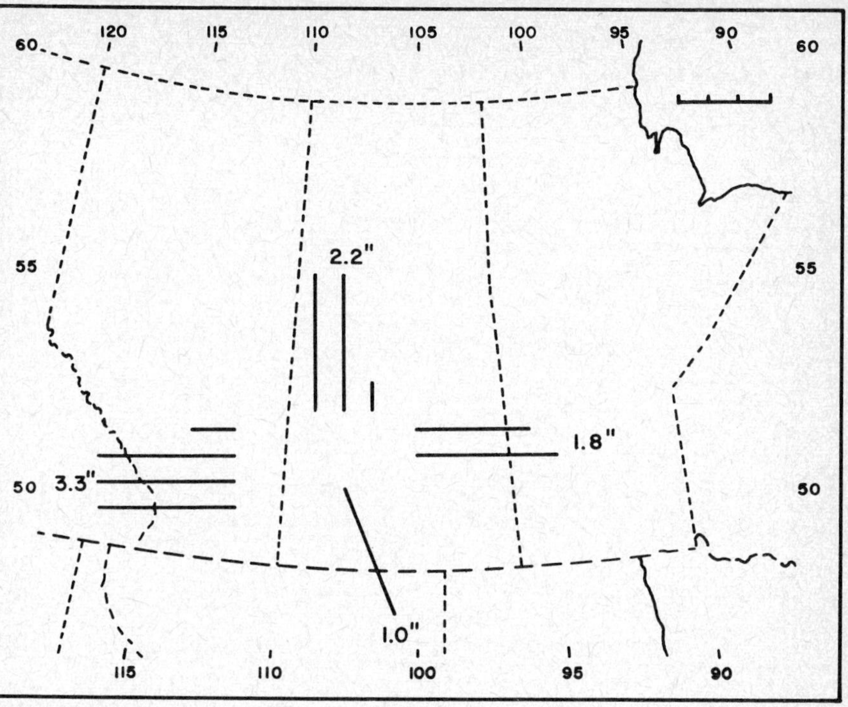

Figure 6. Precipitation gradient per 100 miles from the center of the Dry Belt.

The longest period with less than 0.25 inches of precipitation[13] during the April-through-August span of major dry years shows as clearly as any measure the seriousness of the drought problem in the Dry Belt (Figure 11). Most of the Dry Belt had at least two months without significant precipitation, and the eastern half was "dry" for two and one-half months —all of this during the normally wet season!

Application of the principles and criteria contained in the new Thornthwaite (1948) classification of climate[14] allows mathematical integration of the several factors which combine to produce the world climates. Although the Thornthwaite system is not unique in this respect, it does succeed where other climatic systems fail, viz., in differentiating the Dry Belt from its surroundings.[15] A large area north of the United States-Canada border and lying astride the Alberta-Saskatchewan border in the

[13] Conrad considers 0.25 inches of precipitation the minimum necessary to break a drought. See Victor Conrad and L. W. Pollack, *Methods in Climatology* (2nd ed.; Harvard University Press, 1950), p. 213.

[14] C. W. Thornthwaite, "An Approach Toward a Rational Classification of Climate," *Geographical Review*, Vol. 38 (January 1948), pp. 55–94.

[15] Marie Sanderson, "Drought in the Canadian Northwest," *Geographical Review*, Vol. 38 (1948), pp. 289–99.

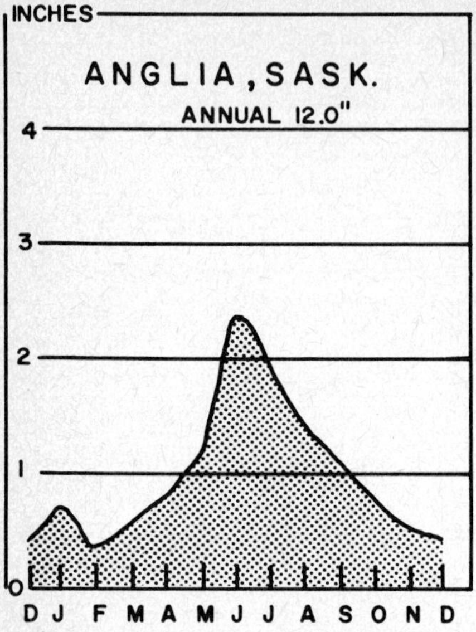

Figure 7. Precipitation profile for Anglia, Sask.

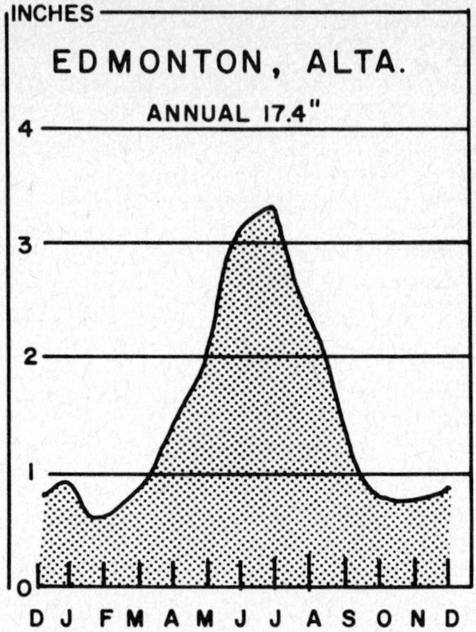

Figure 8. Precipitation profile for Edmonton, Alta.

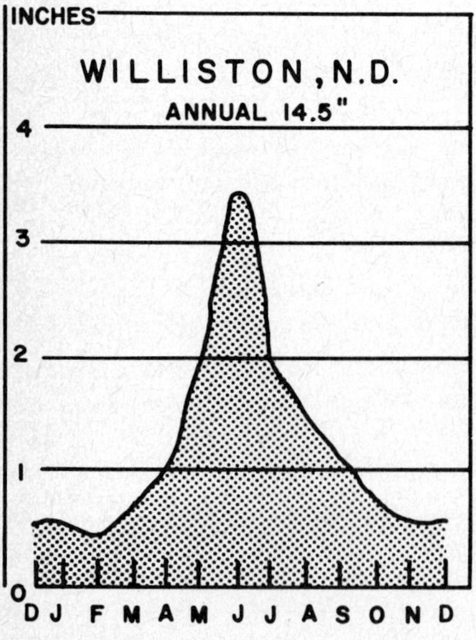

Figure 9. Precipitation profile for Williston, N.D.

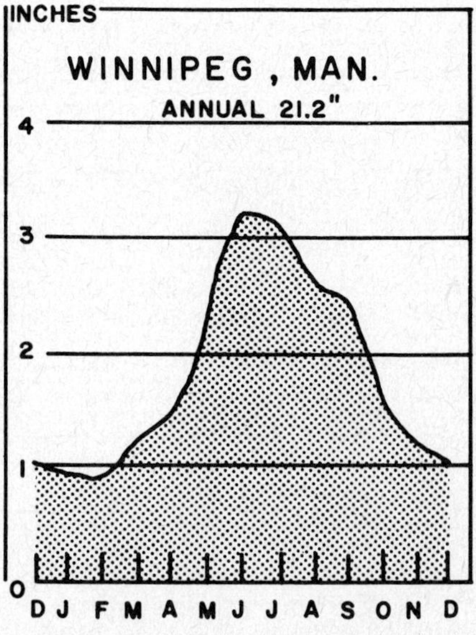

Figure 10. Precipitation profile for Winnipeg, Man.

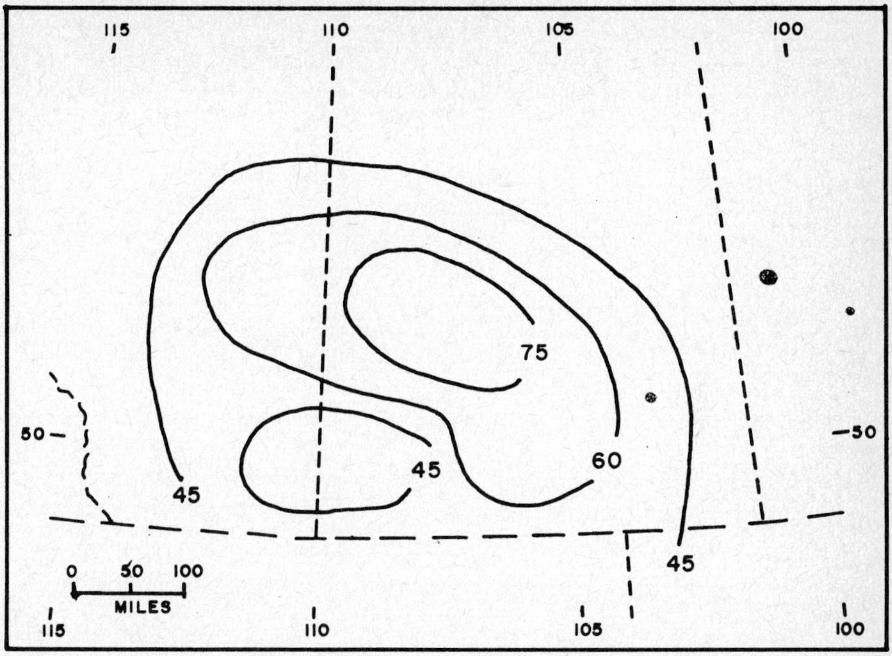

Figure 11. Period of drought – the longest period with less than 0.25″ of precipitation – during the wet season of major dry years (1936, 1945, 1949).

Dry Belt has a moisture index of less than –25 (Figure 12). Since –20 marks the boundary between C_1 (the driest of the humid climates) and D (the moistest of the dry climates), it is apparent that the sector within the isoline of –25 is an intensification of the generally dry conditions east of the Rocky Mountains. This isoline encloses the southern two-thirds of the area of this study and is oriented west-east at latitude 51°N.

For the normally wettest month of the year—June—this area stands out with a marked deficiency of moisture, according to the Thornthwaite system (Figure 13). Within the Dry Belt the following deficiencies of precipitation with respect to potential evapotranspiration occur: Medicine Hat, Alta., 1.8″; Wastina, Alta., 1.5″; Anglia, Sask., 1.7″; Outlook, Sask., 1.9″. Very little or no deficiency during that month occurs in all directions from the Dry Belt; Calgary, Alta., 0.0″; Edmonton, Alta., 0.0″; St. Walburg, Sask., 0.0″; Prince Albert, Sask., 0.0″; Indian Head, Sask., 0.0″; Havre, Mont., 0.4″; Williston, N.D., 0.0″. Much of the Dry Belt thus is characterized by distinctly greater actual deficiencies of precipitation during the wettest month of the year than is true of all areas around it. The magnitude of the June maximum in the precipitation curve typical of the Dry Belt (i.e., Anglia, Sask.) emphasizes its importance to the total moisture picture

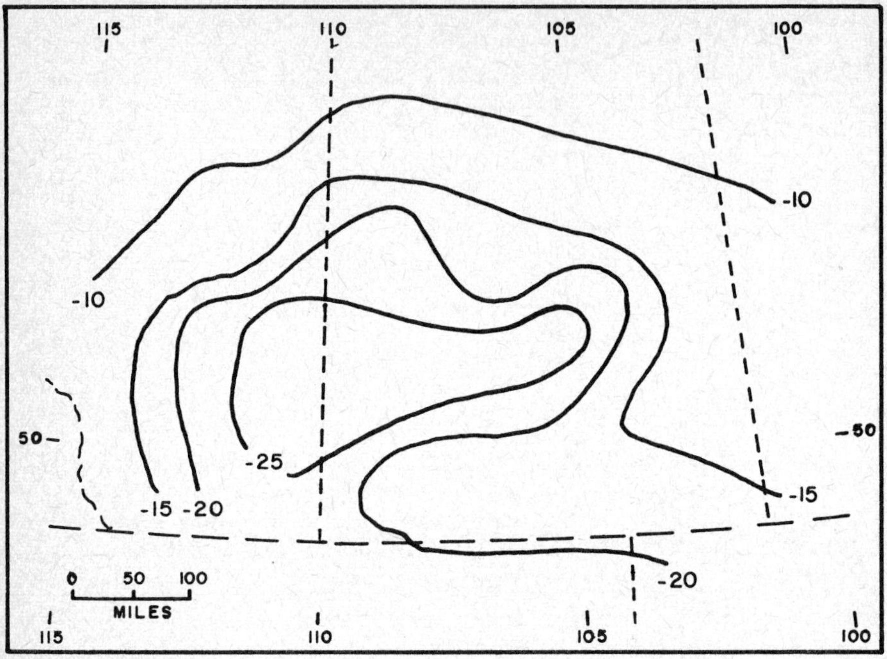

Figure 12. Moisture index according to the new Thornthwaite classification of climate (1948).

here; the Thornthwaite classification reveals that even this strong June maximum is not sufficiently great to overcome normal evaporation.

Almost the same pattern as in the previous instance is produced when the ratio of potential evapotranspiration to annual precipitation is mapped. In a large area of southern Alberta and Saskatchewan, including much of the Dry Belt, the potential evapotranspiration is more than 175 percent of the annual precipitation (Figure 14).

The graphs of potential evapotranspiration and annual precipitation for a number of stations in and about the Dry Belt reveal significant patterns (Figures 15, 16, 17, and 18). The Dry Belt station, Alsask, Sask., has a high water deficiency (9.8 inches), which extends over four months (June-September), and a relatively meager soil moisture recharge (3.0 inches). To the northwest, Edmonton, Alta., has a considerably lower water deficiency (3.5 inches), which also becomes marked somewhat later in summer than to the southeast, and a larger soil moisture recharge (3.7 inches) than Alsask. The pattern to the northeast as evidenced by Prince Albert is very similar in terms of water deficiency and soil moisture recharge (4.8 inches and 3.7 inches, respectively). Estevan, Sask., southeast of the Dry Belt, is characterized by a lower water deficiency (7.1

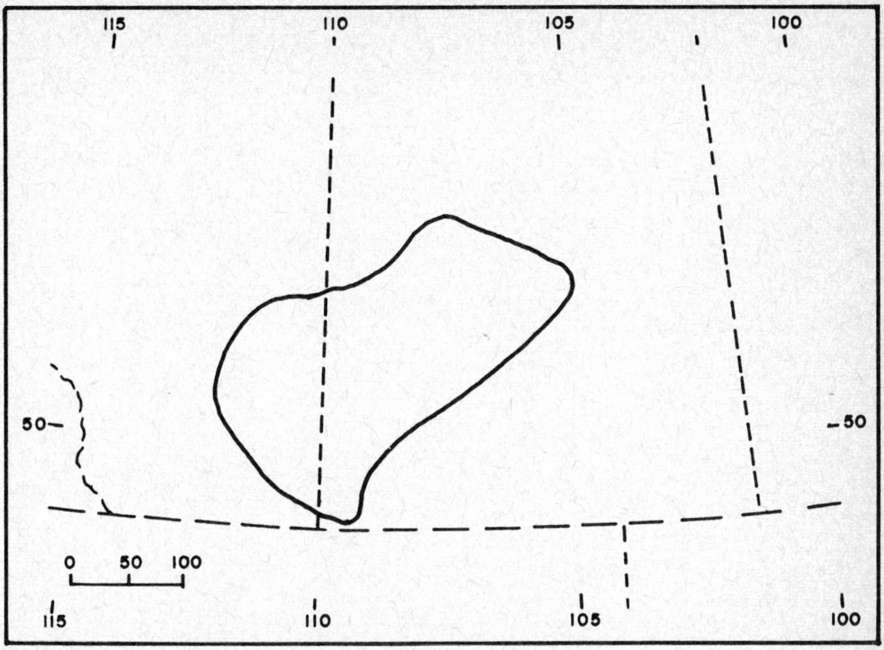

Figure 13. Area with more than 1″ precipitation deficiency in June, according to the new Thornthwaite classification of climate (1948).

inches) and higher soil moisture recharge (3.3 inches) than the Dry Belt station. Moreover, water deficiency does not become apparent until a month later in summer, although it persists somewhat later in autumn.

The new Thornthwaite classification of climate aids significantly in differentiating the Dry Belt from its surroundings in Canada. The moisture isoline of –25 (Figure 12) coincides fairly well with other more direct climatic measures in use to define climatic regions. The Dry Belt has a markedly greater deficiency of moisture per year and a lesser soil moisture recharge than shown by graphs for individual stations outside the Belt. Finally, the wettest month of the year in the Dry Belt, June, has a considerably greater moisture deficit than is true elsewhere.

Distinctive Features of Temperature

During the summers of abnormally dry years (1936, 1945, 1949) the average temperature departure ranges well above the long-time average (Figure 19). According to a recent study the average summer interannual variability of temperature in the vicinity of the Dry Belt is about 2.5°F.;[16]

---

[16] B. W. Currie, *Prairie Provinces and Northwest Territories: Temperature* (Saskatoon: University of Saskatchewan, 1953), Table 7.

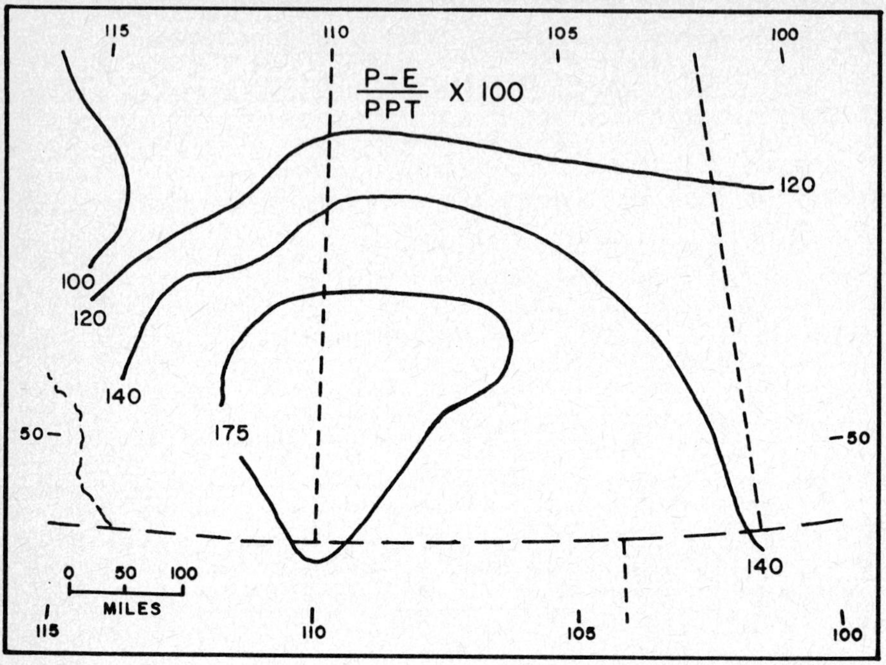

Figure 14. Ratio of potential evapotranspiration to annual precipitation multiplied by 100.

hence temperature departures of +4.0°F. in the Dry Belt during very dry years serve to emphasize the abnormally high evaporating temperatures of these periods. Excessive heat constantly evaporating moisture away from the surface of the earth added to the below-normal return of moisture in the form of precipitation marks the Dry Belt as a region of double misfortune.

Nearly fifty years ago Stupart[17] noted several unusual winter temperature characteristics of the part of Alberta lying east of the foothills of the Rocky Mountains: (1) the area experienced rapid changes of temperature which, in winter, occurred in extremely short periods of time, (2) southern Alberta had milder winters than the land to the east, and (3) there was marked variability in mean winter temperature from year to year. Koeppe[18] and Connor[19] noted much the same picture. Exceptionally rapid changes in temperature in extremely short periods of time[20] along with comparative

17 Stupart, *op. cit.*, p. 298.

18 Clarence Eugene Koeppe, *The Canadian Climate* (Bloomington: McKnight & McKnight, 1931), pp. 117–118.

19 Connor in Köppen and Geiger, *op. cit.*, p. 361.

20 Actually the winter temperature gradient across the Dry Belt (4.2°F/70 miles) is nearly twice as great as the very steep winter gradient in eastern U.S. See Glenn T. Trewartha, *An Introduction to Climate* (3rd ed.; New York, 1954), p. 328.

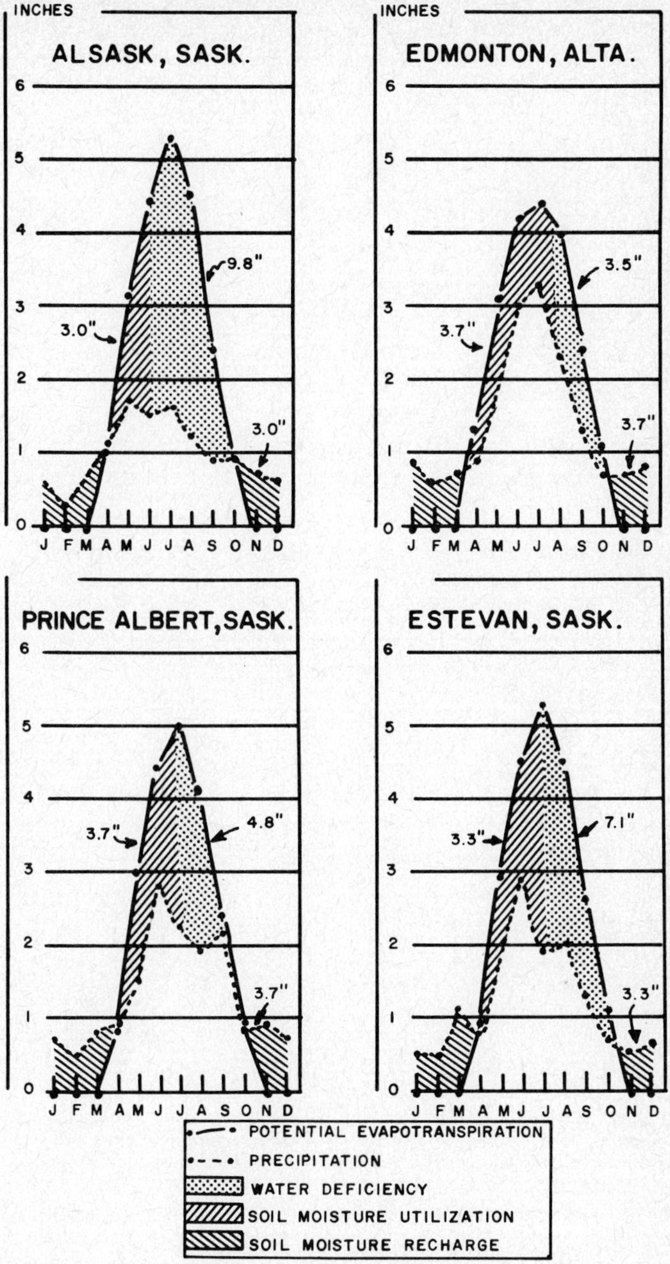

Figures 15, 16, 17, 18. Graphs of potential evapotranspiration and precipitation for Alsask, Sask., Edmonton, Alta., Prince Albert, Sask., and Estevan, Sask.

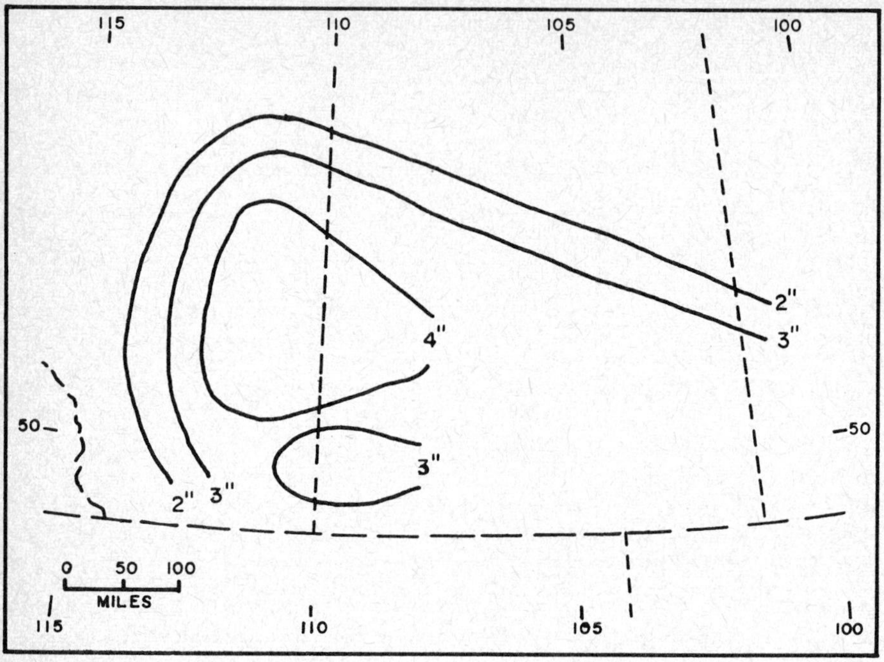

Figure 19. Average positive temperature departure during summers of abnormally dry years – 1936, 1945, 1949.

mildness of winters have led Thomson and Connor[21] to assert that the chinook, or "snow eater," finds its best expression on the North American land mass in southern Alberta.

The interannual variability of mean winter temperature in the Dry Belt (Figure 20) is well marked—averaging over 8.0°F. These statistics indicate a temperature oscillation of three times the average interannual summer variation. Indeed, the pattern of high interannual variability of mean winter temperature comes close to an accurate cartographic representation of frequent chinook activity; for it is the chinook, with its abnormal winter warmth penetrating eastward for a short distance onto the Plains, which causes winter temperatures to exceed the long-time normal. During those winters when cold polar air from northern Canada holds sway westward to the foot of the Rockies, the pendulum swings the other way, and bitterly cold weather dominates the entire Dry Belt as well as much of the vast interior of North America.

---

21 Andrew Thomson and A. J. Connor, *The Climate of Canada* (Toronto: Meteorological Division, Department of Transport, 1949), p. 13.

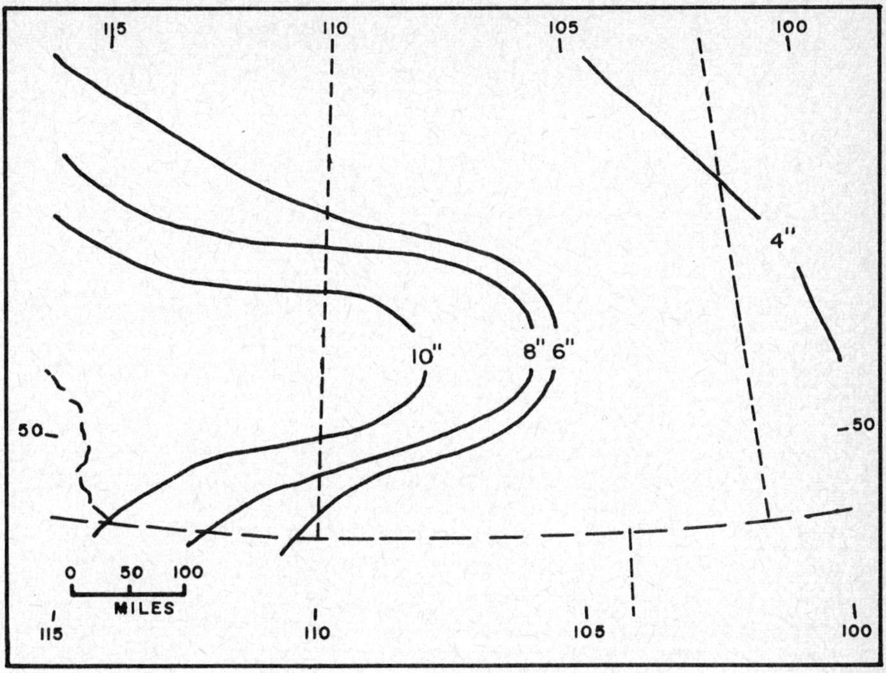

Figure 20. Interannual variability of mean winter temperatures, 1930-1939 incl.

Explanations for the Existence of the Dry Belt Cyclonic and Anticyclonic Frequency

The presence of moist air masses, either at the surface or aloft, is prerequisite to the production of precipitation in a region but does not insure that precipitation. In all cases sufficient lifting of relatively moist air must precede the release of the water content of the air mass. Lifting of air masses can be measured by an enumeration of cyclonic passages. Figure 21, showing the average summer frequency of cyclonic passages near the Dry Belt over the period 1930-49 inclusive, indicates that the Dry Belt is normally a region of infrequent cyclonic action. In northern Alberta lies a heavily traveled track, and to the south, the important North Pacific track. The Dry Belt is frequented by approximately one-half as many cyclonic passages during the summer as areas to the north (compare Figures 21, 7, and 8) and to the immediate south (compare Figures 21, 7, and 9). These comparisons point out the apparent relationship between summer precipitation and frequency of cyclonic passages.

A similar enumeration of frequency of anticyclonic passages (Figure 22) reveals a higher frequency of such occurrences in a northwest-southeast

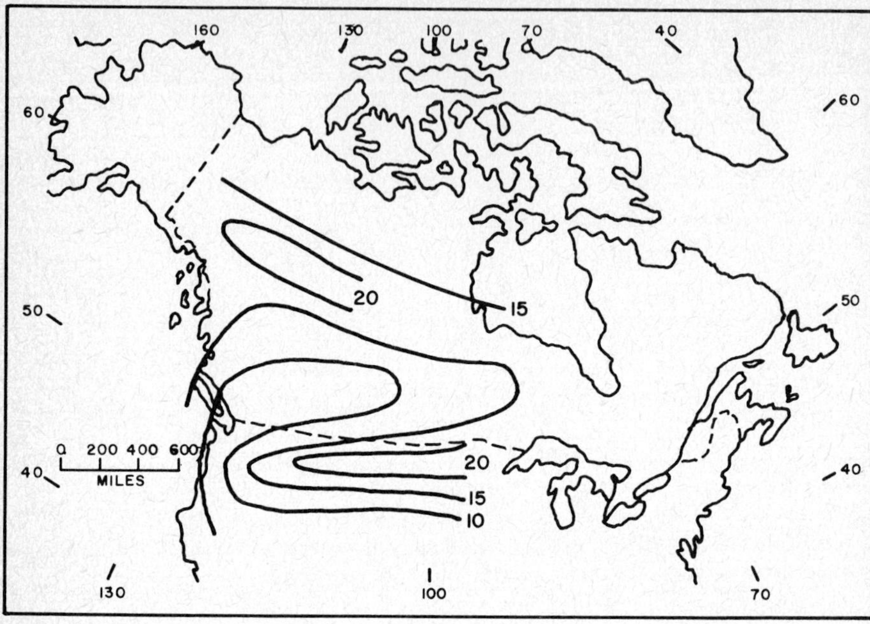

Figure 21. Frequency of cyclonic passages near the Dry Belt, 1930-1949 incl.

line through the Dry Belt. The effect of relatively greater dominance of subsiding air is seen when the relatively dry "wet season" in the Dry Belt is compared with the "wet season" of its surroundings (compare Figure 7 with Figures 8, 9, and 10).

Location Relative To Moist Air Mass Trajectories

In a statistical analysis of surface air mass flow over North America, Brunnschweiler[22] discovered that polar maritime (*mP*) air from the Pacific source region moves onto North America during the summer in a peculiar two-pronged surge. One prong moves southward along the Pacific coast of the United States to southern California. The other crosses the Rocky Mountains and moves southward along the east side of these mountains to the position of the Dry Belt where a sharp gradient occurs in the frequency of *mP* air. This distinctly higher frequency of *mP* air during the summer north of the Dry Belt would help to explain the very sharp maximum of precipitation at 55°N. (Figure 5); conversely, the Dry Belt minimum must, in part, owe its being to the considerably lower frequency of *mP* air here than to the north.

22 Dieter H. Brunnschweiler, "The Geographic Distribution of Air Masses in North America," *Geographische Mitteilungen*, Jahrg. 97 (1952), p. 46.

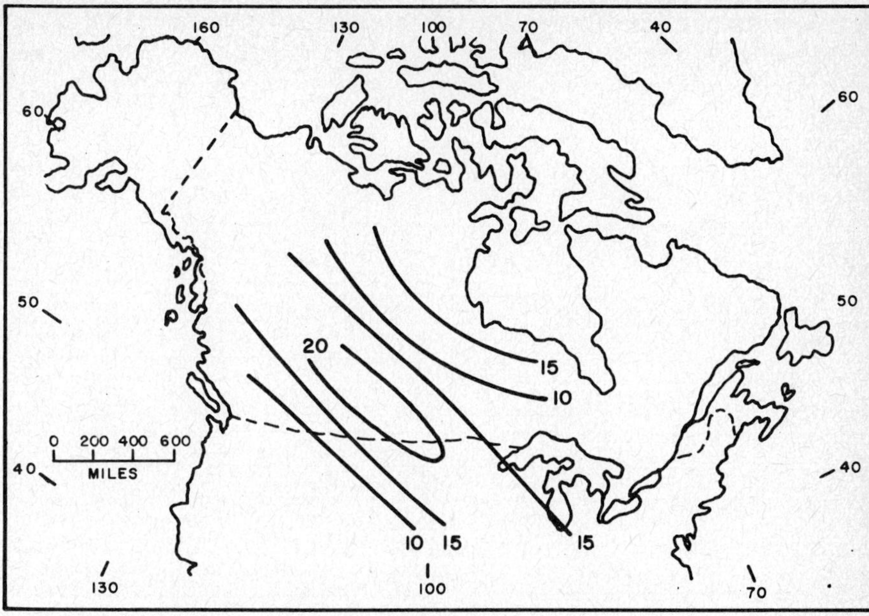

Figure 22. Frequency of anticyclonic passages near the Dry Belt, 1930-1949 incl.

Major Jet Stream Location in North America During July

Figure 23 shows the strength and average position of the jet stream for July,[23] a location near the northern boundary of the United States typical of summer conditions in North America. Numbers indicate geostrophic wind speed in miles per hour at the level of maximum speed.

Rossby indicated a probable relationship between jet streams and precipitation distribution;[24] later, Starrett discovered "when allowance is made for latitudinal variations in the availability of moisture . . . the precipitation tends to be a maximum underneath the jet."[25] However, even in locations where conditions seemed favorable for precipitation, the latter study found 19 percent of the jet streams to be "dry." The Dry Belt's position north of the jet stream means it is in an area of inflow along the left side of the jet. This inflow leads to subsidence aloft, which is probably asso-

---

[23] Jerome Namais and Philip F. Clapp, "Confluence Theory of the High Tropospheric Jet Stream," *Journal of Meteorology*, Vol. 6 (1949), pp. 330–336.

[24] C. G. Rossby *et al.*, "On the General Circulation of the Atmosphere," *Bulletin of the American Meteorological Society*, Vol. 28 (1947), pp. 255–280.

[25] Lloyd G. Starrett, "The Relation of Precipitation Patterns in North America to Certain Types of Jet Streams at the 300-Millibar Level," *Journal of Meteorology*, Vol. 6 (1949), p. 347.

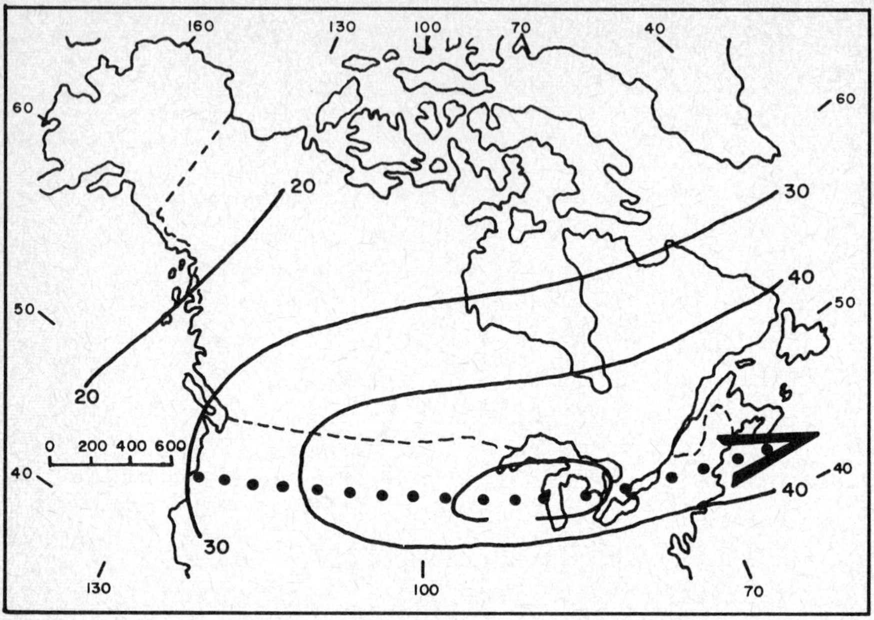

Figure 23. Average position and strength of the jet stream in July; values of geostrophic wind speed in miles per hour at the level of maximum speed.

ciated with the decrease in precipitation during summer north of the jet stream to about lat. 51°N. (Figure 5). That there is no peak in the precipitation profile (Figure 5) beneath the jet seems reasonable on the basis of the position of the jet with respect to the major source of moisture, viz., the Atlantic Ocean or Gulf of Mexico and the average trajectory of the tropical maritime air masses over North America.[26]

Although reaching no maximum beneath the jet, precipitation does increase south of the Dry Belt in the direction of the mean summer position of the jet stream (Figure 5). Moreover, the area to the south of the Dry Belt lies in the path of the important North Pacific low track (Figure 21). Riehl's conclusions clearly apply to this moister area to the south: "When jet stream . . . and low tropospheric disturbance coincide in a favorable sense, ensuing cyclone developments will attain the greatest intensity."[27]

Upper Air Pressure Distribution

Surface pressure charts indicate a probable relationship between low frequency of cyclones and high frequency of anticyclones with the low

---

26 For a summary report on the influence of the jet on surface climates see H. Riehl *et al., The Jet Stream,* Meteorological Monographs, Vol. 2, No. 7 (1954), pp. 34–37.
27 Herbert Riehl, "Jet Stream in Upper Troposphere and Cyclone Formation," *Transactions,* American Geophysical Union, Vol. 29 (1948), p. 186.

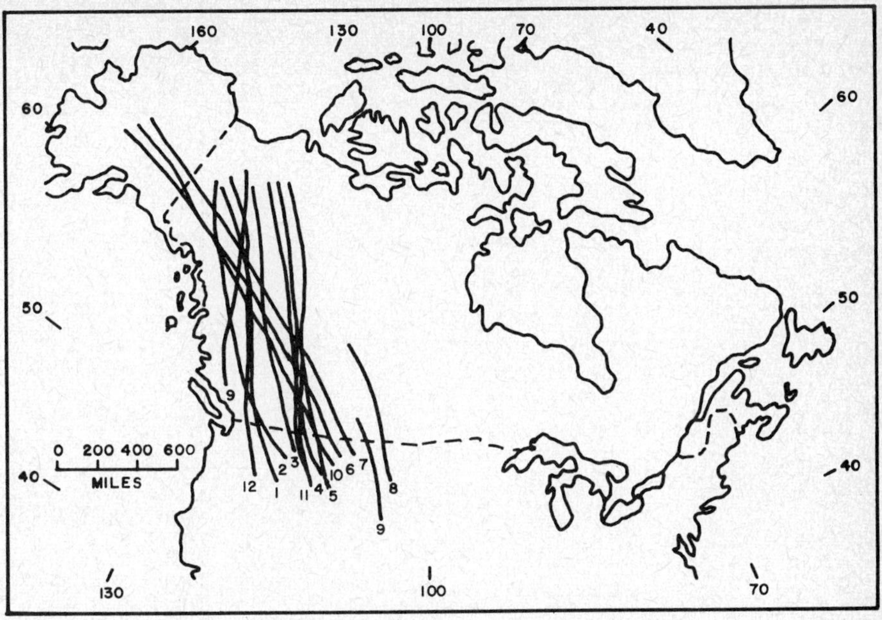

Figure 24. Axis of western anticyclonic ridge at 700 mb height.

precipitation of the Dry Belt. Upper air distribution reveals still further evidence of the subsidence and divergence of air over the Dry Belt. The anticyclonic ridge plotted at a height of 700 mb. (Figure 24) is located over the Dry Belt from June through September, a period of considerably less precipitation in the Dry Belt than in its surroundings. Charts of air pressure distribution at a height of 300 mb. show a ridge over the Dry Belt (Figure 25) during July, indicating subsidence and divergence of considerable depth in the atmosphere.

Westerly Air Flow Over The Rocky Mountains

The effect of a large highland barrier on westerly air flow is a subject of considerable interest. Bolin notes that "the only mountains which may be of real importance in generating planetary waves in the westerlies are the Rocky Mountains and the mountains in the interior of Asia."[28] In a study which bears significance to the Dry Belt, Wexler[29] discusses perturbations in the westerlies which result in the formation of two anticyclonic cells leeward of the Rocky Mountains. The first of these anticyclonic cells, centered in the southern Great Plains, is composed of vast

28 B. Bolin, "On the Influence of the Earth's Orography on the General Character of the Westerlies," *Tellus*, Vol. 3 (1950), p. 192.

29 Harry Wexler, *Some Aspects of Dynamic Anticyclogenesis*, University of Chicago Publications in Meteorology, No. 8 (1943), p. 26.

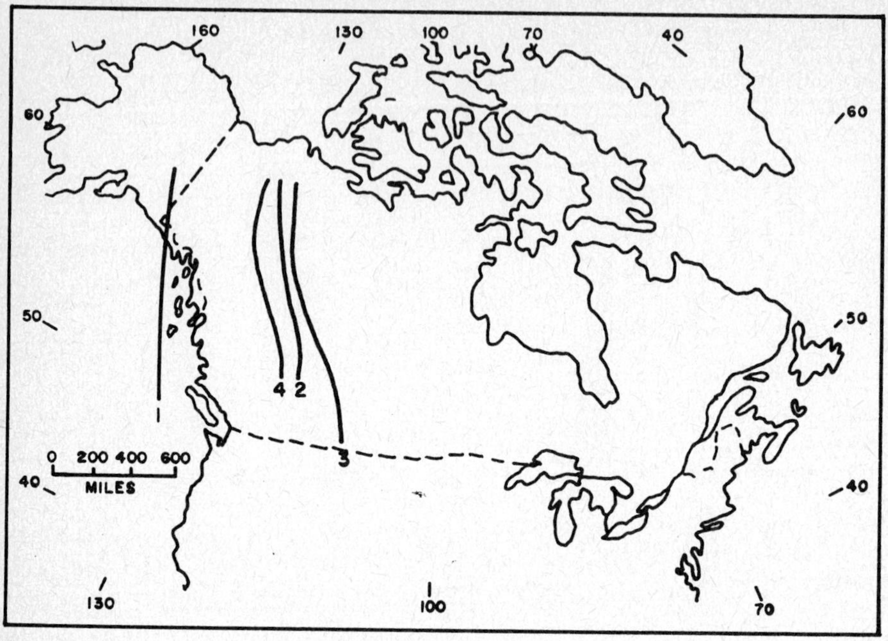

Figure 25. Axis of western anticyclonic ridge at 300 mb height.

moist and dry currents winding about the cell and affecting the surface weather. A normal summer isentropic chart, originally prepared by Namias,[30] is used by Wexler to illustrate certain relationships between the anticyclonic development and precipitation patterns. This chart indicates a dry current of air directly over the Dry Belt, moving eastward along the northern edge of the large anticyclonic cell. Thus the deformation which westerly air flow undergoes as a result of being forced over the highland barrier of western North America results in the development of an anticyclonic flow curvature attended by a dry current of air over the Dry Belt.

Borchert has shown by means of streamlines of resultant mean air flow at the gradient level that the dry continental air flow—or flow directly from the mountains as opposed to flow from the tropical Atlantic—is confined in July to the area bounded by the Alberta storm track, the 100th meridian, and the United States border.[31] Thus the Dry Belt has stronger mean westerlies than other parts of the Great Plains in summer; these westerlies diverging from a northerly flow along the Pacific coast undergo a chinook effect east of the highland barrier during the *summer*

[30] Jerome Namias, *An Introduction to the Study of Air Mass and Isentropic Analysis* (American Meteorological Society, 1940), p. 158.

[31] John R. Borchert, "The Climate of the Central North American Grassland," *Annals*, Association of American Geographers, Vol. 40 (1950), pp. 23–24.

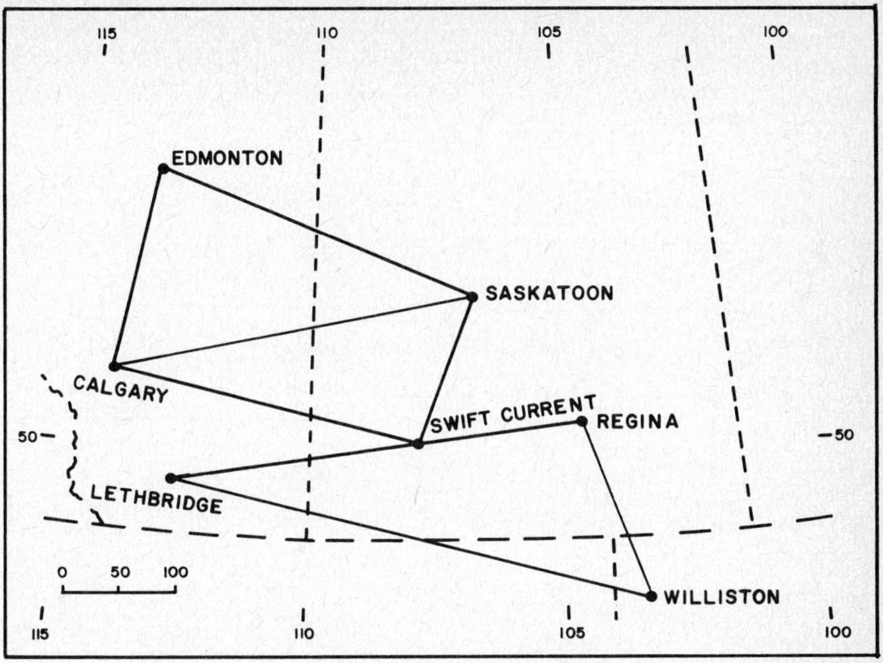

Figure 26. Location of three triangles used in computation of horizontal divergence. (see *Table I.*)

as well as during the winter. In a later study Borchert has demonstrated the dominance of Dry Continental surface wind flow in this area.[32]

Horizontal Divergence

Wind observations lend themselves to another technique by which the Dry Belt may be differentiated from its surroundings in order to explain its unique regional character. The triangular volume method for determining horizontal divergence allows direct computation of horizontal divergence from wind observations.[33] Resultant wind flow for the period April to September was computed, and the partial horizontal divergence for three sets of triangulation stations was determined. The three triangles used in this analysis are plotted on Figure 26. The results are given in the table below, in which negative values correspond to convergence and positive values to divergence.

These calculations clearly indicate that the Dry Belt stands apart from its surroundings in terms of horizontal divergence. The Northern Triangle

32 John R. Borchert, "Regional Differences in the World Atmospheric Circulation," *Annals*, Association of American Geographers, Vol. 43 (1953), p. 19.
33 John C. Bellamy, "Objective Calculations of Divergence, Vertical Velocity and Vorticity," *Bulletin of the American Meteorological Society*, Vol. 30 (1949), pp. 45–49.

has a horizontal divergence value of only +0.1, while the Southern Triangle's horizontal divergence value is 0.0. The Dry Belt, with a value of +1.5, is thus marked in a local sense as a region of relatively strong horizontal divergence.

TABLE I. HORIZONTAL DIVERGENCE

Stations	*Resultant wind*		*Horizontal divergence*	
	Direction	*Velocity*	*Partial*	*Total*
Northern Triangle				
Calgary	303°	10.3 mph	−0.3	
Saskatoon	192	11.6	+0.3	
Edmonton	258	9.4	+0.1	
				+0.1
Dry Belt Triangle				
Calgary	303°	10.3 mph	−0.2	
Saskatoon	192	11.6	+0.8	
Swift Current	221	14.3	+0.9	
				+1.5
Southern Triangle				
Lethbridge	249°	13.4 mph	+0.1	
Williston	135	8.6	+0.3	
Regina	60	13.0	−0.4	
				0.0

Studies of divergence aloft in so small an area as the Dry Belt are severely limited by insufficient coverage in both number of stations and length of record. Namias and Clapp have, however, mapped the distribution of normal fields of divergence and convergence over Anglo-America at 10,000 feet during February, May, August, and November[34] based on station coverage of a comparatively sparse nature. The location of the Dry Belt on these maps is definitely associated with regions of divergence during August and with zero value during the other months studied. Hence, it is clear that the Dry Belt is definitely associated with regions of divergency at the surface and aloft; in the former case, the divergence is quite distinct in a regional sense, differentiating the Dry Belt from its surroundings and clearly helping to explain its regional climatic character.

Summary of Causes For The Existence of The Dry Belt

The Canadian Dry Belt forms a distinct climatic region on the basis of unifying characteristics of precipitation and temperature. Its basic climatic characteristic, deficiency of moisture, is the result of a number of

[34] Jerome Namias and Philip C. Clapp, "Normal Fields of Convergence and Divergence at the 10,000-Foot Level," *Journal of Meteorology*, Vol. 3 (1946), pp. 20–21.

factors which combine to prevent the area from obtaining as much precipitation as its surroundings.

The Dry Belt is frequented by considerably fewer cyclonic passages and more anticyclones than areas to the north and south; further, it lies south of one of the major Pacific air mass trajectories during the summer. The position of the jet stream during the summer, nearly 600 miles south of the Dry Belt in a location very near the axis of maximum frontal activity, places the Dry Belt in a vast feeding area of subsidence and divergence along the left side of the jet stream. Investigation of pressure distribution aloft—at both the 700 mb. and 300 mb. levels—indicates important areas of subsidence and divergence over the Dry Belt during the warmer part of the year when this area is markedly drier than its surroundings.

Studies of air flows across the highlands of western North America reveal that the Dry Belt lies along the northern margin of the first of two standing waves formed in the westerlies. This location along the northern edge of a large anticyclonic cell is in an area dominated at the surface and aloft by a dry current of air which winds eastward around the cell.

Finally, and most conclusively, the Dry Belt is clearly differentiated from areas to the north and south in terms of horizontal divergence. An area to the north has only a minor positive value, while to the south the value is zero; but the Dry Belt stands out with a value of +1.5. Generalized studies at an elevation of 10,000 feet indicate that divergence exists to that height over the Dry Belt part of the year.

Surface Data Sources

Boughner, C. C., and Thomas, M. K. *Climatic Summaries for Selected Meteorological Stations in Canada, Newfoundland and Labrador. Vol. II, Humidity, Wind Speed and Direction*. Toronto, 1948.

Climatic Summaries for Selected Meteorological Stations in the Dominion of Canada. Vol. I, Average Values of Mean and Extreme Temperature, Mean and Extreme Humidity, Sunshine, and Precipitation, Toronto, n.d.

Currie, B. W. *Prairie Provinces and Northwest Territories: Precipitation, Snowfall, and Temperature*. Saskatoon, 1953.

General Summaries of Hourly Weather Observations in Canada and Newfoundland, 1949. Pt. I, British Columbia, Yukon Territory, Alberta, Saskatchewan, Manitoba. Toronto, 1949.

Joint Meteorological Committee. *Daily Synoptic Series, Historical Weather Maps, Northern Hemisphere, Sea Level, 1899–1939*. Washington, n.d.

Monthly Record; Meteorological Observations in Canada. Toronto, 1930–date.

Monthly Weather Review.

Pearce, D. C., and Gold, L. W. *The Canadian Snow Survey, 1947–1950*. Ottawa, 1951.

Schumann, T. E. W., and van Rooy, M. P. "Frequency of Fronts in the Northern Hemisphere," *Archiv für Meteorologie, Geophysik und Bioklimatologie*, Vol. IV (1951), pp. 87–97.

Thomas, M. K. *Climatological Atlas of Canada*. Ottawa, 1953.

Upper Air Data Sources

Mean Monthly Maps of 300-, 200-, 100-, 50-, and 25-mb. Surfaces Over North America. Washington, 1953.

Namias, Jerome, and Clapp, Philip C. "Normal Fields of Convergence of Divergence at the 10,000-Foot Level," *Journal of Meteorology*, Vol. 3 (March, 1946), pp. 14–22.

Normal Weather Charts for the Northern Hemisphere, Sea Level Pressure, 700-mb. Height and Temperature, Thickness (700–1000 mb.), 500-mb. Height. Washington, 1952.

Summaries of Radiosonde Observations in Southern Canada, Stations South of Latitude 55°N., June 1941–December 1949. Toronto, n.d.

4

Agroclimatological Relationships: A Review

W. J. Maunder

Mankind is dependent to a very large extent on the products of the land. These products are, however, closely dependent on the prevailing weather and climatic conditions, and it is clear that climatic variations are an important if not the most important cause of the variations which occur from the over-all trend in per unit production.

The rapid increase in productivity in the United States during the past 35 years has, however, raised the question as to the relative contributions of technology and weather. Bean, in discussing this question asks:[1] "How much of this lift in per acre yields is due to the doings of man,—with what he does with his land, his seeds, his fertilizer, his pest and disease control, his row spacings, and how much may be the result of a fundamental change in weather?"

Commenting further on this question, Thompson[2] says that there has been a growing tendency to believe that technology has reduced the influence of weather on grain production so that we no longer need to fear

---

1 L. H. Bean, "Crops, Weather and the Agricultural Revolution," *Amer. Statist.*, Vol. 21, no. 3 (1967), pp. 10-14.
2 L. M. Thompson, "Weather and our Food Supply," *Center for Agric. and Econ. Development*, Rept. 20 (1964), p. i.

Reprinted in edited form from *Canadian Geographer,* Vol. XII, no. 2 (1968), pp. 73-84.

shortages due to unfavourable weather. Thompson reminds us, however, that "there is increasing evidence . . . that a period of favourable weather interacted with technology to produce our recent high yields, and that perhaps half of the increase in yield per acre since 1950 has been due to a change to more favourable weather for grain crops."[3]

Irrespective of the importance of the over-all trends in the seasonal weather, however, there remains the day to day, week to week, and season to season weather variations. And, it is these variations—the possibility of more accurate long-range weather forecasting and the actual modification of the weather notwithstanding—that are of fundamental importance.

The importance of agroclimatological studies is therefore not lessened by advances in technology. Indeed, increasing pressure of populations on world food supplies suggests that such studies will be of even more value in the future than they are today.

The Setting

The general problem of agroclimatological relationships is complex both in its biology and in its physics, Penman stating:[4]

> Physically, at least five divisions are possible. After a zero order crop-weather interaction, to be defined when it is discussed later, there is the obvious first-order effect in which we know that crop growth depends on weather, and that seasonal and secular changes in yield are caused by weather changes. To be able to demonstrate the dependence may require the absence, avoidance or neutralization of second-, third-, and fourth-order effects, any of which might be dominant. Second-order effects are those arising from pests and diseases, nearly all of which have strong associations with weather in their incidence, in their development, or in their spread. . . . Third-order effects, too, have a biological origin—man. Soil management and crop husbandry frequently have to go on whatever the weather, and though soils and crops will tolerate a lot of ill-treatment, their resilience is not unlimited. . . . The fourth-order effects may be peculiar to a species or even to a variety of plant. They are the effects that involve 'trigger action' or the existence of threshold values; to which for completeness, might be added the effects of meteorological abnormality, sometimes manifest in frost, rain, hail or wind damage.

Agroclimatological relationships can, therefore, be extremely complicated, and as Hogg has said "the answers are not easily come by. . . ."[5] However, Frisby[6] says that Klages[7] voices generally accepted agricultural opinion when he deals unhesitatingly with climatic before physiographic,

---

3 *Ibid.*

4 H. L. Penman, "Weather and Crops," *Quart. Jour. Roy. Meteorol. Soc.*, Vol. 88 (1962), pp. 209-19.

5 W. H. Hogg, "Meteorology and Agriculture," *Weather*, Vol. 19 (1964), pp. 34-43.

6 E. M. Frisby, "Weather-Crop Relationships: Forecasting Spring Wheat Yield in the Northern Great Plains of the U.S.", *Trans. Inst. British Geogrs.*, Vol. 17 (1951), pp. 77-96.

7 K. H. W. Klages, *Ecological Crop Geography* (New York, 1942).

edaphic, or biotic factors; and Frisby further comments that if farming practices, seed, and soil for a given locality are constant, "climate must be the variable resulting in year-to-year fluctuations in crop yield."[8]

Yet, according to Frisby,[9] despite his recognition of the importance of climate, the agriculturalist concerns himself principally with soils, seeds, pests, farming practices, and the intricacies of physiological development of the plant, and shows less willingness even than the climatologist to become involved in questions of climate-agriculture relationships. Frisby continues:[10] "It would seem, therefore, that here there is scope for the imagination of the geographer. His role must be to appreciate and bridge the gaps between climatologist, agriculturist, and statistician."

Similar thoughts are echoed by Reeds who suggested that ". . . clearly . . . a more accurate assessment of world agricultural resources is urgently needed in order that reasonable potentials may be predicted more accurately."[11] These resources include, of course, those of the weather and the climate and its variations. The evaluation of this resource may be made in several ways, some of these ways being discussed in this paper. These evaluations are, however, generally based on the more conventional climatological data and it is appropriate at this point to mention the growing importance of heat and moisture balance studies, as ably reviewed by Tweedie,[12] and their undoubted significance in agroclimatological studies particularly as more and better measurements of this aspect of the weather and climate become available.

The importance of climatic variations on agricultural production is stressed by many observers and Sanderson, who made a detailed study of the methods of crop forecasting,[13] comments that among the various branches of economic activity, agriculture is the only one which seems to be destined forever to be subject to wide and irregular fluctuations of output, the year-to-year variations of agricultural output being largely determined by physical factors. Sanderson further comments:[14]

> We need concern ourselves only with those environmental factors which are variable from year to year. Among these, weather factors are by far the most important. Diseases and insect pests are next in importance, but their year-to-year variations in turn are frequently dependent on meteorological conditions. The same is true of competition by weeds. Economic factors must also be mentioned. Year-to-year variations in the

---

8 Frisby, *op. cit.*
9 *Ibid.*
10 *Ibid.*
11 L. G. Reeds, "Agricultural Geography: Progress and Prospects," *Can. Geogr.*, Vol. 7 (1964), pp. 51-63.
12 A. D. Tweedie, "Challenges in Climatology," *Aust. Jour. Sci.*, Vol. 29 (1967), pp. 273-77.
13 E. H. Sanderson, *Methods of Crop Forecasting* (Cambridge, Mass., 1954).
14 *Ibid.*, p. 181.

acreage planted to a given crop in response to fluctuations in price, are, as a rule, confined to land which is marginal from the point of view of that crop, and thereby affect the average yield. Economic conditions such as the price received in the previous year may also be responsible for variations in the use of input factors, such as fertilizers and sprays, for instance.

Further comments on agroclimatological relationships are made by Smith, who in discussing climatic variations says that the "only constant feature of any climate is its tendency to change."[15] Smith continues:[16] "The weather has an infinite variety of items in its repertoire and although the range may be restricted, the variations within the range are rarely repeated, so that the climate, which is the digest of weather, is always in a state of change."

The *significance* of climatic variations is however difficult to determine. Smith, for example, in looking at the general question of climatic change says that if we are asked the question "Are climatic changes significant?" the answer depends on how we define the word "significant" and as Smith so aptly expresses it:[17] ". . . if . . . we examine recent climatic variations through the eyes of a statistician, we may be forced to conclude that none of them are 'significant'. . . . This may be acceptable to the pure scientist, but it could tend to infuriate the man who has experienced what *appears to him* [italics mine] to be a radical change in the circumstances in which he has to live and work." Similar comments could also be made with regard to the significance of climatological variations, and their association with agricultural production. Some relationships may not be statistically significant, but they may nevertheless be very significant to the plant and the farmer.

Although most of the references in this paper are to American or British Commonwealth authors and experience, it must be emphasized that many other important studies have been made. In this regard it is useful to note (especially if compared with the comments of Tweedie discussed above) that Sinelshikov, in discussing agroclimatological trends in the USSR indicates that:[18] "The great and economically important agroclimatic investigations carried out by Soviet agroclimatologists require a more comprehensive and profound study of the theory of the origin of agroclimatic conditions, on the basis of the radiation, heat, and water balance of agricultural fields."

---

15 L. P. Smith, "Measuring the Effects of Climatic Changes," *New Scientist,* Vol. 12 (1961), pp. 608-11.
16 *Ibid.*
17 *Ibid.*
18 V. V. Sinelshikov, "Agroclimatological Trends in the USSR, Problems, Methods and Scientific and Practical Results," *Agric. Meteorol.*, Vol. 2 (1965), pp. 73-77.

Approaches

There are, broadly speaking, three ways of approaching the issues and establishing whether climate-agriculture relationships are in fact significant. The first is the study of the fundamentals of plant-climate relationships, namely, the radiation and moisture balance for various crops in various climatic environments. Monteith[19] indicates, in fact, that the analysis of the radiation climate is a central problem of agricultural meteorology since the rates of photosynthesis depend on the receipt of visible light and the rates of transpiration depend on the net exchange of radiation by a crop canopy. These views are further explored by Monteith,[20] who discusses the importance of the radiant energy absorbed by a leaf and stored chemically in photosynthesis, the estimation of the potential or maximum photosynthesis of a crop canopy, and the potential rate of transpiration.

The second method of determining climate-agriculture relationships is by studying agricultural data and climatic data for a number of places within a given area, for as long a period as consistent records of both agriculture and climate allow, and deducing agroclimatological relationships from analyses of the data. Despite the many useful agroclimatological relationships obtained using such methods, there are, as pointed out in the following section, many associated problems. In particular, it is very difficult to isolate the contribution made by any one of the various components of the climate such as temperature, day length, light intensity, and precipitation. However, controlled environments, the third method, do permit in many cases an analysis of the importance of the various climatic components, because the individual climatic factors can be varied one at a time. For this method, expensive climate control chambers are necessary, and in many countries these control chambers have only recently been established. Control chambers theoretically can give the optimal and marginal conditions for the growth stages of a plant, nevertheless McWilliam sounds a note of warning about the extrapolation of results derived in the controlled environment to field situations. He states:[21] "It is quite obvious that it is impossible to reproduce natural environments under controlled conditions. For this reason it is not possible to predict whole plant performance."

Statistical Methods

In the past, agroclimatological studies were greatly handicapped by the lack of theoretical insight into the processes of plant growth and the

19 J. L. Monteith, "Radiation and Crops," *Exp. Agric. Rev.*, Vol. 1 (1965), pp. 241-51.
20 J. L. Monteith, "The Photosynthesis and Transpiration of Crops", *Exp. Agric. Rev.*, Vol. 2 (1966), pp. 1-4.
21 J. R. McWilliam, "The Role of Controlled Environment in Plant Improvement," *Aust. Jour. Sci.*, Vol. 28 (1966), pp. 403-407.

responses of plants to weather influences, the inadequacy of the available data, the lack of efficient statistical methods of analysis, and the lack of co-ordination of the efforts made by experts working on the various phases of the problem. However, with the recognition of the need for close co-operation between agronomists, plant ecologists, plant physiologists, meteorologists, climatologists, economists, statisticians, and geographers, recent progress in this field has been encouraging. Further, the advent of electronic computers has made possible the assessment of the importance of a large number of climatic variables in their effect on agricultural production, a task which before was impracticably laborious.

One of the first to apply correlation methods to the problem of the effects of the meteorological factors current during growth was Hooker,[22] who systematically correlated rainfall and accumulated temperature for overlapping eight-weekly periods throughout the year with the yields of a number of farm crops in eastern England. Fisher[23] modified this method, and computed partial regressions of the climate/wheat yield for each period of the year. Many others have also applied statistical methods to agroclimatological relationships.

The application of correlation analysis to geographical areas, rather than to individual experimental stations, however, was not attempted until fairly recently, and one of the first papers in this field was by Rose[24] who assessed climate/corn yield correlations for several counties in the mid-west of the United States. A further paper along similar lines was given by Zacks[25] who discussed the effect of climatic factors on oat production in Southern Ontario, while Frisby[26] assessed the weather crop relationships for wheat in the northern Great Plains of the United States.

In recent years, there have been many more papers published on agroclimatological analyses applied to geographical areas, particularly in the main crop areas of the United States. In particular, multiple regression techniques have been used extensively and Thompson, commenting with regard to analyses of the evaluation of weather factors in the production of grain sorghums in Texas, Nebraska, Kansas, Oklahoma, and Missouri, states that "results from five states analysed simultaneously provide a degree of replication which is desirable and practical in this instance."[27]

22 R. H. Hooker, "The Weather and Crops in Eastern England: 1885-1921," *Quart. Jour. Roy. Meteorol. Soc.*, Vol. 48 (1922), pp. 115-38.

23 R. A. Fisher, "The Influence of Rainfall on the Yield of Wheat at Rothamsted," *Phil. Trans. Roy. Soc. London*, B213 (1925), pp. 89-142.

24 J. K. Rose, "Corn Yield and Climate in the Corn Belt," *Geogr. Rev.*, Vol. 26 (1936), pp. 88-102.

25 M. B. Zacks, "Oats and Climate in Southern Ontario," *Can. Jour. Res.* C 23 (1945), pp. 45-75.

26 Frisby, *op. cit.*

27 L. M. Thompson, "Evaluation of Weather Factors in the Production of Grain Sorghums," *Agronomy Jour.*, Vol. 55 (1963), pp. 182-85.

Problems

There are many difficulties facing the research worker in agroclimatological studies, particularly when statistical techniques are used. This has long been appreciation, Kincer and Mattice stating:[28]

> It is . . . well known that there are critical periods of growth during which certain weather influences are more marked than during other times. These critical periods, in some crops at least, are of comparatively short duration, and consequently, it is necessary for best result to use weather variants based on similar short intervals of time, so that their greater importantance may be reflected in the final result. In most weather and crop correlations the month is used as the basic unit of time, principally by reason of the fact that weather data are usually compiled and published in this way. It is preferable, however, that shorter intervals of time be used in most cases.

The limitations of statistical correlations in studying the influence of weather on crops arise to a considerable extent, because of the large number of weather factors which apparently influence the yield. In addition, the varying importance of different periods of growth necessitates the use of comparatively short time intervals, and Sanderson comments:[29] "Plants pay little attention to the calendar; they germinate, blossom, ripen their seeds according to the seasons not according to the calendar. The progress of the seasons is different from year to year and from locality to locality, so that seeds germinate, flowers bloom and fruits ripen on different dates in different places and in different years at the same place."

The answer to this problem is, according to Sanderson, to select as time intervals stages of equal development of the plant, such as (for wheat): (a) formation of seed, (b) from germination up to formation of first pair of leaves, (c) from formation of first pair of leaves to heading, (d) from heading to flowering, (e) from end of flowering to milky stage, and (f) from milky stage to maturity.[30] However, in many areas the comparative lack of phenological records, together with the paucity of soil moisture and evaporation data, represents a most serious obstacle to further progress in weather-crop research.

This problem is partly overcome by Odell who used two approaches in an investigation of corn and soybean yields in Illinois.[31] The first was to study the relationships within fixed calendar periods each year (this being the usual method and the one followed, by necessity, in many analyses); the second was to study a period which consisted of a specific number of days before and after corn tasseling so that the dates varied from year to year. Odell comments that by using a variable calendar

---

28 J. B. Kincer and W. A. Mattice, "Statistical Correlations of Weather Influence on Crop Yields," *Monthly Weather Rev.*, Vol. 56 (1928), pp. 53-57.

29 Sanderson, *op. cit.*, p. 196.

30 *Ibid.*

31 R. T. Odell, "Effects of Weather on Corn and Soybean Yields," *Illinois Res.*, Univ. of Illinois Agric. Exp. Station (Fall, 1959), pp. 3-4.

period, tied to the date of tasseling each year, we are able to pinpoint more exactly the effects of weather conditions.[32] In addition, Sanderson[33] and others comment that it would also be desirable that the data be grouped by climatically and agriculturally homogeneous regions, rather than by political units.

It can therefore be seen that time units smaller than the month, and regions smaller (or at least more homogeneous) than political or administrative units such as counties, are highly desirable for agroclimatological studies. However, both these ideals may not be met by the available range of data; and Malcolm, in his ecological study of the barley crop in New Zealand, noted that the county is the smallest unit for which reliable and complete crop statistics are available. This, he says, is a serious limitation, especially in large counties where climatic and soil variations may cause yield differences which cannot be determined from district to district within the county.[34]

Equally serious is the small amount of available published climatic data relating to smaller units of time than the month. In most countries climatic data are available on a monthly basis, and although daily data are available in the original records or occasionally in data processing banks, compilation of, say weekly data would in many cases be difficult. Consequently, many agroclimatological analyses are more or less restricted to taking the month as the smallest unit of time. An even more serious problem is that monthly climatic data may not necessarily give a "true" indication of the climate for that particular month. For example, in two consecutive months average rainfalls may be recorded but one month may be completely rainless for (say) the last twenty-five days of the month, while the next month may be completely rainless for the first twenty-five days. There would, therefore, be a period of fifty days without rain, even though the two months had average rainfalls. Such inadequacies are of course well recognized but they are not always taken into account by enquirers, especially when looking at past climatic records.

Complexities

The complexities of the agroclimatological relationship are further emphasized by some recent studies. The advent of electronic computers, for example, enabled Buck[35] to re-examine the results of the Broadbalk ex-

---

32 *Ibid.*, p. 3.
33 Sanderson, *op. cit.*
34 J. P. Malcolm, "An Ecological Study of the Barley Crop in New Zealand" (unpub. M. Agric. Sci. thesis, Univ. of New Zealand, 1947), p. 13.
35 S. F. Buck, "The Use of Rainfall, Temperature, and Actual Transpiration in Some Crop Weather Investigations," *Jour. Agric. Sci.*, Vol. 57 (1961), pp. 355-65.

periment.[36] He calculated multiple regression of yield on seventeen variates —five representing the time trend of yield, together with six rainfall and six temperature variates—for six plots in sixty-seven years, and also repeated Tippett's analysis of sunshine effects.[37] The only significant partial correlation, however, was on total annual rainfall, Watson concluding accordingly that "after eighty years of intermittent but intensive study of a set of data that appears uniquely suited for the purpose, all that has been established with statistical certainty about the dependence of the wheat yield of Broadbalk on weather is that it decreases with increase in annual rainfall above the average."[38] Watson[39] also says that similar studies on long-continued experiments on other crops at Rothamsted have been no more successful. However, he indicates that some studies such as that by Cornish[40] have been more formative, and indeed one could cite many similarly formative agroclimatological analyses.[41] Nevertheless, it is true that many correlation and regression analyses have proved to be unrewarding, some of the reasons for this being given by Watson:[42]

> Past experience . . . does not encourage us to expect that knowledge of how yield depends on weather can come from measurements of yields in naturally varying environments. The fundamental defect of this approach is that the dependence of yield on climatic factors is usually far too complex to be described adequately by linear regressions on a few gross measures of climatic variation, except perhaps when one factor, most likely to be rainfall or lack of it, dominates over all others. At best, any correlations so established are empirical, difficult to interpret reliably in terms of known effects of climatic factors on plant growth, and do not necessarly describe a direct influence of weather on the plants.

A similar view is expressed by Chang[43] who says that the customary

---

36 This experiment relates to the Broadbalk wheat experiment at Rothamsted, England, which was started by Lawes and Gilbert in 1843. In 1880 they published details of their results, and later Fisher (*op. cit.*) made a more comprehensive study. Fairly standardized conditions have prevailed at Rothamsted since 1852 and, hence a long period of agro-climate data is available.

37 L. H. C. Tippett, "On the Effect of Sunshine on Wheat Yield at Rothamsted," *Jour. Agric. Sci.,* Vol. 16 (1926), pp. 159-65.

38 D. J. Watson, "Climate, Weather, and Plant Yield," in L. T. Evans, ed., *Environmental Controls of Plant Growth* (New York, 1963), pp. 337-49.

39 *Ibid.*

40 E. A. Cornish, "The Influence of Rainfall on the Yield of Wheat in South Australia," *Aust. Jour. Scient. Res.*, B3 (1950), pp. 178-218.

41 See, for example, G. C. Doephel, "Regression Analysis of Fleece Weight on Rainfall," *Quart. Rev. Agric. Econ.*, Vol. 12 (1959), pp. 57-60; W. C. Johnson, "A Mathematical Procedure for Evaluating Relationships between Climate and Wheat Yields," *Agronomy Jour.*, Vol. 51 (1959), pp. 635-39; R. J. Millington, "Relations Between the Yield of Wheat, Soil Factors and Rainfall," *Aust. Jour. Agric. Res.*, Vol. 12 (1961) pp. 397-458; and L. M. Thompson, "Evaluation of Weather Factors in the Production of Wheat," *Jour. Soil and Water Conservation*, Vol. 17 (1962), pp. 149-56.

42 Watson, *op. cit.*

43 Chang Jen-Hu, "On the Study of Evaporation and Water Balance," *Erdkunde*, Vol. 19 (1965), pp. 141-50.

procedure of correlating yield with temperature and rainfall is so crude that it contributes little in actual operation and long-range regional planning; and he comments that only through the application of the water balance and other sophisticated concepts can the effect of climate on yield be expressed in exact, quantitative terms.

Prediction of Agricultural Production

Agroclimatological investigations are usually designed to show the association between several climatic elements and various agricultural factors. In particular, emphasis can be placed on the effects of variations from the average climate on variations from the expected (i.e. time trend deviations) agricultural unit production. In such cases, the actual climate or the actual agricultural production is of secondary importance.

Theoretically, however, the prediction (either within the period of record, or beyond the period of record used for assessing the regression coefficients) of agricultural production based on the various regression analyses, is possible by substituting in the various regression equations the specific climatic and time values for a particular season. However, it is doubtful if such prediction through the use of multiple linear regression analyses is justified. Thompson[44] for example, states that although multiple linear regression coefficients are useful in gaining a perspective of the influence of weather variables on crop yields, they should not be used for predicting yields. In this context, Thompson is referring specifically to "prediction" within the period of record—that is, a comparison between the actual yields and the expected (or predicted) yields of the regression equation—and forecasting beyond the period of the regression is not considered. Prediction at least within the period of record is, however, possible with some success, by using multiple curvilinear regression coefficients, as has been demonstrated by Thompson[45] and others. Watson[46] suggests, however, that only when more detailed knowledge has been acquired of how and at what growth stages climatic factors influence yield, will it be possible to derive complex variates that give appropriate weight to the different factors for correlation with yield in naturally varying climates, and eventually use them to predict yields from meteorological records.

A New Zealand Example

In a recent study[47] an "agroclimatological model"[48] was formulated

---

44 L. M. Thompson (ref. 41), p. 153.

45 L. M. Thompson, "Weather and Technology in the Production of Corn and Soybeans," *Center for Agric. and Econ. Development*, Rep. 17 (1963).

46 Watson, *op. cit.*, pp. 347-48.

47 W. J. Maunder, "Climatic Variations and Agricultural Production in New Zealand," *N. Z. Geogr.*, Vol. 22 (1966), pp. 55-69.

48 In the analysis three models were used. See also W. J. Maunder, "An Agroclimatological Model," *Sci. Record*, Vol. 16 (1966), pp. 78-80.

for different agricultural factors and for different areas in New Zealand, based on variations in agricultural production and climate, mainly since the 1930's. Variations from this model were then studied in order that probability assessments could be made of certain variations in agricultural production.

Specifically, three aspects of agroclimatology in New Zealand were examined: (1) an assessment of the variations in climate, and in agricultural production, (2) the agroclimatological associations, and (3) an assessment of the probability of occurrence of significant climatic variations, and the effect upon agricultural incomes of the consequent variations in agricultural production.

A multiple linear relationship was used to ascertain the correlation between agricultural factors and climatic elements, but it must be realized that linear relationships are generally the exception rather than the rule, and that given a sufficient climatic range, optimum values will generally become apparent. Such threshold, optimum, and limiting values are well established in the American mid-west for corn;[49] and it is reasonably apparent that such values apply to other crops in other areas. In New Zealand it is probable, however, that the climatic range is not sufficient to affect seriously the findings of a linear relationship.

The detailed methods and results of the investigation are given elsewhere[50] but, in brief, the coefficients in the multiple regression equations already discussed were used to estimate the effect of specific climatic variations on agricultural production, a specific climatic variation being taken as a positive or negative departure from the average of one standard deviation.

The utilization of such a method enables one to suggest that once in, say, five, six, or seven years[51] the effect on production of a specific climatic variation such as a "dry"[52] October would be in the order of x bushels. Further, the value or economic significance in terms of agricultural income of such a variation in production per acre, can be assessed by multiplying the per acre or per animal variation by the particular total acreage or total livestock population.

One striking example of the application of such a concept is the "economic value" of "dry" and "wet"[53] conditions in the major dairying area of New Zealand—the Waikato. In this case, the analysis showed that a

---

49 Thompson, *op. cit.*
50 Maunder, *op. cit.*
51 Assuming a normal distribution of the climatic data.
52 Defined as a month with a rainfall one standard deviation (or more) below average.
53 Similarly defined as a month with a rainfall one standard deviation (or more) above average.

"wet" January was associated with an "increase" in butterfat production of about 9 ± 2[54] pounds per cow, and this increase, at the 1964 farm-to-factory price of $NZ 0.28[55] per pound butterfat, was valued at about $NZ 2.6 ± 0.6 per cow. Now, the dairy cow population in the Waikato county and surrounding areas has in recent years totalled over 750,000. Accordingly, if the "values per cow" in the Waikato county are taken as an index of the variations in butterfat income per cow for the South Auckland area, it can be seen that a significent climatic variation such as a "wet" January is "worth" about $NZ 2 million to the area (750,000 cows × $2.6 per cow).

The Future

In the past there has been neglect of studies of the dependence of yield on climate and weather, apart from the technical difficulties. One reason for this has been the apparent impossibility of applying the results to improve the productivity of field crops, since farmers cannot control climatic conditions. However, this is surely not a valid reason for it is possible to alter crops or methods of husbandry to suit the prevailing weather and climatic conditions. It is also possible to alter the weather, and although there are claims and counter-claims as to what can and cannot be done, there now seems general agreement that in certain areas, under certain circumstances, and at certain times, weather modification in the form of an increased precipitation of the order of 10 per cent can be achieved.[56]

Recent investigations have re-emphasized the importance of agroclimatological relationships. A comprehensive study by the Center for Agricultural and Economic Development,[57] for example, assessed the effect of recent climatic trends on grain production in the United States. Wadleigh, in an introduction to the study, states that the evaluation of weather in relation to food production is more than a need, it is a must.[58] Further, Wadleigh says that we need to be far more knowledgeable on (1) the phenology of crop plants, (2) weather probabilities, (3) making alternative decisions in management practices in relation to sequential weather events, and (4) weather prediction.[59]

The forecasting of agricultural yields on the basis of meteorological conditions is a problem which has fascinated agricultural climatologists for a long time, although as Bourke has said: "In this type of investigation,

---

54 Standard error.

55 $ N.Z. 1.00 = $ Cdn. 1.22.

56 For a recent appraisal see W. R. D. Sewell, ed.; *Human Dimensions of Weather Modification*, Univ. of Chicago, Dept. of Geography Res. Series (1966), no. 105.

57 Weather and our Food Supply, *Center for Agric. and Economic Development*, Rep. 20 (1964).

58 *Ibid.*, p. 1.

59 *Ibid.*, p. 6

we are groping on the very outside frontiers of our knowledge, and even beyond."[60] However, according to Bourke, the importance of the subject to agriculture is so great that "much of our efforts in the next ten years must necessarily be devoted to finding some reliable basis for extrapolating beyond the short-term weather forecasts adequate for agricultural tactics, to estimates of long-term weather trends which will be applicable to the even more important field of agricultural strategy."[61]

The world is dependent to a very large extent on agricultural production, agricultural production in turn being closely related to climate. The normal climate is however rarely experienced, and what occurs most of the time are variations from the normal. It is reasonably clear that these climatic variations while not random, are not rhythmic enough to produce a prediction equation;[62] nevertheless, significant departures of one or more consecutive seasons from the normal climate are especially important in the world's major agricultural areas; and Thompson, in his most timely study on weather variability and the need for adequate food reserves in the United States,[63] warns that it is important for policymakers to recognize the contribution of favourable weather particularly during the recent 1961-65 period to United States corn and soybean yields. This, he says, is particularly true now that many other countries are becoming dependent on the United States for their grain supply.

That the weather plays an important part in the growth of a crop is well recognized. Further, several recent studies have assessed the more basic physical and biological relationships as, for example, in an analysis of meteorology and grassland farming in Britain by Smith.[64]

The estimation of the precise economic value of climate and weather in their influence on crop and livestock production has not, however, received as much support from researchers as has that of the physical and biological relationships. It is, of course, a particularly difficult problem, and as Curry comments: "All over the world, and as a matter of course, farmers assess weather as a resource but the formal expression of this evaluation is far from easy."[65] Despite these difficulties, however, solutions[66] to

---

60 P. A. M. Bourke, "Agricultural Biometeorology," *Intern. Jour. Biometeorol.*, Vol. 7 (1963), pp. 121-25.

61 *Ibid.*

62 L. M. Thompson, "Weather Variability and the Need for a Food Reserve," *Center for Agric. and Econ. Development*, Rep. 26 (1966), p. 3.

63 *Ibid.*

64 L. P. Smith, "Meteorology and the Pattern of British Grassland Farming," *Agric. Meteorol.*, Vol. 4 (1967), pp. 321-38.

65 L. Curry, "Climate and Livestock in New Zealand – A Functional Geography" (unpub. Ph.D. thesis, Univ. of New Zealand, 1958), p. 252.

66 As shown in the New Zealand example given above. Also see Maunder, *op. cit.*, and W. J. Maunder, "The Effect of Significant Climatic Factors on Agricultural Production and Incomes," *Monthly Weather Rev.*, Vol. 96 (1968), pp. 39-46.

this problem can be found even if only very approximately, and it is imperative that more research along these lines be done, for as Watson states:[67] "Climate determines what crops the farmer can grow; weather influences the annual yield, and hence the farmers' profit, and more important, especially in underdeveloped countries, how much food there is to eat."

[67] Watson *op. cit.*, p. 348.

5

Heat Units and Crop Growth

R. M. Holmes

G. W. Robertson

Introduction

While the advances of science have resulted in mastery of many of the farmer's problems, not all of them have been solved. Climate has not yet been brought under control. Indeed, we have just begun to learn how to "use" it. Because of the weather, yields of many crops fluctuate enormously each year; most crops are adapted to very specific localities; and all crops suffer periodically. Many of these effects are very subtle and difficult to explain. However, they have to do with things like sunlight, wind, rain, and temperature.

The effects of temperature on crop growth are perhaps the most obvious. Although only very slight artificial temperature modifications are now possible, much can be done to "work with" temperature. One of the simplest means of "making use of" the weather is through heat units, which have been variously referred to as degree days, day degrees, and growing degree days.

Plants and Temperature

Temperature is one of the primary factors controlling the growth of plants and their distribution over the earth. The rate of many physiological processes occurring in plants, including growth, is markedly influenced by

---

Reprinted from *Canada Dept. of Agric. Pubn.* 1042 (1959).

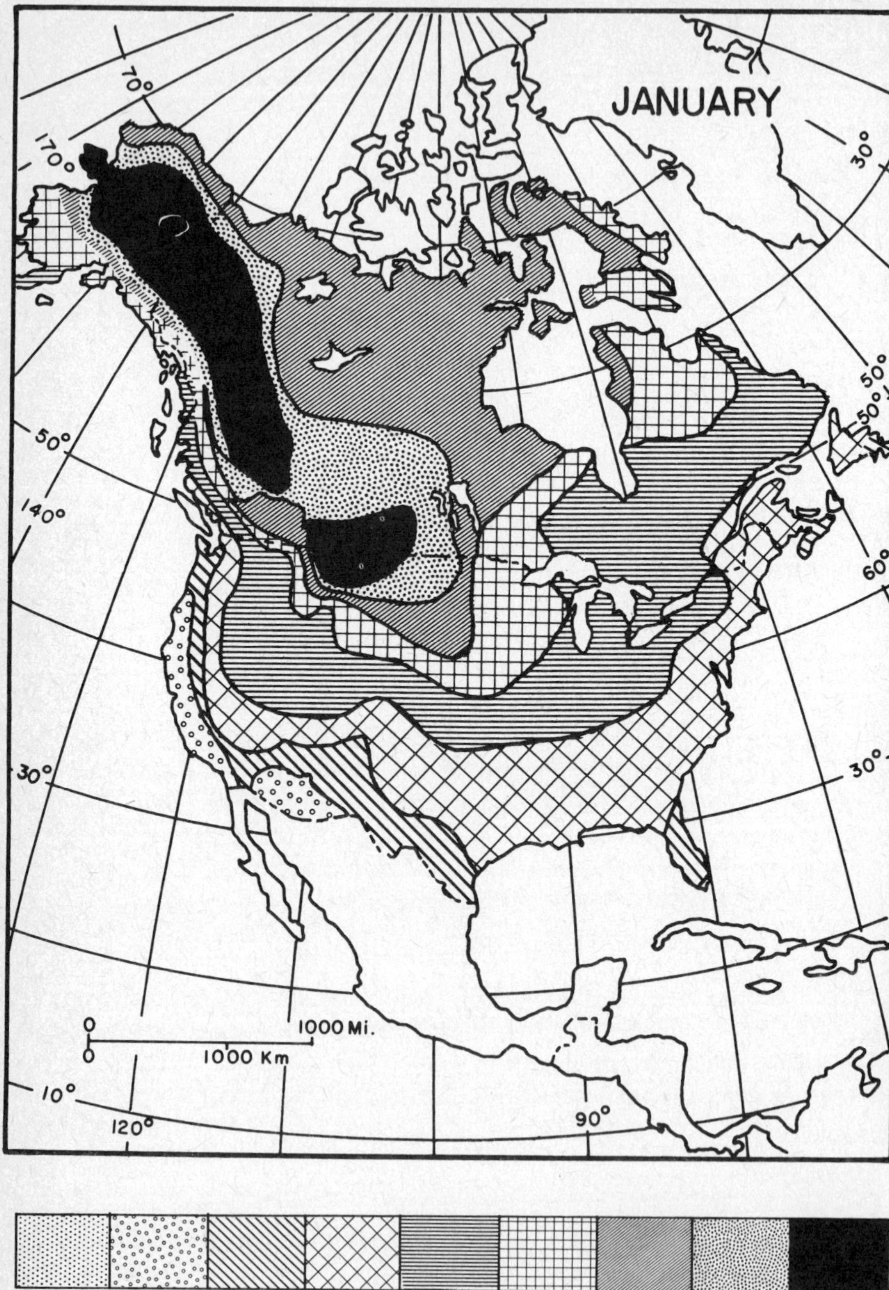

Figure 1. Deviation (σ) of temperatures in North America.

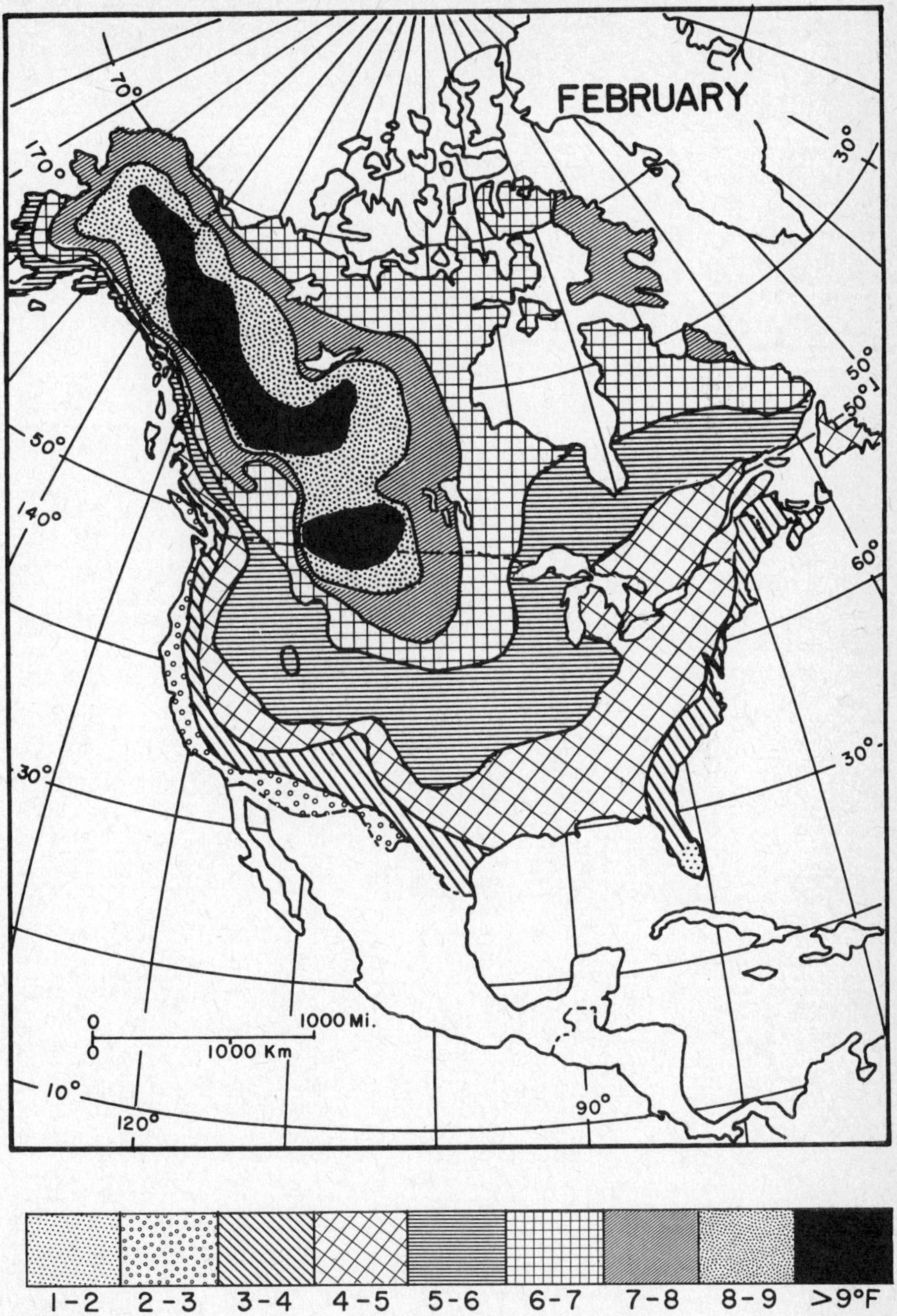

Figure 2. Deviation (σ) of temperatures in North America.

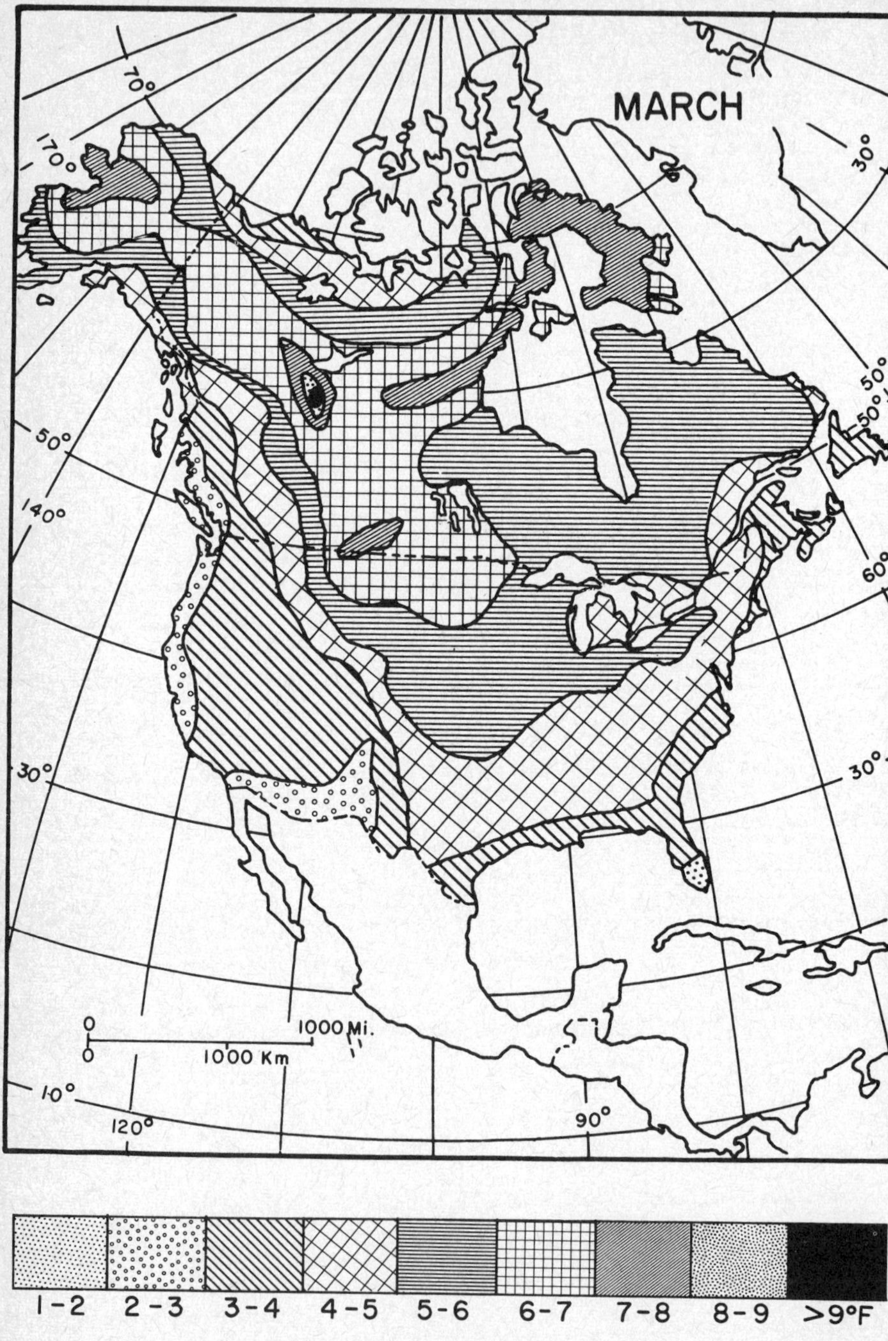

Figure 3. Deviation (σ) of temperatures in North America.

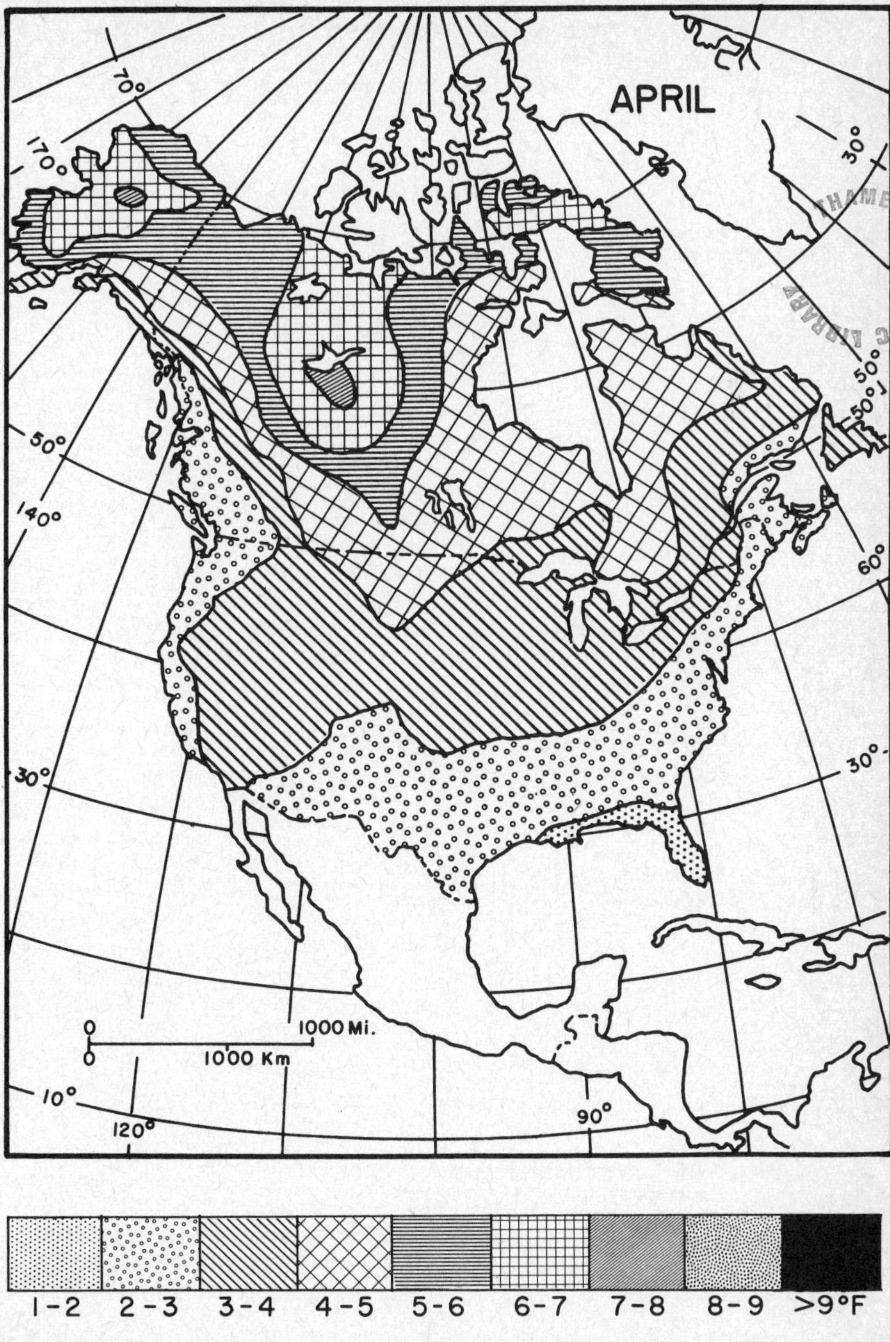

Figure 4. Deviation (σ) of temperatures in North America.

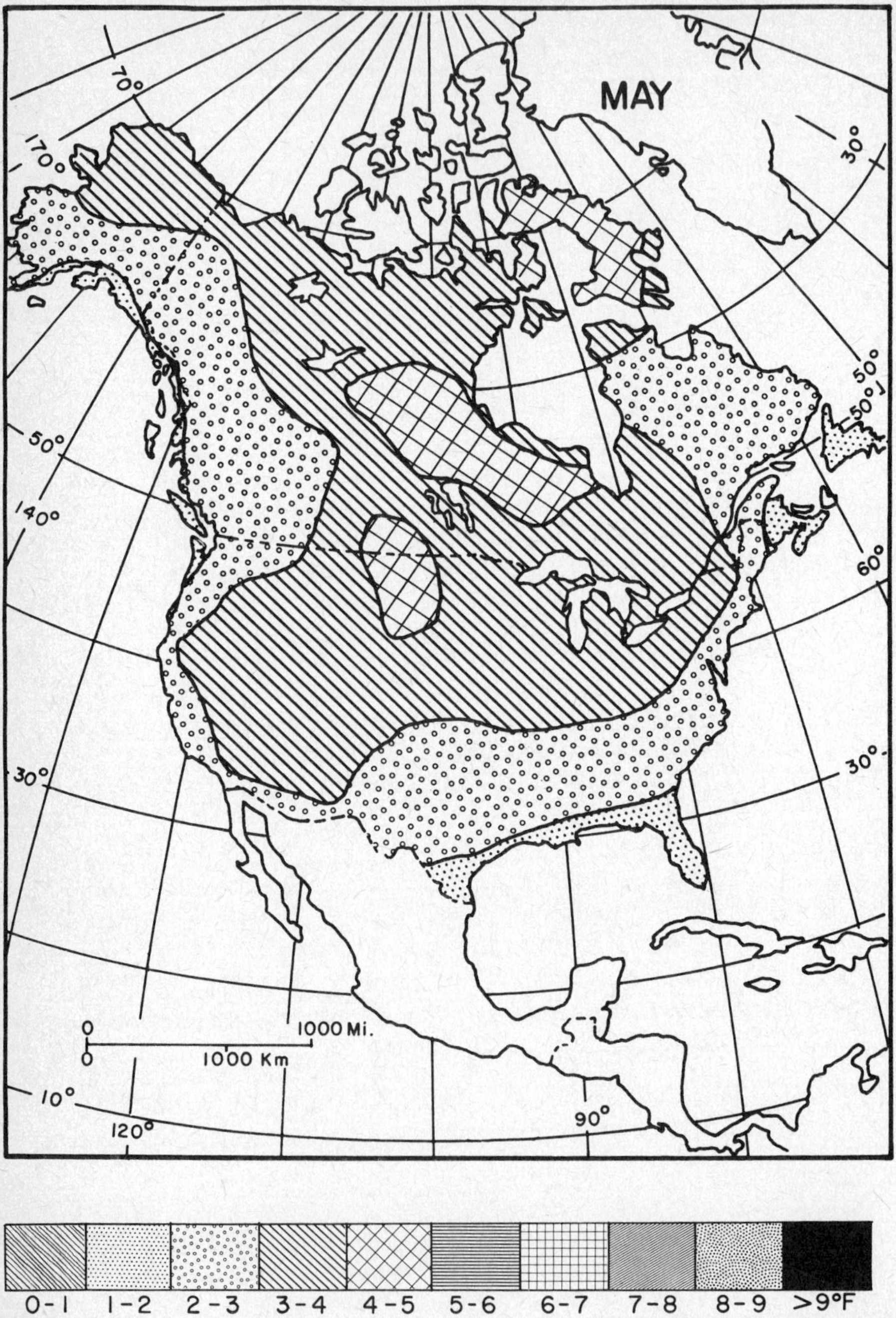

Figure 5. Deviation (σ) of temperatures in North America.

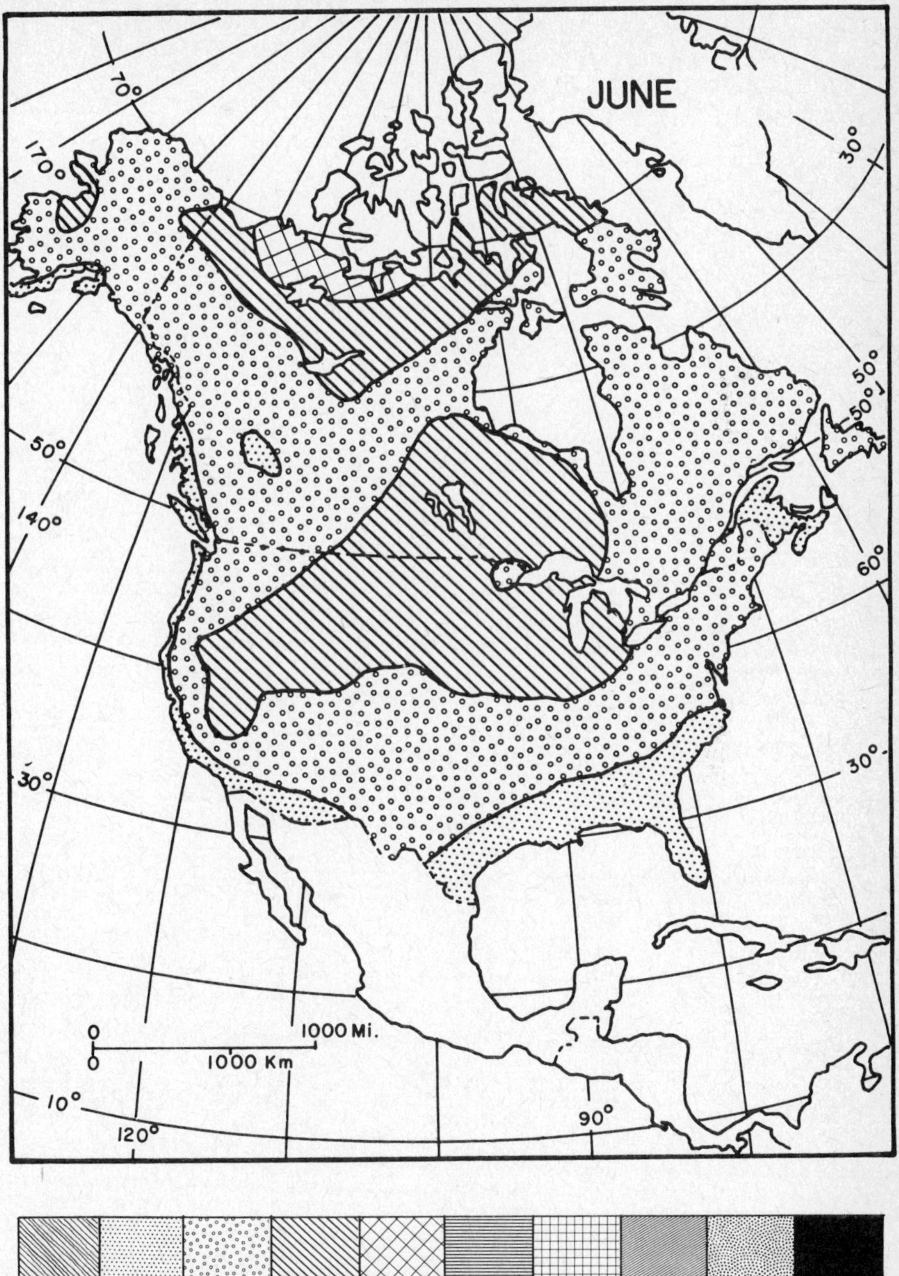

Figure 6. Deviation (σ) of temperatures in North America.

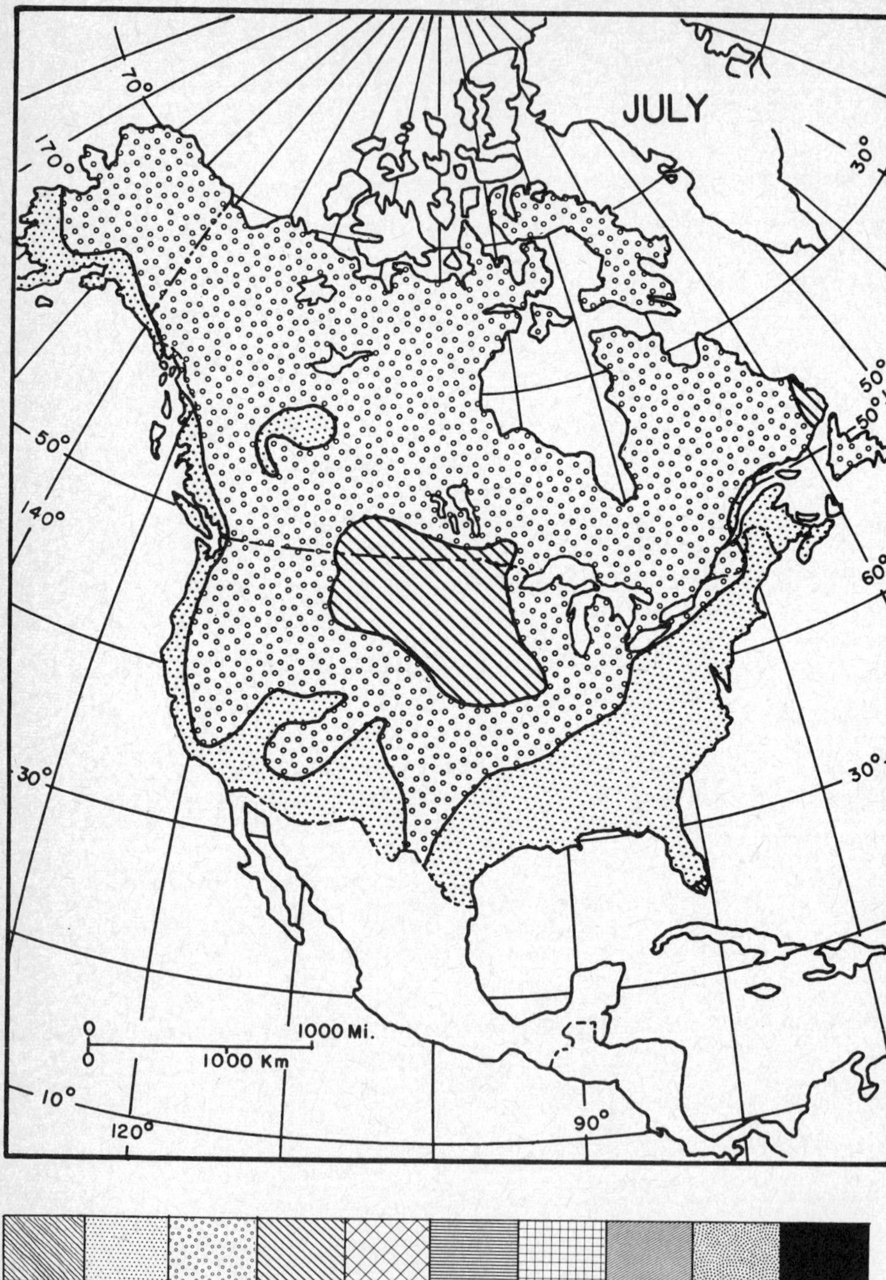

Figure 7. Deviation (σ) of temperatures in North America.

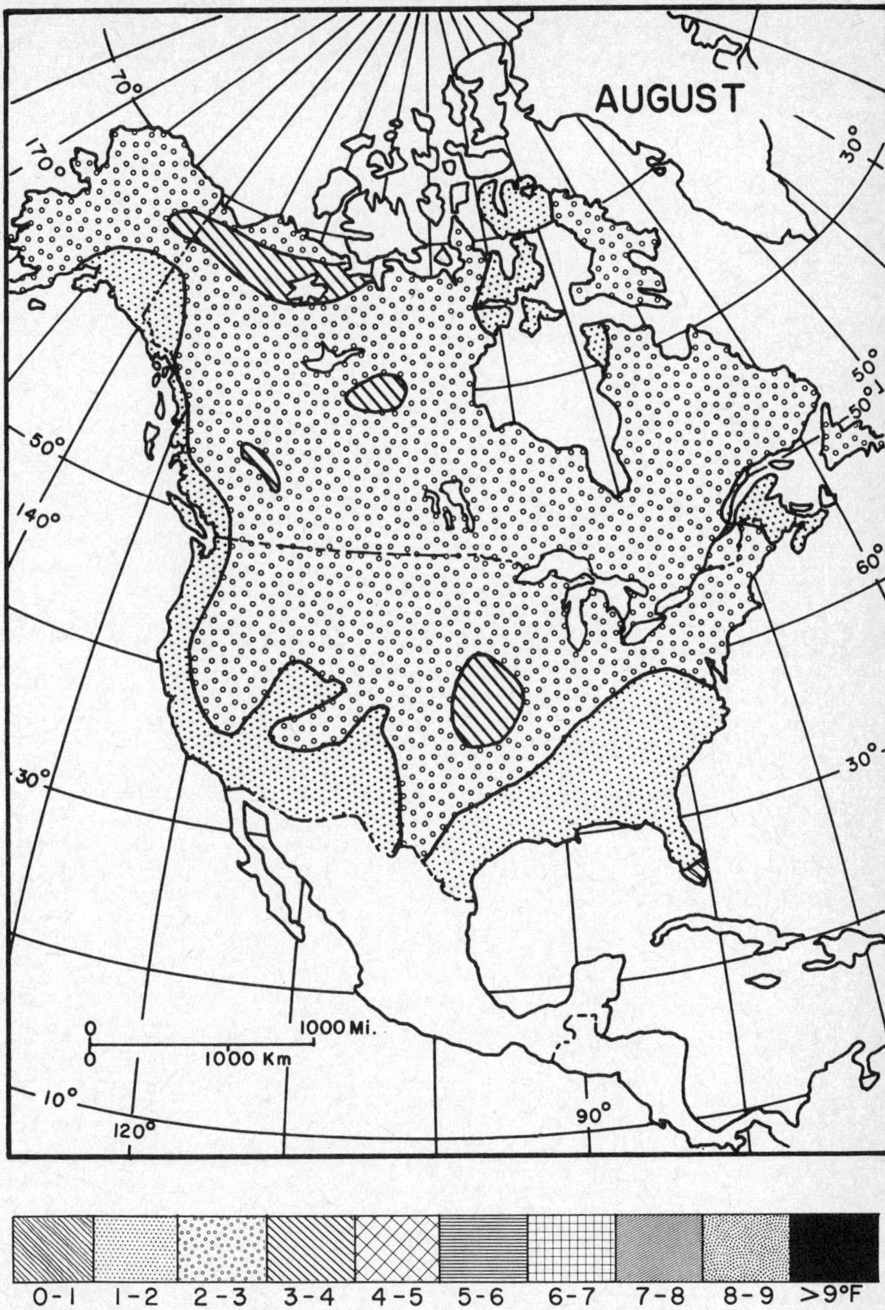

Figure 8. Deviation (σ) of temperatures in North America.

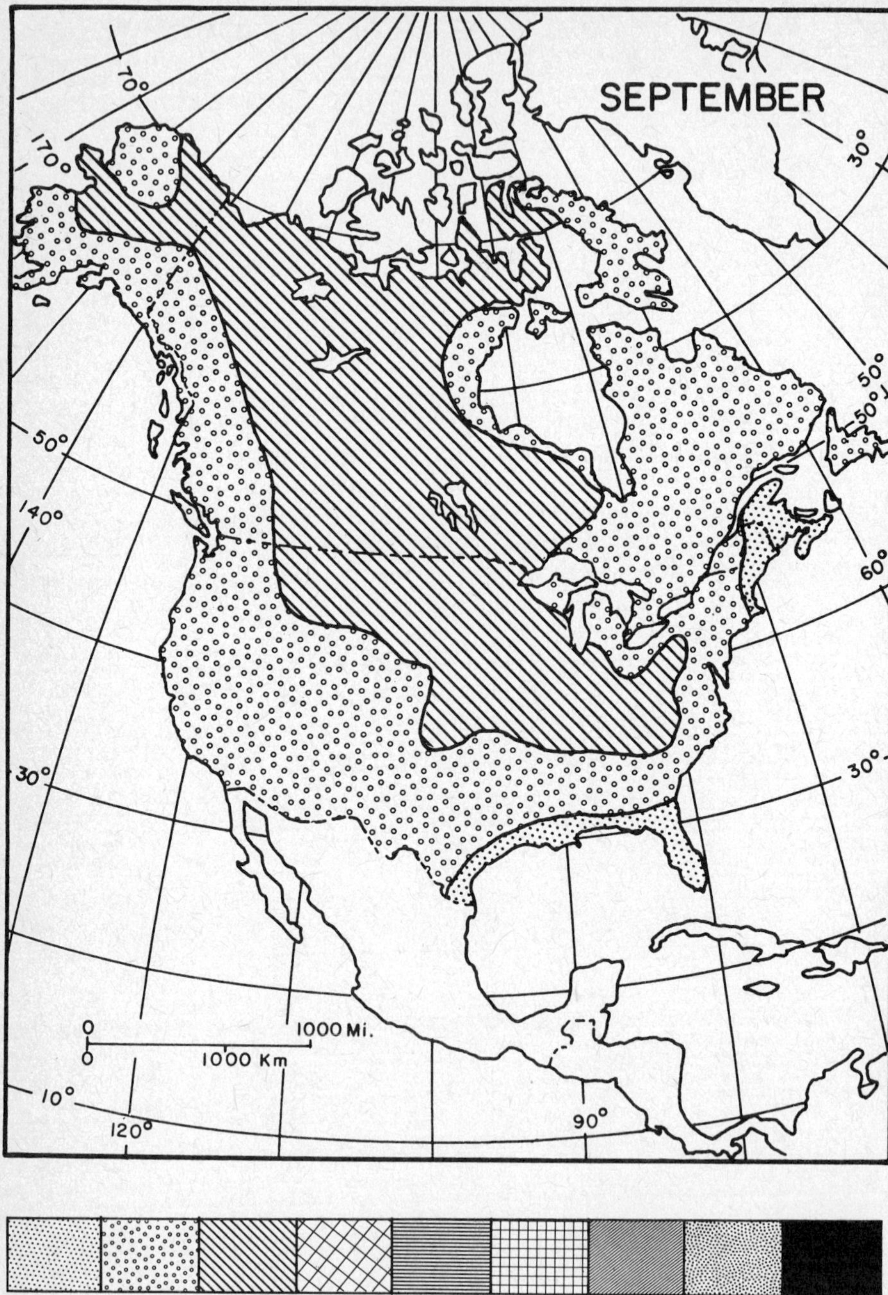

Figure 9. Deviation (σ) of temperatures in North America.

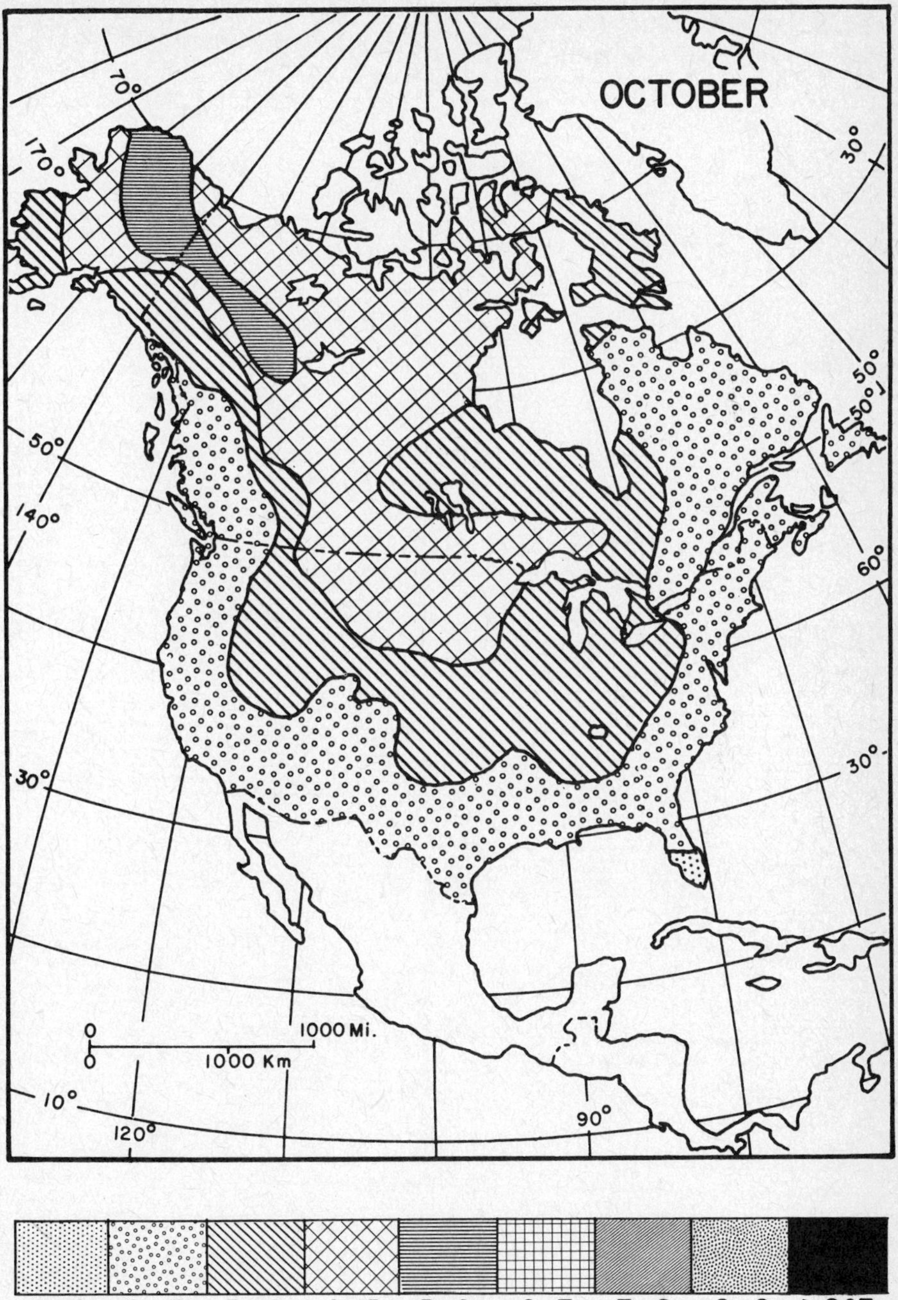

Figure 10. Deviation (σ) of temperatures in North America.

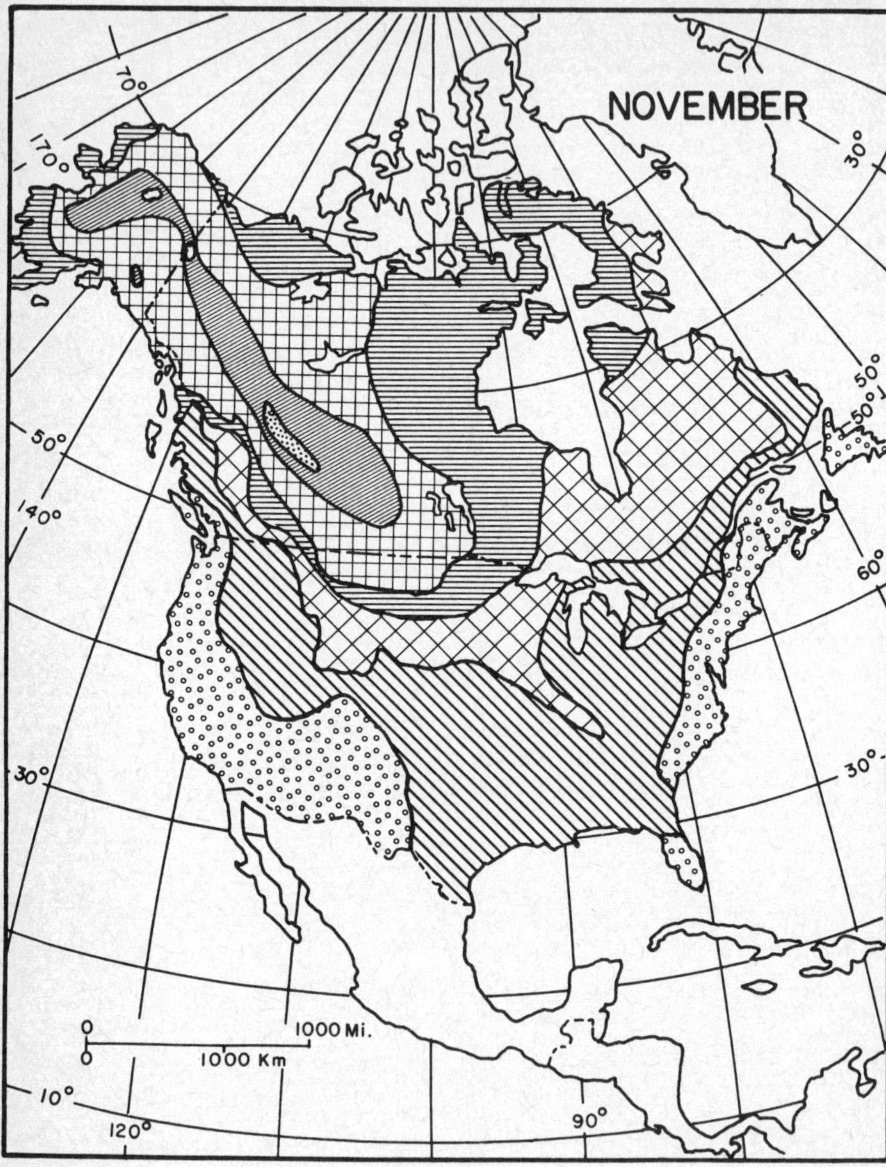

Figure 11. Deviation (σ) of temperatures in North America.

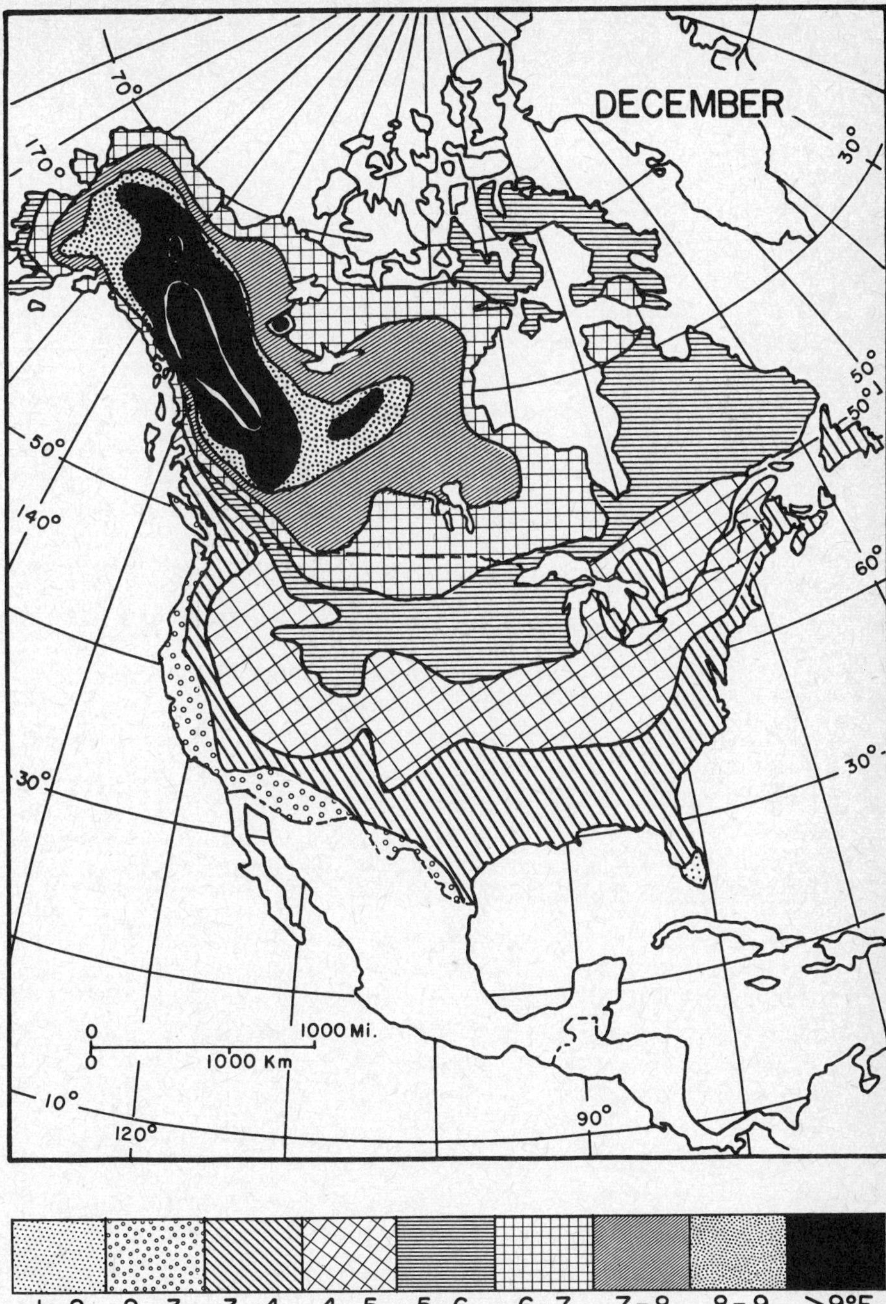

Figure 12. Deviation (σ) of temperatures in North America.

temperature. Qualitative as well as quantitative effects of temperature are manifest in the development of plants, or in other words, the structural and physiological development of a plant may vary greatly, depending upon the temperature pattern of the environment.

Most agricultural crops grow in conditions where the temperature fluctuates widely. Cool nights and warm days are usually favorable (54, 55). For example, certain varieties of tomatoes will set fruit only when the night temperature is near 64°F (54). Consequently these tomatoes are not grown commercially in the tropics where nights are usually warmer. On the other hand, in more northern areas night temperatures may be too cool. Tomato yields are therefore reliable in areas where maritime-type conditions keep temperatures consistently in a favorable range during the growing season. In other areas production may be very variable. Peas do best when day-time temperatures stay below 80°F. Above 85°F growth is markedly reduced (54). The onion will flower only under low night temperatures, but higher temperatures are required for other growth processes.

Optimum tuber set of potatoes occurs only if the night temperatures during set fall between 50° to 57°F. Therefore, more northern regions like Maine, Idaho, and Ireland are best suited for potato production. Most fruit trees will not open their leaf or flower buds until they have gone through a chilling period (54). For strawberries to develop optimum flavor, they must ripen under day-time temperatures of about 45° to 50°F. Therefore strawberries harvested in the summer when it is warmer are practically without aroma, while the flavor of strawberries from Alaska is good at any time. Actually this is important because we pay 40¢ a basket for flavor while the cost of the acid, vitamins, and sugar in that quantity of strawberries amounts to less than one cent.

It has been customary to consider minimum, optimum, and maximum temperatures for plant growth. These cardinal points vary greatly from one stage of growth to another. In many temperate-zone plants, the optimum temperature for germination of seeds is less than that for vegetative growth, which in turn is often lower than the temperature most suitable for flowering and fruiting.

From the above, it may be difficult to see how any relationship could be formulated to express the overall growth of a plant from planting to maturity. However, chemical and physical processes whether occurring without or within the plant are subject to the same laws. Many chemical reactions are doubled in rate for every 18°F rise in temperature. This would occur regardless of whether the reaction took place in a palisade cell or a beaker. Any apparent inconsistencies manifested in the gross growth of plants are probably the result of the overlapping of many varied reactions. However, the overall development of many plants follows a remarkably close relationship to temperature. Variations within the lifetime of any one plant either are of too small a magnitude or they cancel each other.

The Heat Unit Concept

That plants have a "temperature requirement" was first suggested in the eighteenth century (40). It was not until as late as 1916 that any practical use was made of the idea (31). Abbe has published a complete review of the early work in degree days and crop-weather relationships (1). It was suggested that a base temperature exists below which plants will not develop, each plant having its own base temperature (5, 9, 33, 37). The amount of "effective growth-heat" accumulated during the day was obtained by subtracting the plant base temperature from the daily mean temperature (or hourly mean as the case may be). This figure was called the "growing degree-day" accumulation, (or "growing degree-hour"), or "heat" accumulation for the day. The amount accumulated each day was added to that of the previous day until the plant matured. The total "accumulated heat", at maturity was called the "summation constant", or the "remainder index". This figure was nearly the same each season for any one crop grown at a particular location.

The theory assumes that there is only one base temperature throughout the life of the plant; that day and night temperatures are of equal importance to plant growth; and that the plant response to temperature is linear over the entire temperature range. Some investigators have sought to overcome these weaknesses by using an exponential expression [which favors the various growing temperatures depending upon their position in the temperature scale (25)]. Following the rule of Van't Hoff., plant growth should double in rate for every 18°F rise in temperature above the base temperature. Hence, if a plant grows at the rate of unity at 40°F it supposedly would grow twice as fast as 58°F and four times as fast at 76°F. Some investigators have assumed that a maximum temperature exists beyond which plants do not grow.

Livingston's physiological index is yet another modification of the heat unit idea (31). It is based on the work of Lehenbauer (30), who measured the growth rate of corn seedlings under controlled temperatures. According to Livingston, to obtain the physiological index for any temperature, the growth rate at that temperature is divided by the growth rate at 40°F (Livingston's base temperature). The index was further modified to include a factor for rainfall and another for the evaporating power of the atmosphere. (Several atmometers were devised to obtain the evaporation factor). The application of the physiological index has been very limited.

Of all the various indices and modifications described above, the remainder index has been most enthusiastically received, in spite of the fact that it is based on simplified assumptions. This may be because growth is virtually linear for many agricultural crops over the temperature range existing in the temperate regions.

When the remainder index is used it is common practice to assume that the daily minimum temperature does not fall below the plant base temperature. For example, on a cool day, the actual daily mean temperature may be below the plant base temperature. According to the theory, under these conditions no heat units are accumulated. In fact, however, there is usually a period during such days when the temperature is high enough for the plants to grow. Therefore, by assuming that the daily minimum temperature never falls below the plant base temperature, there will never be a day when heat units are not accumulated, as long as the daily maximum is above the plant temperature.

In actual practice it has been found that the summation constant for any particular crop varies depending upon planting day and environment (51). Heat accumulation for early plants are usually less than those for later plantings. This can be explained by the fact that as the season progresses, the daily mean temperature increases, and there is a higher frequency of occurrence of temperatures that are too high for good growth. This may also be partially accounted for by the fact that the daylength shortens as the season progresses(41). For some plants, short day lengths are compensated for by a larger number of heat units.

Environmental Factors and Heat Unit Accumulation

Environmental factors that may cause fluctuations in the summation constant of a crop are:

Soil fertility level: Low fertility causes slow growth, high nitrogen and consequently heavy stem growth delays maturity, and high phosphorous tends to hasten maturity.

Plant population: A low plant population will mature slightly earlier than a more dense population, provided weeds do not make up the difference.

Soil type: Sandy soils warm up earlier than clay soils. Other variables such as fertility status and moisture characteristics are associated with soil type.

Soil temperature: During the spring warm-up, soil temperature lags appreciably behind air temperature. Hence heat units accumulated on the basis of air temperature may be too high. Soil temperature may be used until emergence (42). (Some workers reduce the plant base temperature by 50°F and use soil temperature instead of air temperature in computing heat accumulation). Southern slopes warm up sooner than northern slopes.

Seeds planted deep are cooler and usually emerge later than those planted at shallow depths provided moisture is not lacking.

Moisture: Poorly drained soils are cold and also give rise to a variety of nutritional problems. Drought during the latter part of the life span of the plant usually hastens maturity. The plants may even die before maturity is achieved. Many investigators are of the opinion that lack of moisture does not affect maturity unless it causes more than day-time wilting. If this happens, maturity is usually delayed because plant growth is retarded or ceases. Maturity will be delayed if moisture is lacking at seeding time or during the early growth period, even though heat units have been accumulating. A correction factor for heat units depending on the percentage emergence might be useful in this connection.

Further Considerations Regarding the Heat Unit Theory

Usually the temperatures used in the calculation of heat units are based on measurements of air temperature in a Stevenson screen which is supported 4½ to 6 feet above the ground. Often there is an appreciable difference between this temperature and the temperature of the plant tissue or the immediate plant environment. Fluctuations in solar energy cause concomitant fluctuations in the temperature of plant tissues (40). However, if the summation constants are based on these (Stevenson screen) measurements, the difference does not cause a great error. There is agreement between heat units accumulated at various levels above and below the surface as long as summation constants and base temperatures are based on measurements of temperature at the particular level (17).

Topography affects the local crop temperature. If the temperature obtained from the Meteorological Service is not applicable to an area, maximum and minimum thermometers may be set out, and private records kept.

The summation constant for a given crop may vary from one region to another. This is particularly true when the photo-periods vary. Longer periods of daylight reduce the heat requirement of many crops, particularly those that thrive in cool weather (13, 39). Therefore in order to make the summation constant apply over many widely separated areas, additional refinements are necessary. This can be partially accomplished by incorporating a threshhold value for daylength into the calculations (41). Below a certain base number of "light units" (hours of daylight) a plant may not mature regardless of the temperature. However, suitable relationships between daylength and temperature for a given crop are not easy to obtain, and they have been worked out for only a few crops (35, 41). Heat units seem to work well without light units at any one location because during the life span of most crops in the temperature zone, daylight hours vary little from one season to another. In other zones or in fringe areas in the north, light duration may have to be taken into consideration. Some plants mature more rapidly in the north where days are long in the summer than would be expected from temperature accumulation alone.

Use of Heat Units in Scheduling Plantings

Normal Heat Accumulation

Long-term temperature records can be used to estimate heat accumulation to predict the probable date of maturity of a crop. The normal heat unit accumulation curve based on these records is a gradual one being nearly horizontal in the spring during cool weather and becoming nearly vertical during the summer when heat is rapidly accumulated (Figures 15, 16, and 17). To predict the harvest date of an individual crop the normal curve is projected forward from the day of planting. The expected harvest date may have to be adjusted as the season progresses because of seasonal variation of temperature from normality. In any case the actual temperature is used to compute heat accumulation up to the day in question, and the normal is expected for the remainder of the season. A surplus or deficiency of heat units will either hasten or delay the predicted harvest date (57, 58, 59).

Simple Planting Ratio

It is difficult to plant a series of fields so that the crops will not blossom, and consequently mature, at the same time. If planting is done in proportion to heat unit accumulation, however, a fairly orderly harvest can be arranged. For example, if the capacity of a pea cannery in acres per day is known in proportion to the normal heat units accumulated per day at harvest time, then planting may be done in the same proportion as heat units are accumulated in the spring (44). If the normal heat accumulation per day at harvest is 30, and the capacity of the cannery and viners at harvest is 60 acres per day, then the ratio of heat accumulated at harvest to the number of acres harvested per day is 30:60 or 1:2. Therefore, after the accumulation of each heat unit at planting time, two acres should be planted; after 10 heat units have accumulated, 20 acres may be planted. Strict adherence to this ratio, however, might result in plantings being seriously delayed in the case of a cool spring. In that case more acres would have to be planted than heat accumulated. Then the method would forecast the time when "bunching" would occur at the cannery.

Planting Ratio Applied to a Combination of Crops

If a number of fields scattered over an area were planted to peas on the same day, they would not all mature at the same time. Experience has shown that harvest would normally be spread over a two- or three-day period because of the factors previously mentioned. Therefore it is common practice to plant two or three times as many acres on the first day as would normally be done according to the heat unit accumulation.

Let us suppose four varieties of peas are to be planted and each has a different acreage and summation constant as follows:

Variety	*Acreage*	*Summation constant range*
Alaska	400	1225 - 1275
Sweets	200	1300 - 1350
Perfection	300	1500 - 1550
Superior	300	1600 - 1650

Assume that the normal daily heat accumulation at harvest time is 30 heat units and that the cannery has a capacity of 40 acres per day.

The planting ratio is 4 acres for every 3 heat units accumulated. Since 400 acres of Alaska peas are desired, the planting would normally be spread over 300 heat units. But because of the fluctuation in summation constants between different fields, three times the normal acreage should be planted during the first 30 heat units accumulated. Then, 120 acres of Alaska peas are planted during the first 30 heat units, and 280 acres planted during the next 210 heat units. The rate of planting of the last 280-acre portion is done according to the planting ratio.

After the accumulation of 1225 heat units, the first Alaska peas will enter the cannery and continue for 10 days since the capacity of the cannery is 40 acres per day. Three hundred heat units have normally accumulated during this time. The total heat unit accumulation from the first plantings to the last harvest of Alaska peas than is 1225 plus 300 = 1525.

At this time, early Sweets peas must be ready for harvesting. They require 1300 heat units to mature so planting must begin 1525 minus 1300 = 225 heat units after the first Alaska planting. Since 200 acres are required, and since the planting ratio is 4:3, planting will be distributed over 150 heat units. Late Sweets variety peas require 1350 heat units to mature so the processing time of Sweets will be prolonged 50 heat units. Therefore total heat accumulation from first plantings of Alaska peas to the last processing date of Sweets peas is 1525 plus 150 plus 50 = 1725 units.

At this time Perfection peas must be ready for processing. The earliest peas of this variety require 1500 heat units to mature, therefore planting must begin 1725 minus 1500 = 225 heat units after the first Alaska plantings. Since 300 acres are required, the planting must be distributed over 225 heat units. Late Perfection peas require 1550 heat units to mature so the processing time of this variety will be prolonged 50 heat units. The total heat accumulated up to this point is then 1725 plus 225 plus 50 = 2000 units since the first planting of Alaska peas.

Superior peas must enter the cannery at this time and since the early portion matures after 1600 heat units, the plantings must start 2000 minus

1600 = 400 heat units after the first Alaska planting. Since 300 acres are required, the planting will be distributed over 225 heat units. Late Superior peas require 1650 heat units to mature so the processing time for Superiors will be prolonged 50 heat units.

Therefore the total heat accumulated since the first Alaska variety was planted is 2000 plus 225 plus 50 = 2275. The entire planting schedule would be spread over 400 plus 225 = 625 heat units. This plan is briefly summarized below.

Variety	*Heat Unit Distribution of Plantings*	*Heat Unit Distribution of Harvest*
		(from first Alaska planting)
Alaska	0 - 240	1225 - 1525
Sweets	225 - 375	1525 - 1725
Perfection	225 - 450	1725 - 2000
Superior	400 - 625	2000 - 2275

After some preliminary experimentation and adjustment for local peculiarities the heat unit theory can be a very reliable tool for farmer and processor alike. This theory is not meant to completely replace any practice already in use in field operations. It is rather an attempt to express mathematically the influence of temperature on crop growth. Many commercial canners have found this method of scheduling plantings very beneficial.

A survey of processing companies using the heat unit system was made during the winter of 1949-1950, by the Continental Can Company (3). Their report indicated that of 75 companies operating 145 pea canneries, and 47 companies operating 93 sweet corn canneries only two companies had experimented with the system before 1947. About 10 companies started using the system in 1947, three in 1948, and the remainder in 1949. It was suggested that several other companies would use the system for the first time in 1950. Seaton and Huffington (45) state that 39 per cent of the acreage planted to peas and 32 per cent of the acreage planted to sweet corn in the United States in 1949 were scheduled by the heat unit system.

HEAT UNITS APPLIED TO CROPS

Peas

The literature contains more information on heat units and the growth of peas than on any other single crop (7, 8, 9, 10, 19, 25, 31, 36, 37, 39, 42, 43, 44, 55). The growth of peas has been found to be almost linear over the range of temperatures occurring during the growing season in most temperature regions (7, 9, 10, 44). It is important to note that the summation constant for any one variety of pea will vary with geographic location because of factors discussed previously. In addition there is a variation from season to season in any one locality (44). However, the spread between varieties, and between plantings of the same variety, has been found to be essentially the same (for any one locality) regardless of seasonal variation. Users of the heat unit theory must of necessity establish their own summation constant values. It is not feasible to set a definite value for any one variety and expect it to be usable under all conditions.

The base temperature for peas is usually taken at 40°F although a few investigators prefer 42°F (31, 44, 19, 10). Several published reports deal with the summation constants of several pea varieties, and the reader is referred to the above papers for more detailed information. At Morden, Man., the following summation constants were found for various pea varieties (base at 40°F).*

	Year			
Variety	*1954*	*1955*	*1956*	*3-year average*
Early Wisconsin	1239	1483	1234	1319
All 3	1301	1457	1234	1331
Little Marvel	1392	1552	1327	1424
D.P. Perfection	1488	1635	1559	1560
Lincoln	1517	1732	1607	1619
Wyola	1549	1732	1583	1621
Mean of 6 varieties	1414	1598	1424	

The high values for 1955 are attributed in part to early seeding followed by high air temperatures and low soil temperatures. In southern Ontario the following is reported for the same three-year period (base of 40°F).**

	Year			
Variety	*1954*	*1955*	*1956*	*3-year mean*
Wisconsin	1450	1426	1380	1418
Surpass	1365	1485	1412	1411
Wyola	1727	1738	1753	1739
Perfection	1772	1731	1745	1749
Mean of 6 varieties	1581	1591	1575	

Another report from north-central United States lists the following summation constants for various pea types (44).

*A. L. Shewfelt, Experimental Farm, Morden, Manitoba (1957). Private Communication.

**H. Fried, Canadian Canners Ltd., Petrolia, Ontario (1957). Private Communication.

Type	*Heat Units above 40°F*
Alaska	1200 - 1250
Early sweets	1200 - 1300
Pride	1400 - 1500
Perfection	1500 - 1575
Late sweets	1625 - 1725

In Chatham, Ont., the Libby, McNeil, & Libby Co., use the following summation constants for peas(*).

	Heat Units above 40°F
Very early	1165 ± 42
Early	1250 ± 42
Mid-season	1420 ± 65
Late	1540 ± 65

Soil differences were found to cause an 80- to 125-unit variation in summation constants. High temperatures lowered, and low temperatures raised, these values. In southern Alberta 1232 to 1300 heat units above 40°F are required to mature Alaska peas(**).

Corn

Corn is a crop that does well in warm latitudes where heat unit accumulation is high. The warm mid-west states of the United States are noted for the crops produced there. Moisture variations often complicate the "heat unit picture" for this crop. Corn subjected to drought does not mature so rapidly as corn receiving sufficient moisture. Therefore the amount of heat required to bring droughty corn to maturity is increased.

The base temperature for corn is usually taken as 50°F (some prefer 49°F) (5, 7, 15, 17, 21, 24, 28, 33, 36, 40, 44, 52). At Morden, Man., the following summation constants were obtained during 1954, 1955, 1956(†).

	Year			
Variety	*1954*	*1955*	*1956*	*Average*
Golden Rocket	1444	1401	1346	1397
Sugar Prince	1432	1401	1362	1398
Carmel Cross	1529	1495	1467	1497
Golden Bantam	1582	1495	1462	1514
Golden Rush	1512	1551	1481	1515
Seneca Arrow	1605	1600	1481	1562

*R. J. Hall, Libby, McNeil and Libby Co., Chatham, Ontario, (1957). Private Communication.

**Alex Metcalfe, Broder Canning Co., Lethbridge, Alberta (1957). Private Communication.

†A. L. Shewfelt, *op. cit.*

The Libby, McNeil, & Libby Company in Chatham, Ont., has found that many seasonal variations are removed by addition of heat units from planting to general silk, then again from general silk to harvest. The following are the values received(††).

	Heat Units above 50°, planting to maturity
Early	1790 ± 40
Late	2000 ± 40
	Heat Units above 50°, planting to silk
Early	1270 ± 20
Late	1440 ± 20

Beans

Maturity of snap beans is perhaps the most important factor of quality in the processed product. The heat unit theory has been used with much more success in predicting maturity of snap beans than has the number of growing days alone (20, 44, 35). A base temperature of 50°F is usually used for snap beans, and about 1125 heat units are required to mature many varieties (20). The following is a list of several varieties of snap beans and their summation constants when grown in the North-Central States (20).

Beans are hand picked through a large part of the growing season. If mechanical pickers were available, heat units would be of greater advantage in harvesting operations.

Variety	*Summation constant*
Bountiful	1100 - 1200
Tendergreen	1100 - 1300
Logan	1150 - 1300
Fruitful	1100 - 1150
Burpee St.	1000 - 1300
Commodore	900 - 1200
Rival	1150 - 1550

The base temperature for lima beans is approximately 50° F (8). Further information on the use of heat units and the growth of lima beans would be very desirable.

††R. J. Hall, *op. cit.*

Other Vegetable and Canning Crops

Lettuce

Madariaga and Knott (32) used the heat unit theory in 389 fields of commercially grown lettuce. Their data indicate that when planted at varying dates, lettuce did not utilize the same amount of heat to reach maturity. However all fields of lettuce planted on the same date were fairly uniform in their heat requirement. The base temperature was 40°F and all temperatures above 70°F were eliminated. This procedure made the data more uniform. These investigators also utilized a finding of Nuttonson (35) and multiplied the summation constant by the average daylength. This did not reduce the variation in summation constants at any one location but it did reduce variability in summation constants between fields in the Salinas and Imperial Valleys. The average range of heat unit accumulation for lettuce in the Salinas Valley was 1450 to 1580.

Asparagus

Culpepper and Moon (14), determined that the growth of asparagus does not become appreciable until a temperature of 42°F is reached. They noted that it took approximately 120 degree days for a stalk to grow from the base to a height of 20 cm., and about 410 degree days to grow 150 cm. in height. They found these summation constants to be altered by changes in soil fertility, moisture, and disease.

Spinach

Boswell (11) has described the growth of spinach in relation to temperature. He tabulated the variation in summation constants for various places by the direct and exponential methods. At any one location he achieved reasonable consistency of summation constants by the direct method using 36°F as the base temperature. The coefficient of variability was greatest using the exponential index method.

Pumpkin and Tomatoes

It would be of great commercial importance if it could be shown that the heat unit concept could be applied with reasonable accuracy to other vegetable crops. Several vegetables have had base temperatures tentatively assigned to them: pumpkins 55°F (18); and tomatoes 59°F (44). Kitchin used three methods to determine the length of time between field setting of Rutgers tomatoes and date of first harvest. Degree-day accumulation above 50°F seemed to be most accurate. However, none of the methods gave a very reliable index by which harvest could be predicted from date of field set (7).

Fruits

Tree Fruits

The time of full bloom of fruit trees marks a definite stage in the development of fruit and has often been used as an index of future fruit development. The interval between full bloom and fruit maturity is specific for many varieties, with variations from year to year due to various factors. The forecasting of date of fruit maturity is different from that of the common vegetable crops in that high temperatures at some stages of development have been found to hasten maturity more than at other times (12).

One investigator found with Elberta peaches that the mean maximum temperature for the first 50 days from bloom gave reliable estimates of date of first picking ($r = .93$). Increasing the interval to 60 or 70 days did not increase the accuracy of the method (53).

Another worker reported that elapsed time from full bloom to maturity was found to be fairly constant for many individual varieties under differing climatic and cultural conditions and constituted the most reliable index for estimating maturity (23). Baker and Brooks (6) used three different statistical methods for predicting the time of apricot maturity. Apricot harvests were predicted within three days of the actual harvest 80 per cent of the time, and prunes 50 per cent of the time.

The validity of the heat unit approach has been shown in many other studies with various fruits. Most of the observations have shown that, while methods used in one area may be applicable to another, the data usually are not. Deviations occur because of local climatic and cultural variations. It remains for the individual grower or researcher to establish relationships based on conditions existing in his own area.

Cantaloupe

The effect of temperature on cantaloupe production has been reviewed by Rauchenstein (38). In the Imperial Valley of California, cantaloupes are set out in December and early January and then usually covered with hot caps. The base temperature for cantaloupes is approximately 50°F. The heat accumulation above 50°F in the Imperial Valley from January 1 to maturity of fruit has been found to be between 1900 and 2100. The mean accumulation of day degrees was 2000, and the correlation between heat accumulation and days to maturity was $r = 0.76$.

Cereals

While the heat unit theory has been shown to be useful in dealing with results from a single location, serious errors may occur if these relationships are extrapolated to other areas. A more comprehensive relationship is necessary to predict maturity of crops in a number of different locations. No universal expressions have been worked out for vegetables and only a few in the case of cereals. Nuttonson's work is a notable example (35). He studied the variation in summation constant, at various stations using Marquis and Thatcher wheat, with base temperatures of 32 and 40°F respectively. The variations in the summation constants were found to be related to daylength. The variation between stations was markedly reduced by introducing the mean daylength into the calculations.

RELATIONSHIP BETWEEN DEGREE DAYS AND MEAN MONTHLY TEMPERATURE

Degree-day (heat unit) normals are often difficult to obtain and tedious to compile. Temperature means, however, are usually available from many stations or are easily computed from published data. Thom has worked out a relationship between monthly mean temperature and degree days which can be widely used for calculating degree days above or below any base temperature, provided the monthly mean temperature and the standard deviation of the monthly mean are available (50).

A Generalized Formula

The generalized formula is as follows:

$$D = N[(t - b) + L\sigma\sqrt{N}]$$

D is the normal degree days accumulated for the month concerned, b is the base temperature, σ is the standard deviation of the monthly mean temperature, and L is a proportionality coefficient obtained from Table II.

For example, the normal temperature for Ottawa during the month of May is $t = 54.7°$ F and the standard deviation $= 3.0$ (taken from Figure 5). H is then calculated in order to obtain the value of L from the following: $H = (t - b) / \sigma\sqrt{N}$. Proper substitution in this equation yields the value of .76 for H if our base temperature is 42° F. From Table 1, L is found to equal .01. Therefore, the normal degree-day accumulation above 42° F in May at Ottawa is:

$$D = 31\,[(54.7 - 42) + .01\,(3.0)\,(5.56)] = 394 \text{ degree days.}$$

This is a monthly value, but daily averages can be obtained by plotting the calculated monthly values on the mid day of each month and joining the points. The daily values are then read from the graph. Similarly degree days above or below any base temperature can be calculated for any station in Canada from the tables and charts in this review, provided the monthly mean temperatures are known.

Standard Deviation (σ) of Mean Monthly Temperatures

The standard deviation of monthly mean temperature is an expression of the probability of departure of the monthly means from the long-term values. For example, at the eastern foot of the Rockies particularly in Alberta and Montana, in the winter months, it is probable that the temperature will vary considerably each year above or below the long-term means (Figures 1 and 12). On the other hand, the low sigma value (standard deviation) in El Paso in January shows that no radical departure from the January average can be expected. That is to say, one January in El Paso is much the same as any other as far as temperature is concerned. (See Figures 1 to 12).

The highest values of sigma are associated with the chinook effect which finds its greatest development at the eastern base of the Rocky Mountains in Canada, especially in Alberta. Since the chinook causes great changes in temperature during the winter months in this area, it is presumable that chinooks also will cause considerable change in the mean monthly temperature of a month from year to year.

On the basis of the normal distribution it can be assumed that 68 per cent of the time, the monthly mean temperature will not vary more than sigma; 95 per cent of the time, it will not vary more than 2 sigma; and 99.7 per cent of the time it will not vary more than 3 sigma. This means then that at Lethbridge or Winnipeg the monthly mean for May will vary little more than 6° to 8°F above or below the long-term mean. During August, however, the monthly mean temperature will vary little more than 4° to 6°F above or below the long-term mean at any point in southern Canada.

DEGREE-DAY ACCUMULATION AT SELECTED SITES IN CANADA

Degree-day data were calculated from seven years (1950 to 1956) of hourly temperature data taken at nine meteorological stations at the following airports located in or near principal horticultural and cropping areas of Canada: Patricia Bay, B.C.; Penticton, B.C.; Lethbridge, Alta.; Medicine Hat, Alta.; Winnipeg, Man.; Windsor, Ont.; Trenton, Ont., Ottawa, Ont.; and Greenwood, N.S. All calculations are based on a 42°F threshold temperature. Growing degree-day accumulation based on threshold values between 35°F anad 50°F would be very similar, relatively speaking, to accumulations with a base temperature of 42°F.

Rate of Accumulation of Degree Days

The influence of the temperature air masses from the Pacific is felt at Patricia Bay. It is the earliest of the nine stations to accumulate heat units above 42°F (Figures 14 and 18). Accumulation rapidly falls behind, however, and during June, July, and August, the seasonal total at Patricia Bay is nearly the lowest of the nine stations. Degree-day accumulation at

Windsor begins soon after Patricia Bay and continues rapidly through the season, achieving the highest value of the nine stations listed by November 30. In descending order of total seasonal accumulation the stations rank as follows: Windsor, Trenton, Ottawa, Penticton, Medicine Hat, Greenwood, Winnipeg, Patricia Bay, Lethbridge, (Figures 15, 16 and 17). In order of earliness of beginning of accumulation in the spring they rank as follows: Patricia Bay, Penticton, Windsor and Trenton, Lethbridge, Greenwood, Ottawa and Medicine Hat, Winnipeg (Figure 18). By the end of June the stations have nearly assumed an order that continues till the end of November when accumulation of heat units above 42°F has virtually ceased.

Figures 16 and 17 also give some understanding of the differences in climate between areas that are agriculturally similar. While fruit growing is

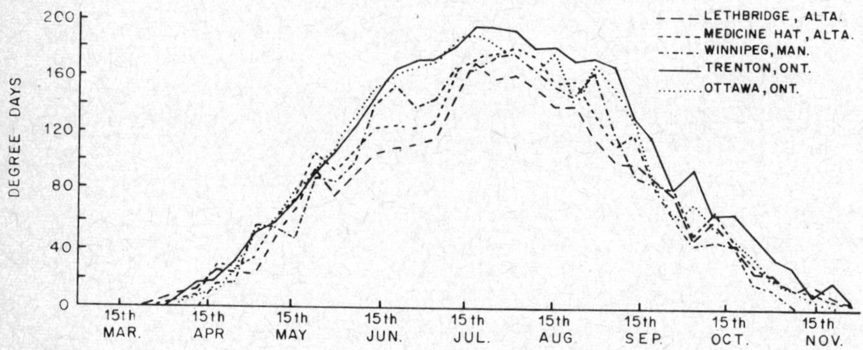

Figure 13. Number of degree days above 42°F accumulated per week (7-year average).

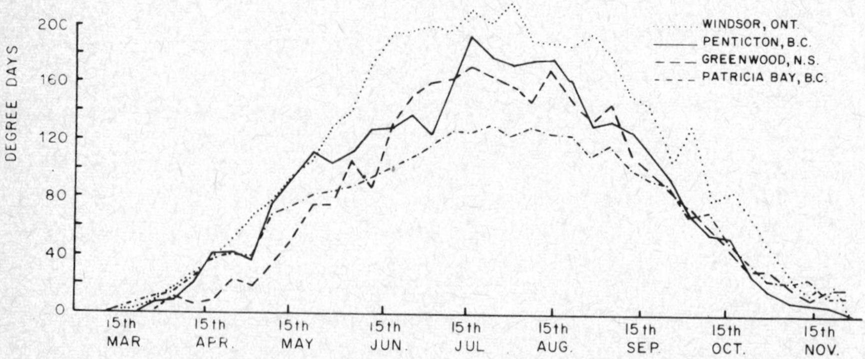

Figure 14. Number of degree days above 42°F accumulated per week (7-year average).

a large part of the agriculture of Penticton, Windsor, and Greenwood, the accumulation of heat at these locations is very different. This is the main reason for growing the varied fruits in these areas, e.g., grapes in southern Ontario, peaches and apricots in the Okanagan Valley and southern Ontario, apples in the Annapolis Valley and small fruits, strawberries, etc. on Vancouver Island. The Prairie stations are fairly similar in heat accumulation. Crops less sensitive to heat (some cereals) can do well at all three locations. However where vegetables (heat sensitive) are grown, greater care should be used in matching the area with the proper varieties. Caution should be exercised in trying to explain differences in agriculture between various locations on the basis of heat accumulation alone. While temperature markedly affects the growth of all crops, degree-day accumulation is one phenomenon in many that can be considered. Other important factors of climate that need clarification are diurnal fluctuation of temperature, day-length, rainfall, frost frequency, etc., not to mention a host of soil factors.

In Tables I to X are presented data on the weekly accumulation of degree-days above 42°F at the stations listed above. These data are the high, low, and seven-year mean of the weekly accumulation of heat. Graphic representations of these data are shown in Figures 13 and 14.

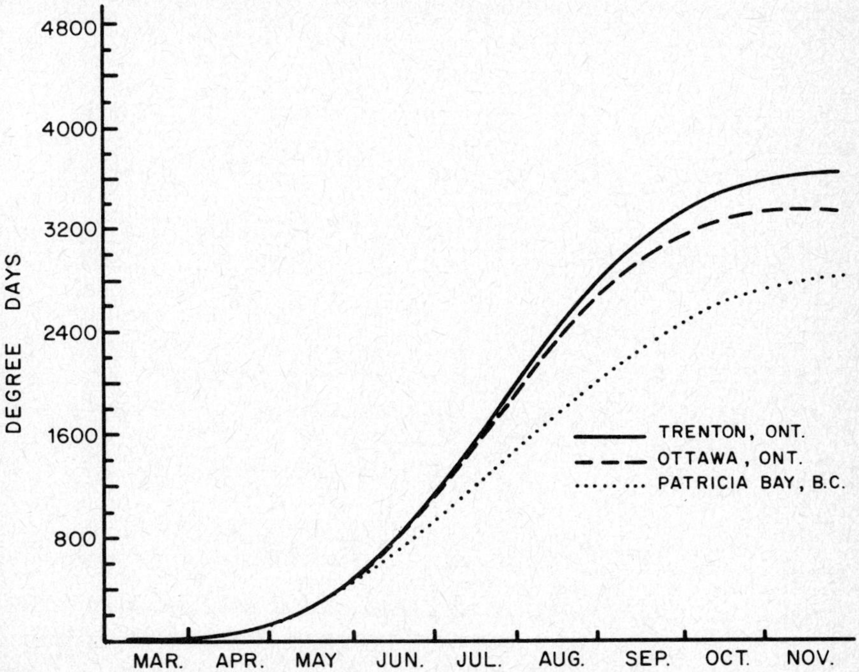

Figure 15. Number of degree days above 42°F accumulated per week (7-year average).

Temperatures Above 72°F and 90°F

The growth of cool weather crops such as peas is retarded when temperatures exceed 72° to 80°F (56). Predictions of harvest dates have been improved by discounting temperatures above a maximum (32). In Table XI are data showing the number of hours during the growing season when temperatures were above 72°F at the above nine stations. The location accumulating the largest number of heat units above 72°F is Windsor, followed in order by Trenton, Penticton and Ottawa, Medicine Hat, Winnipeg, Greenwood and Patricia Bay.

By discounting temperatures of 72°F at Lethbridge, the minimum improvement in the heat unit method of predicting harvest date for peas is 25 per cent; and at Windsor the improvement would be 36 per cent. If the daily temperatures were much above 72°F the improvement might be considerable. The above figures are calculated on an entire season rather than a per season basis. Nevertheless the indications are that this modification is worthy of note.

Often, however, with extreme high temperatures the refinement is no greater than the inherent accuracy of the method. In Figure 19 is presented the accumulated number of hours during the year when the temperature is

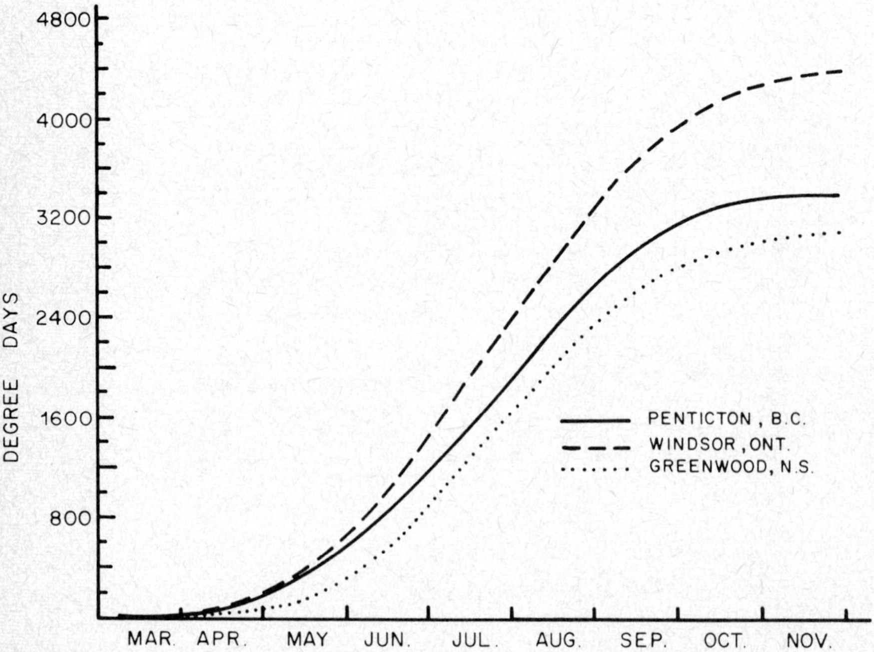

Figure 16. Number of degree days above 42°F accumulated (seasonal) (7-year average).

above 90°F. If it is assumed that 40 hours per growing season is the maximum amount of time the temperature is above 90°F, at Medicine Hat for example, the improvement in the method by discounting such temperatures will be 2½ per cent in the case of a crop of peas. Obviously application of the theory is not so accurate. However in other areas where very high temperatures occur frequently, the heat unit method may be greatly improved by the above procedure.

In decreasing order of number of hours above 90°F, the nine stations being considered are arranged as follows: Medicine Hat, Windsor, Penticton, Ottawa, Winnipeg, Lethbridge, Trenton, Greenwood, Patricia Bay. The arrangement according to beginning date of occurrence of temperatures above 90°F is as follows: Winnipeg, Medicine Hat, Lethbridge and Trenton, Windsor and Ottawa, Penticton, Greenwood. Temperatures above 90°F did not occur at Patricia Bay. Admittedly, a seven-year sample is not a large one, and the above arrangements might be slightly changed if a greater number of years had been considered.

SUMMARY

The degree day has many agricultural uses. Even in its simplest

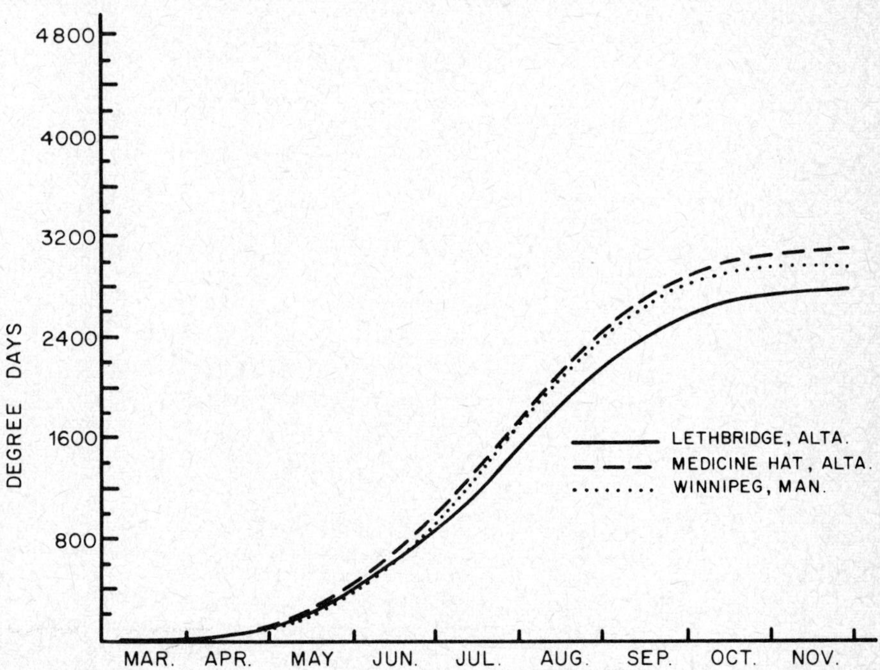

Figure 17. Number of degree days above 42°F accumulated (seasonal) (7-year average).

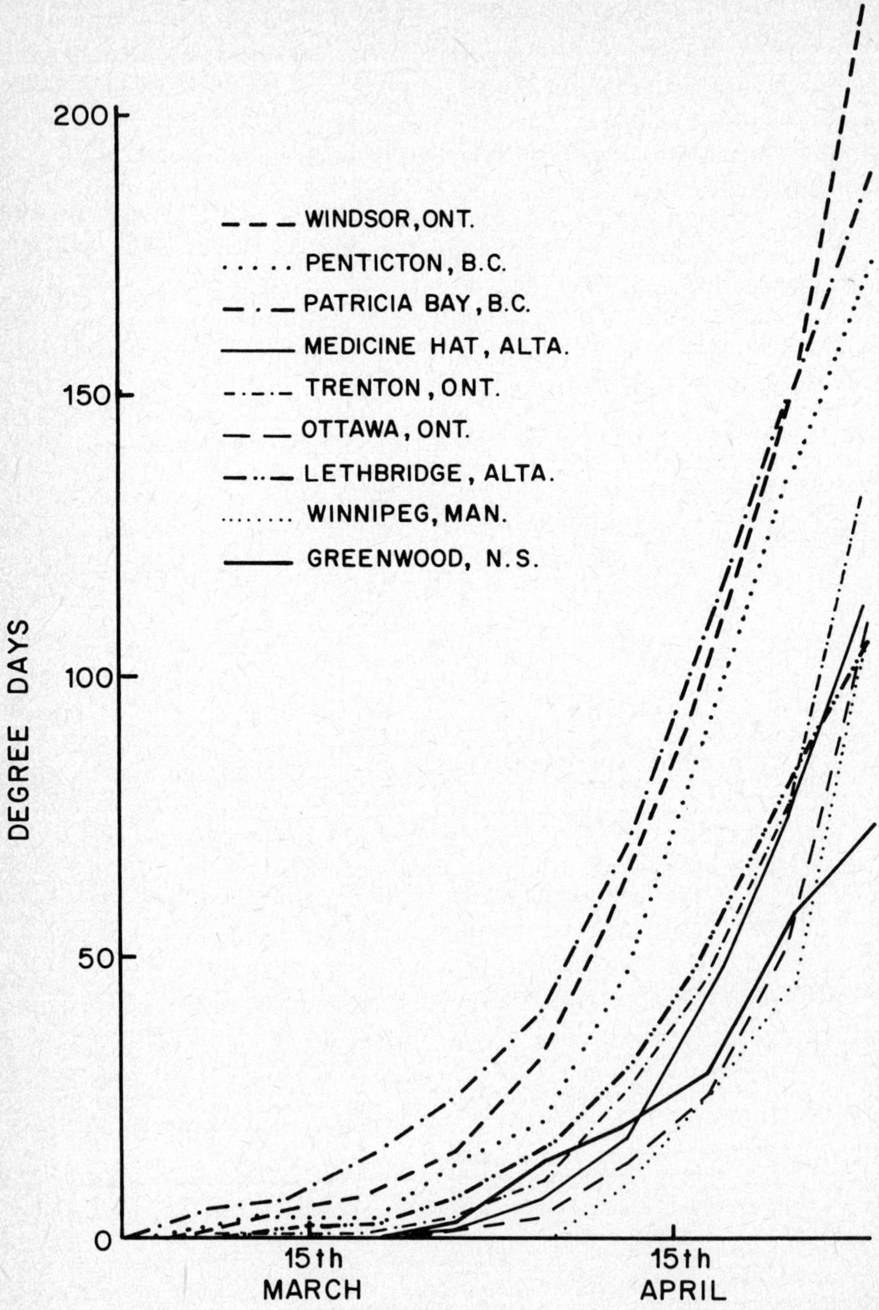

Figure 18. Early spring accumulation of degree days above 42°F (7-year average).

form, the heat unit concept has been a boon to many vegetable food processors and growers, not only in predicting data of harvest, but in arranging for an orderly one. The potentialities for use of the basic idea of the heat unit concept are great. It has possibilities for being refined for wider use in predicting date of maturation and harvest of fruit and vegetables, not only in single locations but over wide geographical areas. Canada has large expenses of potentially arable land yet untouched. If sufficient information were available concerning the heat requirements of crops, agriculturalists could more easily determine the probability of maturing various crops in our northern areas. Since soil genesis is markedly influenced by temperature, relationships might be found in temperature data that would explain some apparent incongruities in our soils. Undoubtedly one of the

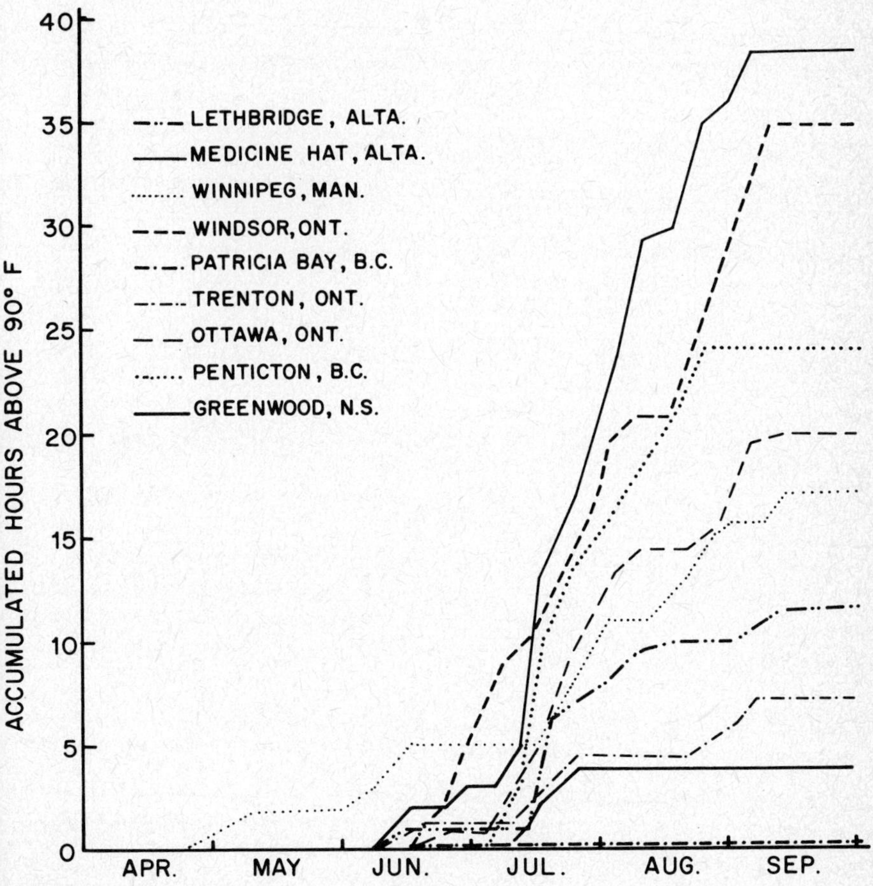

Figure 19. Total accumulated hours above 90°F (7-year average).

greatest contributions to be made by the heat unit or degree-day theory will be that of crop zonation. Comparatively speaking, this subject remains untouched.

For more readily comparative purposes, some of the pertinent information in the figures and tables is summarized in Table I.

APPENDIX

TABLE I. DEGREE-DAY COMPARISONS AT NINE STATIONS IN CANADA
(based on hourly recorded temperatures 1950-56 inclusive)

	(1)	(2)	(3)	(4)	(5)	(6)
Patricia Bay, B.C.	March 6	39	660	2900	150	0
Penticton, B.C.	" 16	85	865	3400	780	24
Lethbridge, Alta.	" 17	51	770	2825	560	12
Medicine Hat, Alta.	" 26	74	855	3140	769	38
Winnipeg, Man.	April 15	166	860	3020	670	18
Windsor, Ont.	March 19	93	1010	4380	1265	35
Trenton, Ont.	" 17	45	970	3720	840	7
Ottawa, Ont.	April 8	152	940	3440	780	20
Greenwood, N.S.	March 22	52	865	3100	480	4

(1) Average date on which accumulation of day degrees above 42°F begins.
(2) Average accumulation of day degrees during first 30 days following above date.
(3) Average accumulation of day degrees during July.
(4) Average accumulation of day degrees by end of season (November).
(5) Average number of hours with temperature 72°F or above during the growing season.
(6) Average number of hours with temperatures 90°F or above during the growing season.

TABLE II. THE FACTORS H, AND L, FOR USE IN COMPUTING DEGREE DAYS

H	L	H	L	H	L	H	L
−0.70	0.70	−0.32	0.39	0.05	0.17	0.42	0.05
− .69	.70	− .31	.38	.06	.17	.43	.05
− .68	.69	− .30	.38	.07	.16	.44	.04
− .67	.68	− .29	.37	.08	.16	.45	.04
− .66	.67	− .28	.36	.09	.15	.46	.04
− .65	.66	− .27	.36	.10	.15	.47	.04
− .64	.65	− .26	.35	.11	.14	.48	.04
− .63	.64	− .25	.34	.12	.14	.49	.03
− .62	.63	− .24	.34	.13	.13	.50	.03
− .61	.62	− .23	.33	.14	.13	.51	.03
− .60	.61	− .22	.32	.15	.13	.52	.03
− .59	.60	− .21	.32	.16	.12	.53	.03
− .58	.59	− .20	.31	.17	.12	.54	.03
− .57	.58	− .19	.30	.18	.11	.55	.03
− .56	.58	− .18	.30	.19	.11	.56	.02
− .55	.57	− .17	.29	.20	.11	.57	.02
− .54	.56	− .16	.29	.21	.10	.58	.02
− .53	.55	− .15	.28	.22	.10	.59	.02
− .52	.54	− .14	.27	.23	.10	.60	.02
− .51	.53	− .13	.27	.24	.09	.61	.02
− .50	.53	− .12	.26	.25	.09	.62	.02
− .49	.52	− .11	.25	.26	.09	.63	.02
− .48	.51	− .10	.25	.27	.08	.64	.02
− .47	.50	− .09	.24	.28	.08	.65	.01
− .46	.50	− .08	.24	.29	.08	.66	.01
− .45	.49	− .07	.23	.30	.07	.67	.01
− .44	.48	− .06	.23	.31	.07	.68	.01
− .43	.47	− .05	.22	.32	.07	.69	.01
− .42	.47	− .04	.22	.33	.07	.70	.01
− .41	.46	− .03	.21	.34	.06	.71	.01
− .40	.45	− .02	.20	.35	.06	.72	.01
− .39	.44	− .01	.20	.36	.06	.73	.01
− .38	.44	− .00	.19	.37	.06	.74	.01
− .37	.43	.01	.19	.38	.06	.75	.01
− .36	.42	.02	.18	.39	.05	.76	.01
− .35	.41	.03	.18	.40	.05	.77	.01
− .34	.41	.04	.17	.41	.05	.78	.00
− .33	.40						

For $H > 0.78$, L = O.
For $H < -0.70$, L = h.

TABLE III. DEGREE DAYS ABOVE 42°F
Accumulated Weekly at Patricia Bay, B.C. (1950-56)

Date		*High*	*Low*	*7-year Average*	*Date*		*High*	*Low*	*7-year Average*
March	7	15	0	5	August	1	1,811	1,284	1,579
	14	25	0	7		8	1,945	1,417	1,715
	21	30	3	14		15	2,069	1,537	1,847
	28	52	11	25		22	2,219	1,660	1,976
						29	2,323	1,781	2,094
April	4	63	18	40					
	11	90	45	68	September	5	2,442	1,920	2,218
	18	159	58	107		12	2,544	2,028	2,325
	25	212	71	149		19	2,670	2,160	2,424
						26	2,761	2,181	2,519
May	2	267	105	190					
	9	353	156	261	October	3	2,854	2,258	2,592
	16	438	204	339		10	2,942	2,305	2,669
	23	567	272	425		17	3,000	2,360	2,724
	30	683	335	512		24	3,036	2,404	2,765
						31	3,053	2,434	2,793
June	6	772	413	603					
	13	852	549	706	November	7	3,064	2,461	2,820
	20	961	650	813		14	3,094	2,481	2,849
	27	1,063	747	921		21	3,100	2,481	2,864
						28	3,117	2,485	2,879
July	4	1,214	832	1,042					
	11	1,355	934	1,175					
	18	1,516	1,059	1,311					
	25	1,675	1,179	1,449					

TABLE IV. DEGREE DAYS ABOVE 42°F
Accumulated Weekly at Penticton, B.C. (1950-56)

Date		*High*	*Low*	*7-year Average*	*Date*		*High*	*Low*	*7-year Average*
March	7	17	0	2	August	1	2,190	1,792	1,970
	14	23	0	4		8	2,365	1,960	2,155
	21	24	0	4		15	2,564	2,122	2,341
	28	43	0	13		22	2,756	2,261	2,509
						29	2,922	2,378	2,649
April	4	48	6	20					
	11	65	20	46	September	5	3,034	2,504	2,789
	18	125	60	90		12	3,160	2,623	2,927
	25	188	88	134		19	3,313	2,734	3,044
						26	3,420	2,838	3,146
May	2	245	117	173					
	9	324	174	252	October	3	3,496	2,878	3,216
	16	409	230	347		10	3,559	2,938	3,275
	23	562	322	462		17	3,626	2,979	3,330
	30	712	403	570		24	3,645	3,004	3,359
						31	3,646	3,004	3,375
June	6	850	513	685					
	13	966	688	818	November	7	3,648	3,020	3,385
	20	1,095	812	954		14	3,657	3,041	3,396
	27	1,241	979	1,098		21	3,657	3,091	3,404
						28	3,657	3,102	3,407
July	4	1,386	1,080	1,230					
	11	1,556	1,239	1,399					
	18	1,767	1,468	1,602					
	25	2,000	1,624	1,788					

TABLE V. DEGREE DAYS ABOVE 42°F
Accumulated Weekly at Lethbridge, Alta. (1950-56)

Date		*High*	*Low*	*7-year Average*
March	7	3	0	0
	14	7	0	2
	21	9	0	3
	28	22	0	7
April	4	40	1	16
	11	47	2	28
	18	93	34	55
	25	179	34	80
May	2	264	34	104
	9	313	72	152
	16	409	123	222
	23	521	169	322
	30	607	232	402
June	6	748	323	498
	13	851	441	608
	20	955	581	724
	27	1,050	740	841
July	4	1,192	832	962
	11	1,331	950	1,113
	18	1,464	1,170	1,289
	25	1,656	1,345	1,454
August	1	1,807	1,513	1,622
	8	1,950	1,681	1,781
	15	2,104	1,833	1,927
	22	2,241	1,961	2,072
	29	2,385	2,029	2,193
September	5	2,439	2,077	2,299
	12	2,541	2,183	2,402
	19	2,675	2,288	2,496
	26	2,769	2,300	2,578
October	3	2,872	2,353	2,627
	10	2,927	2,438	2,689
	17	2,950	2,467	2,735
	24	3,015	2,467	2,761
	31	3,057	2,475	2,784
November	7	3,074	2,476	2,797
	14	3,091	2,477	2,814
	21	3,092	2,477	2,820
	28	3,092	2,477	2,824

TABLE VI. DEGREE DAYS ABOVE 42°F
Accumulated Weekly at Medicine Hat, Alta. (1950-56)

Date		*High*	*Low*	*7-year Average*	*Date*		*High*	*Low*	*7-year Average*
March	7	0	0	0	August	1	1,988	1,674	1,823
	14	2	0	0		8	2,161	1,865	2,004
	21	2	0	0		15	2,331	2,042	2,165
	28	47	0	2		22	2,478	2,195	2,329
						29	2,649	2,319	2,468
April	4	14	0	7					
	11	54	2	17	September	5	2,729	2,403	2,585
	18	100	7	48		12	2,852	2,519	2,709
	25	191	34	76		19	2,954	2,621	2,798
						26	3,073	2,635	2,882
May	2	309	34	112					
	9	376	94	177	October	3	3,174	2,695	2,933
	16	484	180	256		10	3,215	2,800	3,001
	23	608	251	369		17	3,228	2,849	3,048
	30	710	350	467		24	3,292	2,855	3,077
						31	3,331	2,855	3,099
June	6	864	447	579					
	13	977	556	708	November	7	3,347	2,855	3,112
	20	1,087	659	840		14	3,357	2,855	3,122
	27	1,200	827	969		21	3,357	2,855	3,126
						28	3,357	2,856	3,128
July	4	1,348	959	1,102					
	11	1,497	1,148	1,272					
	18	1,641	1,343	1,452					
	25	1,817	1,502	1,636					

TABLE VII. DEGREE DAYS ABOVE 42°F
Accumulated Weekly at Winnipeg, Man. (1950-56)

Date		*High*	*Low*	*7-year Average*	*Date*		*High*	*Low*	*7-year Average*
March	7	0	0	0	August	1	2,195	1,507	1,811
	14	0	0	0		8	2,397	1,699	1,982
	21	0	0	0		15	2,599	1,850	2,141
	28	0	0	0		22	2,804	1,948	2,291
						29	2,992	2,064	2,460
April	4	2	0	0					
	11	58	0	0	September	5	3,154	2,218	2,587
	18	91	0	26		12	3,215	2,316	2,682
	25	163	0	44		19	3,338	2,403	2,771
						26	3,367	2,515	2,835
May	2	345	0	103					
	9	442	0	161	October	3	3,397	2,546	2,880
	16	484	34	210		10	3,480	2,584	2,929
	23	592	97	304		17	3,507	2,612	2,974
	30	674	199	394		24	3,512	2,616	2,991
						31	3,524	2,680	3,003
June	6	800	328	500					
	13	958	412	646	November	7	3,524	2,680	3,004
	20	1,095	572	806		14	3,524	2,680	3,006
	27	1,245	713	950		21	3,524	2,680	3,006
						28	3,524	2,680	3,006
July	4	1,420	840	1,100					
	11	1,590	1,032	1,274					
	18	1,776	1,143	1,445					
	25	1,992	1,303	1,627					

TABLE VIII. DEGREE DAYS ABOVE 42°F
Accumulated Weekly at Windsor, Ont. (1950-56)

Date		*High*	*Low*	*7-year Average*	*Date*		*High*	*Low*	*7-year Average*
March	7	10	0	1	August	1	2,843	2,202	2,456
	14	15	0	5		8	3,091	2,400	2,655
	21	23	0	8		15	3,295	2,588	2,854
	28	46	0	15		22	3,552	2,766	3,046
						29	3,753	2,975	3,250
April	4	48	25	33					
	11	126	28	62	September	5	3,923	3,145	3,436
	18	219	58	101		12	4,080	3,304	3,596
	25	308	78	154		19	4,290	3,432	3,745
						26	4,414	3,521	3,857
May	2	410	93	226					
	9	535	153	308	October	3	4,535	3,634	3,993
	16	645	278	406		10	4,677	3,700	4,077
	23	783	366	516		17	4,758	3,827	4,167
	30	931	487	650		24	4,811	3,924	4,241
						31	4,860	4,010	4,292
June	6	1,106	583	797					
	13	1,243	798	977	November	7	4,874	4,076	4,322
	20	1,442	1,013	1,179		14	4,904	4,083	4,339
	27	1,617	1,225	1,386		21	4,908	4,099	4,366
						28	4,911	4,099	4,370
July	4	1,862	1,412	1,594					
	11	2,114	1,607	1,798					
	18	2,353	1,800	2,018					
	25	2,589	2,005	2,230					

TABLE IX. DEGREE DAYS ABOVE 42°F
Accumulated Weekly at Trenton, Ont. (1950-56)

Date		*High*	*Low*	*7-year Average*	*Date*		*High*	*Low*	*7-year Average*
March	7	6	0	1	August	1	2,519	1,785	2,088
	14	6	0	1		8	2,765	1,975	2,274
	21	6	0	1		15	2,973	2,171	2,462
	28	22	0	4		22	3,222	2,333	2,639
						29	3,506	2,506	2,819
April	4	34	1	10					
	11	54	5	27	September	5	3,550	2,660	2,992
	18	99	13	46		12	3,671	2,791	3,127
	25	150	16	80		19	3,822	2,890	3,249
						26	3,908	2,935	3,335
May	2	238	34	133					
	9	329	58	193	October	3	4,000	3,027	3,434
	16	415	138	272		10	4,121	3,074	3,501
	23	527	188	370		17	4,212	3,150	3,568
	30	676	373	481		24	4,260	3,219	3,621
						31	4,303	3,265	3,657
June	6	844	381	612					
	13	1,008	557	764	November	7	4,322	3,296	3,684
	20	1,207	736	937		14	4,332	3,311	3,693
	27	1,371	911	1,113		21	4,332	3,321	3,711
						28	4,332	3,321	3,718
July	4	1,596	1,079	1,292					
	11	1,832	1,249	1,480					
	18	2,056	1,422	1,683					
	25	2,281	1,614	1,887					

TABLE X. DEGREE DAYS ABOVE 42°F
Accumulated Weekly at Ottawa, Ont. (1950-56)

Date		*High*	*Low*	*7-year Average*	*Date*		*High*	*Low*	*7-year Average*
March	7	0	0	0	August	1	2,508	1,655	2,040
	14	0	0	0		8	2,718	1,843	2,209
	21	0	0	0		15	2,926	2,032	2,393
	27	13	0	2		22	3,160	2,182	2,540
						29	3,320	2,314	2,716
April	4	25	0	4					
	11	42	0	13	September	5	3,474	2,470	2,873
	18	50	6	26		12	3,571	2,527	3,008
	25	98	6	53		19	3,721	2,646	3,099
						26	3,781	2,672	3,163
May	2	179	21	106					
	9	269	36	165	October	3	3,859	2,722	3,237
	16	398	115	251		10	3,959	2,765	3,296
	23	517	151	349		17	4,017	2,845	3,344
	30	679	225	464		24	4,038	2,896	3,382
						31	4,062	2,919	3,401
June	6	851	362	603					
	13	1,006	523	760	November	7	4,070	2,980	3,415
	20	1,214	693	927		14	4,073	2,982	3,420
	27	1,362	858	1,099		21	4,073	2,982	3,429
						28	4,073	2,982	3,432
July	4	1,598	1,015	1,276					
	11	1,860	1,177	1,467					
	18	2,082	1,334	1,662					
	25	2,281	1,507	1,851					

TABLE XI. DEGREE DAYS ABOVE 42°F
Accumulated Weekly at Greenwood, N.S. (1951-56)

Date		*High*	*Low*	*7-year Average*	*Date*		*High*	*Low*	*7-year Average*
March	7	0	0	0	August	1	1,770	1,426	1,679
	14	2	0	0		8	1,958	1,569	1,835
	21	2	0	0		15	2,148	1,729	2,013
	28	11	0	3		22	2,313	1,872	2,169
						29	2,473	1,988	2,308
April	4	48	0	13					
	11	65	1	19	September	5	2,620	2,135	2,460
	18	70	7	30		12	2,728	2,240	2,578
	25	89	16	55		19	2,825	2,338	2,681
						26	2,927	2,373	2,768
May	2	113	18	74					
	9	200	33	110	October	3	3,017	2,419	2,842
	16	280	66	162		10	3,092	2,475	2,905
	23	345	118	238		17	3,141	2,511	2,955
	30	407	154	318		24	3,152	2,557	2,989
						31	3,183	2,557	3,025
June	6	484	314	431					
	13	586	398	523	November	7	3,197	2,597	3,046
	20	744	536	661		14	3,229	2,618	3,059
	27	900	668	817		21	3,254	2,634	3,078
						28	3,265	2,647	3,097
July	4	1,080	824	985					
	11	1,250	960	1,156					
	18	1,435	1,123	1,337					
	25	1,602	1,281	1,510					

References

Abbe, C., "A first report in the relations between climates and crops," U.S.D.A. Weather Bureau Bul. 36, W.B. No. 342, P.O. 167-294 (1905).

Anonymous, "Relation of temperature to maturity in peas and sweet corn," Crop Prod. Memo., Continental Can Co. (1949).

Anonymous, "Report of survey of companies using the heat unit system – 1949," Crop. Prod. Memo., Continental Can. Co. (1950).

Appleman, C. O., "Forecasting the date of best canning stage for sweet corn," Md. Exp. Sta. Bul. 254 (1923).

Appleman, C. O. and S. V. Eaton, "Evaluation of climatic temperature efficiency for ripening processes in sweet corn," *Jour. Agric. Res.* Vol. 20 (1921), pp. 795-805.

Baker, G. A. and R. M. Brooks, "Effect of temperature on number of days from full bloom to harvest of apricot and prune fruits," *Proc. Am. Soc. Hort. Sci.*, Vol. 45 (1944), pp. 95-103.

Barnard, J. D., "Heat units as a measure of canning crop maturity," *The Canner* (April 17, 1948).

Bomalaski, H. H., "Growing degree days; how to apply this unit to measure maturity of crops," *Food Packer*, Vol. 29, no. 8, pp. 51-59; Vol. 29, no. 9, pp. 47-51.

Boswell, V. R., "The influence of temperature upon the growth and yield of garden peas," *Proc. Am. Soc. Hort. Sci.*, Vol. 22 (1924), pp. 178-87.

Boswell, V. R., "Factors influencing yield and quality of peas – biological and chemical studies," Maryland Agric. Exp. Sta. Bul. 306 (1929).

Boswell, V. R., "A study of the temperature, daylength, and development interrelationship of spinach varieties in the field," *Am. Soc. Hort., Sci. Proc.*, Vol. 32 (1934), pp. 549-57.

Brooks, R. M., Effect of daily temperatures on the date of picking apricots and pears," *The Blue Anchor*, Vol. 22, no. 1, (February 1945).

Carder, A. C., "Growth and development of some field crops as influenced by climatic phenomena at two diverse latitudes," *Can. Jour. Pl. Sci.*, Vol. 37 (1957), pp. 392-406.

Culpepper, C. W. and H. W. Moon, "Effect of temperature on the rate of elongation of the stems of asparagus grown under field conditions," *Pl. Phys.*, Vol. 14 (1939), pp. 255-70.

Culpepper, C. W. and C. A. Magoon, "Studies upon the relative merits of sweet corn varieties for canning purposes and the relation of maturity of corn to the quality of the canned product," *J. Agr. Res.*, Vol. 28 (1924), pp. 403-43.

Daubenmire, R. F., *Plants and environment*, (New York, John Wiley & Sons, Inc., 1947) pp. 186-97, 212-19.

Davis, J. F., "Some factors and their interactions affecting the yield of sweet corn on organic soil," *Mich. Quart. Bull.*, Vol. 34, no. 3, (1952), pp. 253-61.

Erwin, T. S., "Climatic and varietal factors in pumpkin production," *The Canner* (Convention number, 1929) pp. 71-72.

Fetcher, R. F., "Soil type, air temperature affect heat units for peas," *Food Packer*, Vol. 31, no. 4 (1950), 33 pp.

Gould, W. A., "Here's a heat unit guide for 47 varieties of snap beans," *Food Packer* (March, 1950).

Gould, W. A., F. A. Krantz, and J. Mavis, "Quality evaluation of fresh, frozen, and canned yellow sweet corn," *Food Technol.*, Vol. 5 (1951), pp. 175-79.

Hardenburg, E. V., "Ecological factors affecting tuber set in potatoes," *Potato Assoc. Amer. Proc.* Vol. 10 (1924), pp. 165-72.

Haller, M. H., "Evaluation of indexes of maturity for apples," U.S.D.A. Tech. Bul. No. 1003 (January, 1950).

Huelsen, W. A., "Sweet corn 1954," Interscience Publications Inc., New York, N.Y. (1954).

Katz, Y. H., "The relationship between heat unit accumulation and the planting of canning peas," *Jour. Am. Soc. Agron.*, Vol. 44 (1952), pp. 74-78.

Kitchin, J. T., "The heat unit theory for vegetables," *Hort. News*, Vol. 34, no. 5 (May, 1953).
Kitchin, J. T., "Evaluation of the heat unit theory in measuring the interval from field setting to first harvest of Rutgers tomatoes," M.Sc. Thesis, Rutgers University, New Brunswick, N.J. (1953).
Lana, E. P. and E. S. Haber, "Seasonal variability as indicated by cumulative degree hours with sweet corn," *Proc. Am. Soc. Hort. Sci.* Vol. 59 (1952), pp. 389-91.
LeCalvex, L., "Climate Calendar gives crops on time," *N.J. Farm and Garden*, Vol. 22, no. 12 (1951), pp. 18-21.
Lehenbauer, P.A., "Growth of maize seedlings in relation to temperature," *Physiol. Res.*, Vol. 1 (1914), pp. 247-88.
Livingston, B. E., "Physiological temperature indices for the study of plant growth in relation to climatic factors," *Phy. Res.*, Vol. 1 (1916), pp. 399-420.
Madariago, F. J. and J. E. Knott, "Temperature summations in relation to lettuce growth," *Am. Soc. Hort. Sci. Proc.*, Vol. 58 (1951), pp. 147-52.
Magoon, C. A. and C. W. Culpepper, "Response of sweet corn to varying temperatures from time of planting to canning maturity," *U.S.D.A. Tech. Bul.* 312 (1952).
Nuttonson, M. Y., "Some preliminary observations of phenological data as a tool in the study of photoperiodic and thermal requirements of various plant material," Vernalization and Photoperiodism. A symposium (1948), pp. 217-43.
Nuttonson, M. Y., "Wheat climate relationships and the use of phenology in ascertaining the thermal and photothermal requirements of wheat," *Amer. Inst. Crop. Ecol.* (1955).
Phillips, E. E., "Heat unit summation theory applied to canning crops," *Canner*, Vol. 110, no. 10 (1950), pp. 10-26.
Price, H. L., "The application of meteorological data in the study of physiological constants," *Va. Agric. Exp. Sta. Ann. Rep.* (1914), pp. 1708-1714.
Rauchenstein, E., "Economic aspects of the cantaloupe industry," *Univ. Cal. Agric. Exp. Sta.* Bul. 419 (1927).
Reath, A. N. and S. H. Wittwer, "The effects of temperature and photo period on the development of pea varieties," *Proc. Am. Soc. Hort. Sci.*, Vol. 60 (1952), pp. 301-10.
Reaumur, R. A. F. de 1735, Observations du thermomètre, faites à Paris pendant l'année 1735, comparée avec celles qui ant été faites sous la ligne, à l'Isle de France, à Alger et en quelques-unes de nos isles de l'Amérique. Paris Memoirs, Acad. Sci.
Robertson, G. W., "Some Agro-meteorological problems in Canada," *Roy. Met. Soc. Can. Br.*, Vol. 4, no. 2 (1935).
Sayre, C. B., "Heat unit method of forecasting maturity of peas," *Ann. Rep. N.Y. Agric. Exp. Sta.*, Geneva, N.Y. (1949), p. 35.
Scott, G. C., "Quality control in canning peas and corn," *Food Packer*, Vol. 27, no. 13 (1946), pp. 40-42.
Seaton, H. L., "Scheduling plantings and predicting harvest maturities for processing vegetables," Continental Can Co. Research Department (1955).
Seaton, H. L. and J. M. Huffington, "Raw product quality control," Continental Can Co., Bul. 26 (1951).
Smith, O., "Relation of temperature to anthesis and blossom drop of the tomato, together with a histological study of the pistils," *Jour. Ag. Res.*, Vol. 44 (1932) pp. 183-90.
Stevens, N. E. and C. H. Higgins, "Temperature in relation to quality of sweet corn," *Jour. Ag. Res.*, Vol. 17 (1919), pp. 275-84.
Sumner, A. R., "Standard deviation of monthly mean temperatures in Anglo-America," *Geo. Rev.*, Vol. 43 (1953), pp. 50-59.
Thom, H. C. S., "The rational relationship between heating degree days and temperature," *Mo. Weather Rev.*, Vol. 82 (1954), pp. 1-6.
Thom, H. C. S., "Standard deviation of monthly average temperature," Climatological Services Div., U.S. Weather Bureau, Washington, D.C. (1954).

Thornthwaite, C. W. and J. R. Mather, "Climate in relation to Crops," *Met. Mono.*, Vol. 2, no. 8 (1954).

Wallis, E. P., "Predicting maturity dates from temperature records," *Canning Trade* (April 10, 1950), pp. 7-8.

Weinburger, J. H., "Influence of temperature following bloom on fruit development period of Elberta peach," *Am. Soc. Hort. Sci. Proc.*, Vol. 51 (1948), pp. 175-78.

Went, F. W., "Plant growth under controlled conditions," "Thermoperiodicity in growth and fruiting of the tomato," *Amer. Jour. Bot.*, Vol. 31 (1944), pp. 135-50.

Went, F. W., "The response of plants to climate," *Science,* Vol. 112 (1950), pp. 489-94.

Went, F. W., "Climate and Agriculture," *Sci. Am.* Vol. 196 (June 1957), pp. 82-94.

Winkler, A. J. and W. O. Williams, "The heat required to bring Tokay grapes to maturity," *Proc. Am. Soc. Hort. Sci.*, Vol. 37 (1939), pp. 650-53.

Winkler, A. J. and W. O. Williams, "Relation of heat summation to time of maturing and palatability of table grapes," *Cal. Agric.* (March, 1948).

6

Drought Patterns in the Canadian Prairies

A. H. Laycock

Many procedures have been employed in mapping drought patterns. Some of the most useful are those described by C. W. Thornthwaite and his associates in 1948 and subsequent years. We have used these and other procedures as bases for mapping drought frequency, intensity, duration and timing patterns for the Prairie Provinces Water Board.

One of our major objectives was to define regional drought patterns more closely than had been done previously. To do this, we employed the data of the 575 station locations listed in the Canadian Monthly Record with one or more complete years of record in the period 1921-1950. Only 91 stations had over 20 years of record and there were relatively few data for some regions but the major patterns appear to be well established.

Average annual precipitation (Figure 1) ranges from less than 12″ (305mm) in the driest areas to well over 50″ in some of the higher mountain areas. The major agricultural regions receive between 12 and 20″. We have made conservative allowances for the effects of topography upon precipitation patterns. We might reasonably assume that the moisture deficiency, or drought, for any year will be equal to potential evapotranspiration minus precipitation plus or minus allowances for water surpluses and storage change. Water surpluses are small—less than 3″ in most agricultural

Reprinted from Publication No. 51, Symposium of the International Association for Scientific Hydrology (Helsinki, 1960), pp. 34-47.

Figure 1. Average Precipitation.

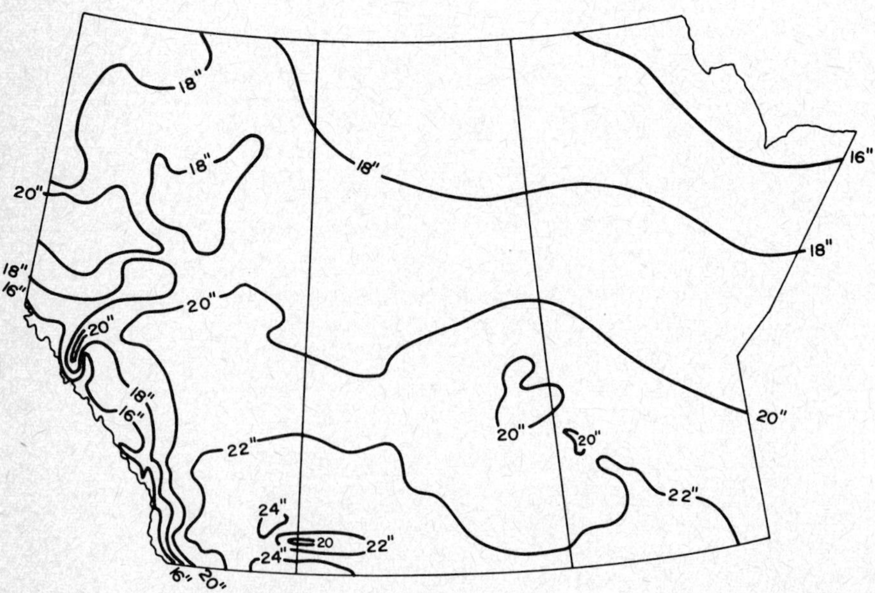

Figure 2. Average Potential Evapotranspiration.

regions—thus effective precipitation is not much smaller than actual precipitation.

Potential evapotranspiration (Figure 2)—the amount of water that would be evaporated or transpired if the surface were completely covered with vegetation and there was sufficient moisture in the soil at all times for its use — ranges from 20 to just over 24″ in the agricultural regions.

The average deficit (Figure 3) ranges from about 12″ in the driest areas to between 4 and 6″ in the more humid areas. The deficit patterns in the drier areas are similar to those mapped by Sanderson in 1948 but the more humid area deficits are larger. If certain months had deficits in all years, we would find that deficits based on average monthly temperature and precipitation data would be the same as average deficits based on calculations for all years of the period. Since many of the months have surpluses in some years and deficits in others, particularly in the more humid parts of the Prairies, we must calculate water balance values for all years if we wish to determine average deficits. In making these calculations, we have obtained abundant information for establishing many different frequency-intensity and other deficit patterns. The more detailed studies are essential if we are to obtain adequate record of moisture deficits in non-irrigated areas and water needs in irrigated and irrigable areas.

In most sub-humid and semi-arid regions there is a significant variation in drought intensity from one year to the next (Figure 4). The agricultural areas experienced deficits of from 8 to 16″ in the driest year of the period 1921-1950. If we assume that crop yields will be marginal (little or no return is provided to the farmer for his work) if a deficit of 8″ is experienced, almost all of the agricultural areas with average soil moisture storage capacities suffered from crop failure in at least one year of the period.

In the most humid years (Figure 5) moisture deficits were small or absent. Areas with deficit of over 2″ were widely scattered in the drier parts of the Prairies and owed their distribution largely to chance since adjoining areas that are normally as dry experienced no deficit.

The regional patterns of drought are not necessarily the same from one year to the next. In 1927 (Figure 6), most of the southern regions that are normally dry were relatively moist and the Peace River District in the north was moderately dry. In 1936 (Figure 7) a large part of the southern plains was very dry. The Peace River District was relatively moist. The severity of this drought in the south was almost matched in 1937 and much of the reputation of the 1930's as dry years can be attributed to this two year combination. In 1950 (Figure 8), the major drought area was centred well to the west in Alberta and the northern and eastern areas were moist. We can find little support for different theories of drought cycles in this region for the period studied.

Figure 3. Average Deficit.

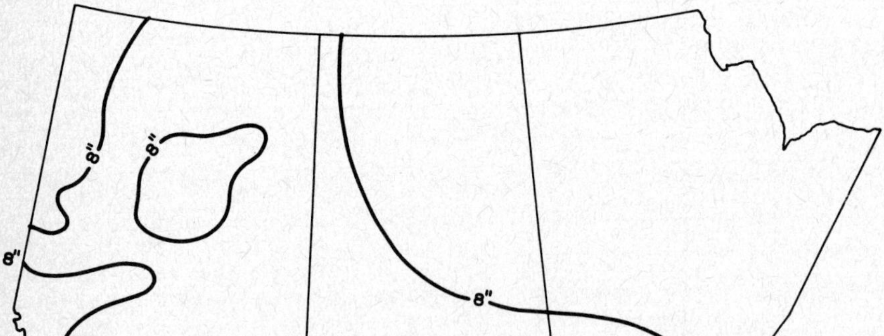

Figure 4. Maximum Deficit.

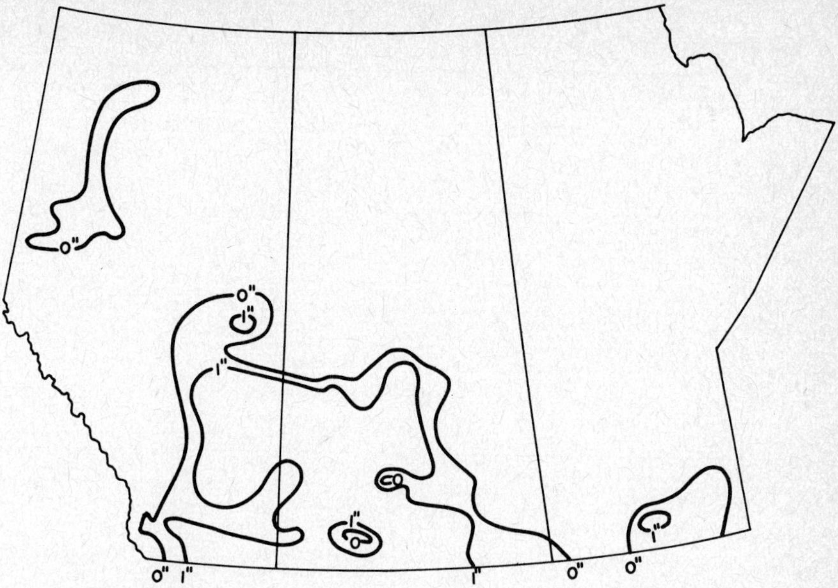

Figure 5. Minimum Deficit.

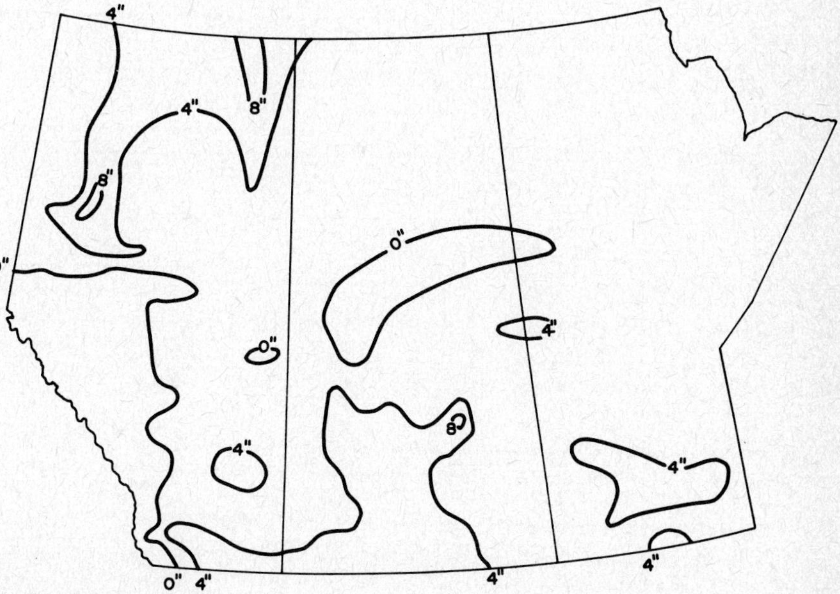

Figure 6. Deficit – 1927.

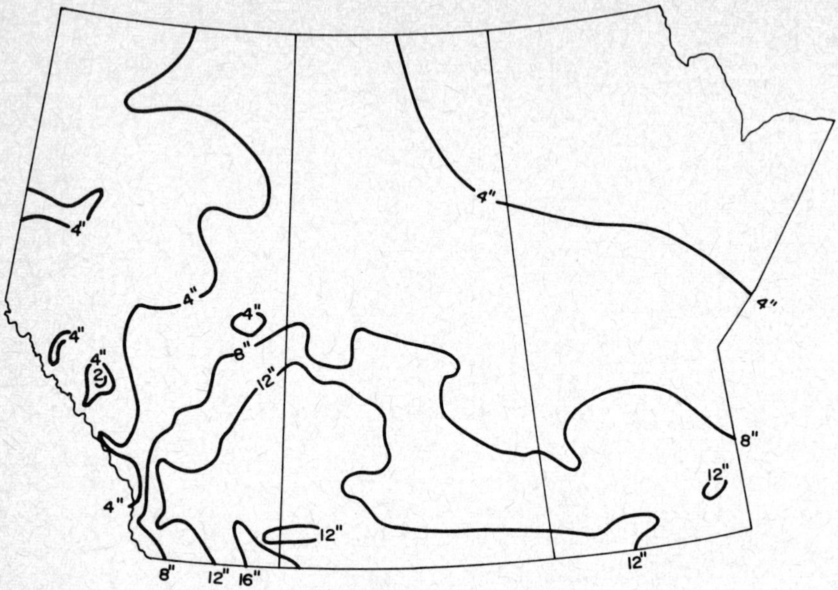

Figure 7. Deficit – 1936.

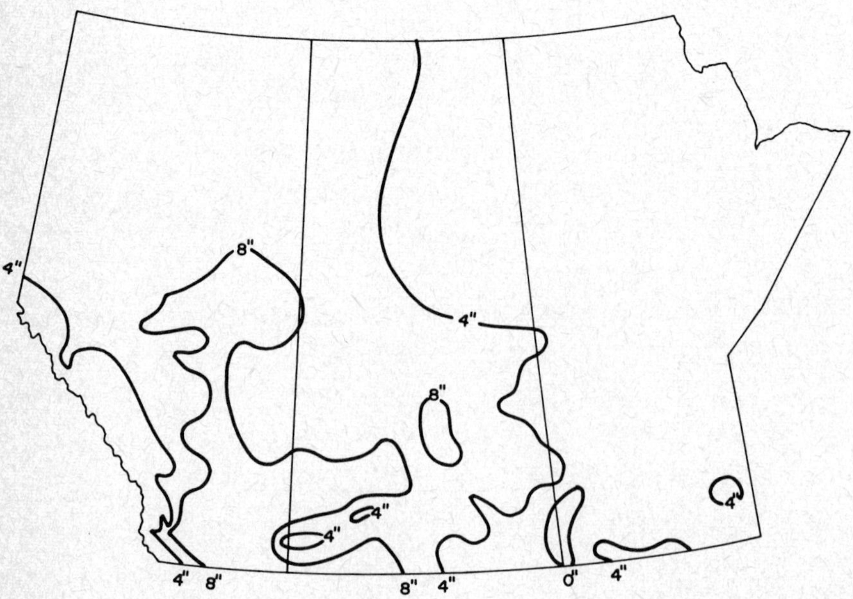

Figure 8. Deficit – 1950.

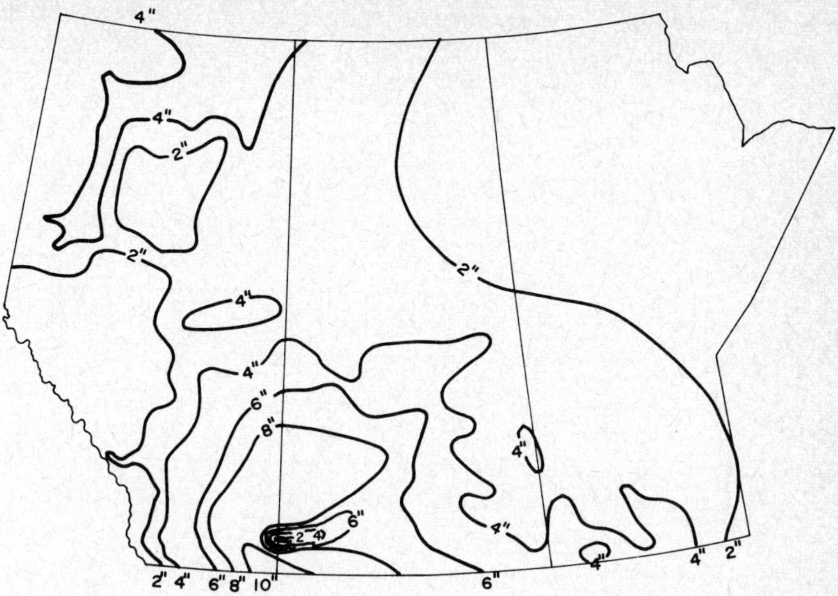

Figure 9. Moisture Deficit – Lower Quartile.

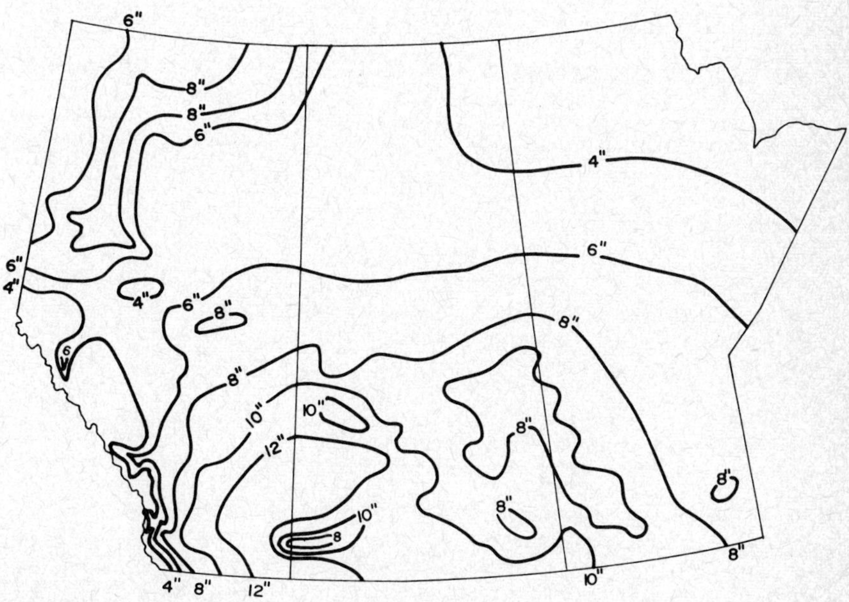

Figure 10. Moisture Deficit – Upper Quartile.

Figure 11. Average Deficit.

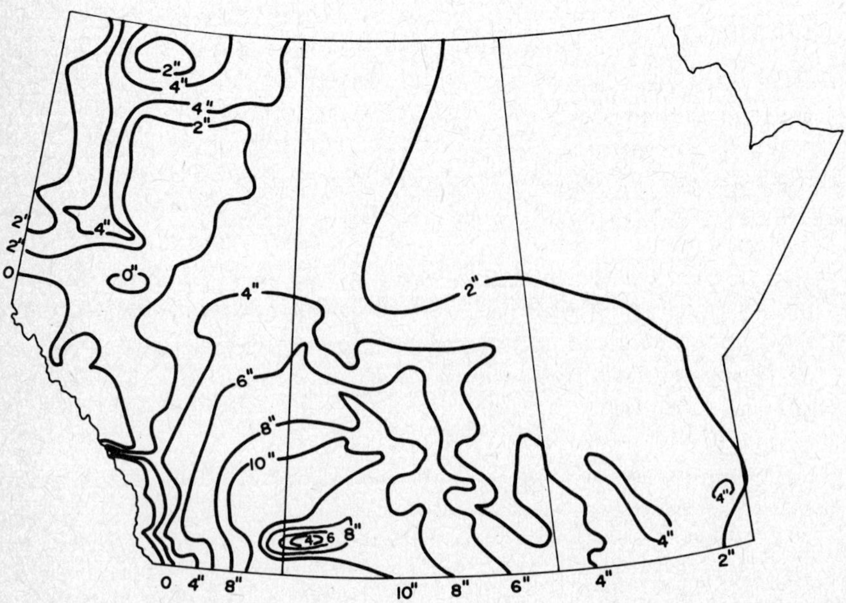

Figure 12. Average Deficit.

Information concerning the extremes of drought intensity is useful in planning, but agriculturists are most interested in frequency patterns of intermediate drought intensities. These can be shown in various ways. The wettest quarter of the years (Figure 9) have deficits of less than 4″ in the more humid agricultural areas and over 8″ in the driest areas. If we assume that at least one quarter of the years should have droughts of less than 8″ if dry farming is to be practised successfully, we might suggest that crops should not be grown in areas drier than this unless supplementary water supplies are available. Irrigation is most likely to be successful in these dry regions, at least partly because farmers are least likely to defer irrigating in their hope for rain.

The driest quarter of the years (Figure 10) have quite intense droughts and crops may be quite marginal in many of the more humid regions. Supplementary irrigation and moisture conservation practices can be of value in almost all agricultural areas in almost all years. The low intensity of drought in the drier years in west-central Alberta was an attraction to settlers forced out of the driest regions in the 1920's and 30's.

One of the major problems in mapping moisture deficit patterns is that the water supply available for plant use varies appreciably in small areas because of variations in soil moisture retention storage capacity and in the capacity of plants to utilize the stored moisture. It is conceivable that detailed maps of soil moisture storage capacities might be prepared, but such maps are not yet available. It is possible for us to determine approximately local deficiency patterns if we draw maps for each of a number of storage capacities and apply the one that is appropriate for local soil and growth conditions. The values that we mapped were ½″, 1″, 2″, 4″, 6″, 8″ and 12″. The maps discussed previously are based on an average storage capacity of 4″—approximately the amount of moisture that light loam soils might hold for use by cereal crops with moderate root development.

Plants in coarse textured sandy soil areas with limited moisture retention storage capacities (Figure 11) are subject to severe droughts in even the more humid regions in average years. These patterns indicate that the greater yields of much of Southern Manitoba can be attributed largely to the high moisture storage capacities and the appreciable filling of these capacities before the growing season begins.

Fine textured clays and clay loams have large capacities for moisture retention storage (Figure 12). If alfalfa or other crops with deep and extensive rooting habits are grown on these soils and storage capacities are filled at the start of the growing season the moisture deficiencies will tend to be small. Under these conditions the average annual drought in the more humid regions is only 3 to 4″. In drier regions, the fine textured soils can retain better moisture supplies than coarse soils, but they are rarely filled to capacity during fall, winter and spring and the storage advantage is frequently not used.

Irrigated crops in the drier regions require approximately 4″ more water in sandy soil areas than in areas with clay soils. We might also conclude that supplemental irrigation might be utilized to a much greater degree in sandy than in clay soil areas, and for shallow than deep rooted plants in the more humid regions.

The percentage of years with over 8″ deficit in areas with sandy soils (Figure 13) is very high in all regions. In contrast, the percentage of years with this deficit in areas with clay soils (Figure 14) is much smaller. The advantages of crop cultivation in the better soil areas are apparent.

We have drawn numerous maps to show drought conditions for specific crops by varying the length of growing season and moisture storage capacity in our calculations. For example, in the south-western plains, the germinating wheat seeds and seedlings have very limited access to soil moisture storage in May (Figure 15). Drought may occur after ½″ of moisture has been withdrawn from storage and deficiency calculations based on this capacity might be useful. In June, the plants will be able to utilize more of the soil moisture and droughts might be based on a 2″ capacity. (Figure 16). In July the wheat root systems are well developed and droughts may be calculated using a 6″ capacity (Figure 17). Potential moisture deficiencies in subsequent months may not be important because head filling and ripening may be complete by early August.

Some of you have worked with the Thornthwaite procedures of 1948 and 1957 and may wonder why we have not used the latter. They do have some advantages and we have checked some of our patterns with them but our major study is based on the 1948 procedures for several reasons: (1) we started our study in early 1956 and had processed data for many of the station years by late 1957; (2) the 1948 procedures could be employed more easily and quickly; (3) our information was to be used largely for irrigation planning and a changing rate of storage water use was less needed than it would be if major storage depletion were to take place; (4) over 70% of the precipitation of the Prairies is received in the growing season—current rainfall is used at close to optimum rates by plants no matter what moisture storage conditions might be present; (5) we wished to introduce variations in storage to explain some of the drought patterns present; (6) our comparative studies showed few significant pattern changes; and (7) our study is being compared with daily calculations for the 30 year period for at least two dozen stations in the Prairies—these data are being made available to us by George Robertson, Meteorologist with the Experimental Farms Service and the records for 8 stations have

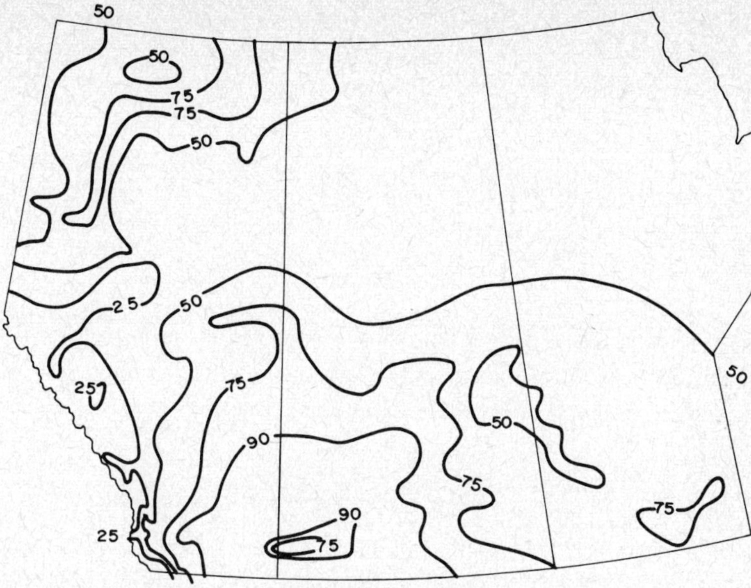

Figure 13. Percentage of years with over 8″ deficit.

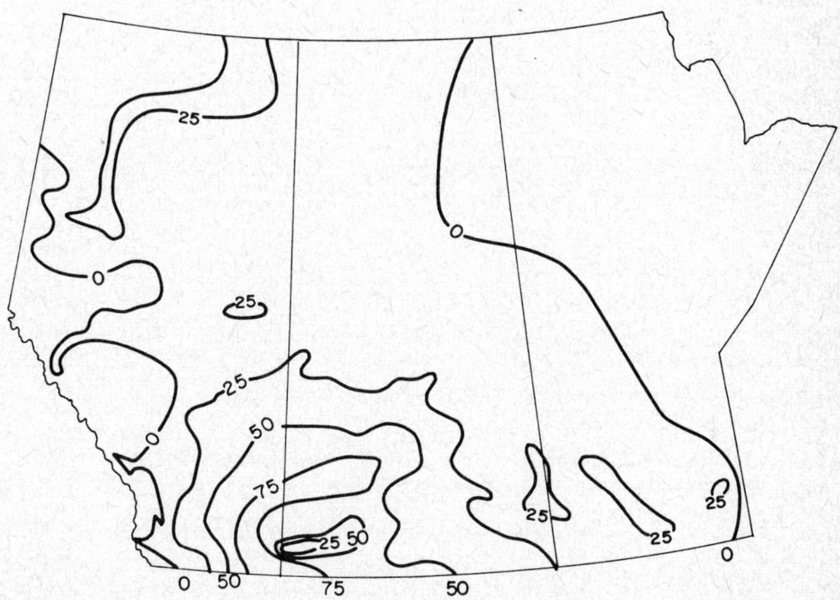

Figure 14. Percentage of years with over 8″ deficit.

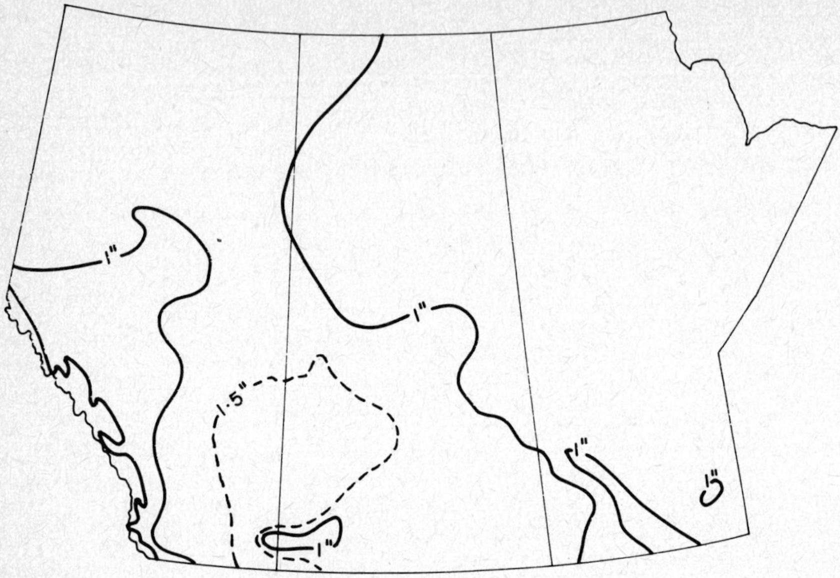

Figure 15. Average Deficit in May.

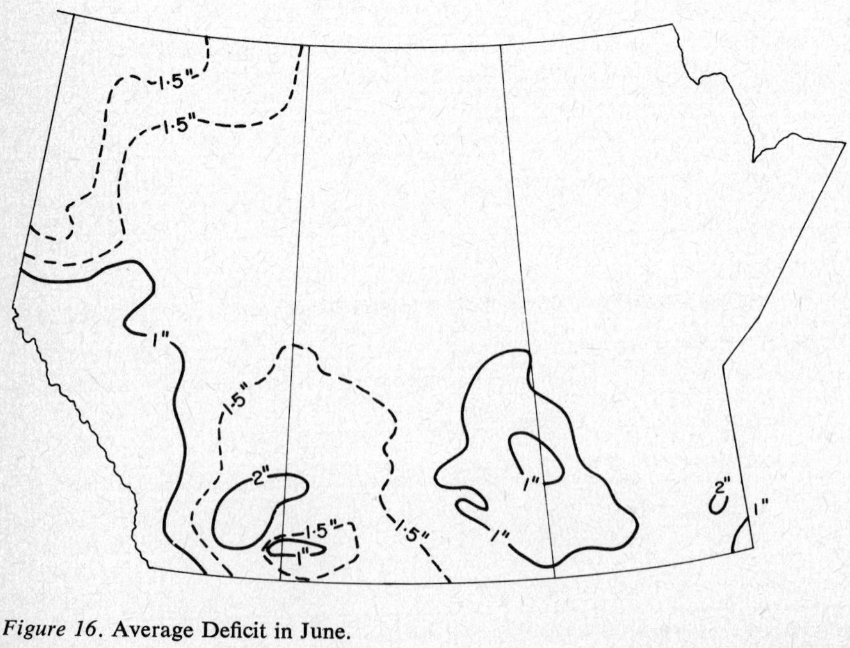

Figure 16. Average Deficit in June.

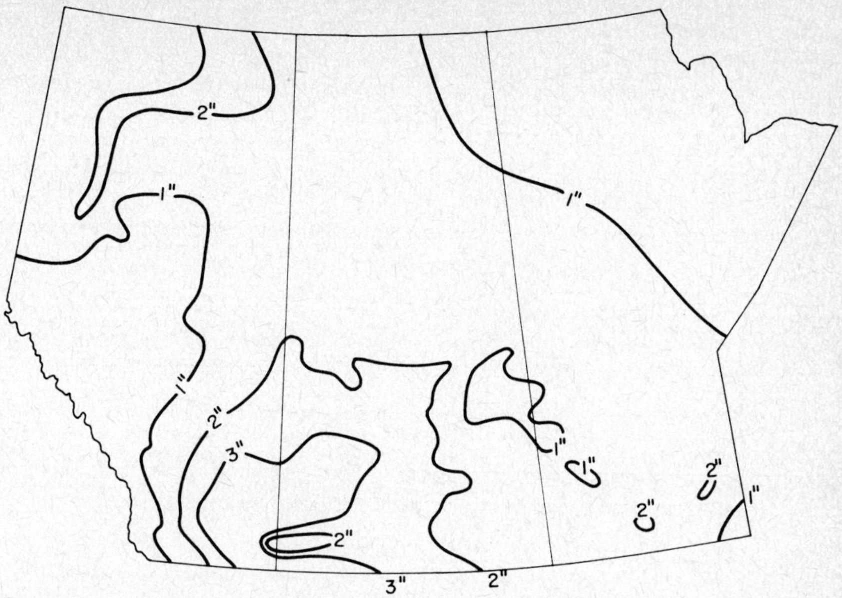

Figure 17. Average Deficit in July.

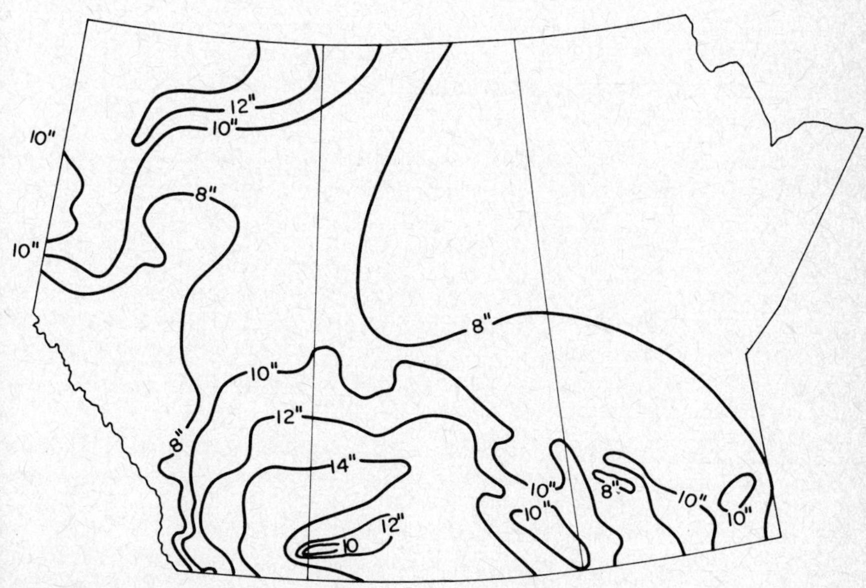

Figure 18. Moisture Deficit.

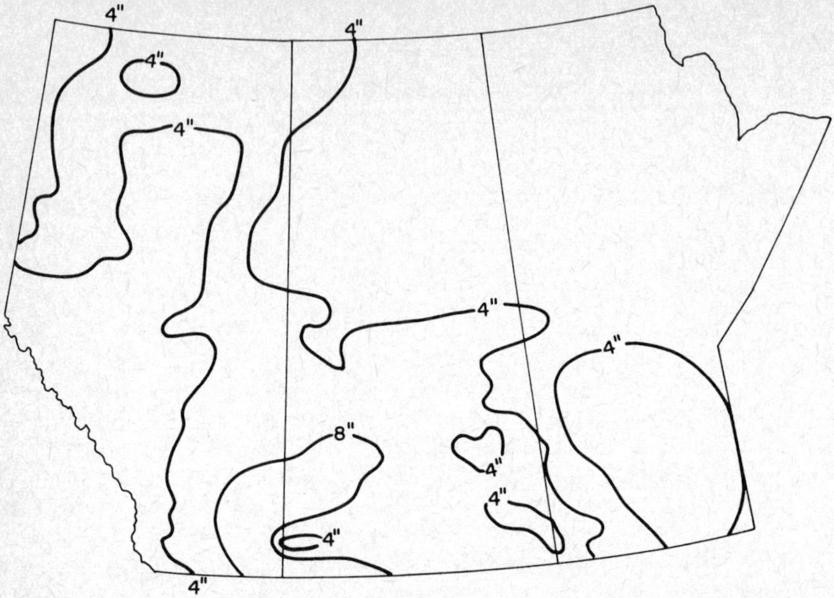

Figure 19. Average soil deficit for wheat.

been received. Mr. Robertson has found the 1948 procedure to be better adapted to I.B.M. programming and as useful for his needs. We will be able to use the daily data and suggest refinements in our patterns but there is no indication that changes will be significant.

We have recognized that the Thornthwaite procedures are empirical and that studies employing other procedures would help to confirm the major patterns we have established. The Lowry-Johnson procedures of 1941 have been employed for 43 stations and the Blaney & Criddle procedures of 1952 have been used for 89 locations in the Prairies. Comparative studies have been made employing methods outlined by Penman, Turc and others. Local studies by the Experimental Farms Service, the Prairie Farm Rehabilitation Administration, Currie and others are also being used in confirmation and correction of values for different needs.

The deficit values based on Lowry-Johnson procedures (Figure 18) are a bit larger than those based on Thornthwaite procedures, especially in the more humid regions. The differences may be attributed in part to smaller allowance for soil moisture storage. When we used the storage values obtained with Thornthwaite procedures to reduce the deficit values mapped, we found that the two procedures provided almost identical results.

The Blaney and Criddle procedures were used for calculating moisture deficits for wheat and alfalfa. The results for wheat (Figure 19) are similar to those based on Thornthwaite procedures. The major difference appears to be the greater allowance for day length variations in the Blaney and Criddle procedures. This difference and the greater deficits in general, may also be noted in deficits for alfalfa.

Though amounts may differ, the major and local patterns are sufficiently alike that we may consider them to be valid. It would appear that improved measurements of the local water balance patterns will now be more useful than additional attempts to develop or apply other empirical procedures designed to define regional drought patterns.

References

Blaney, H. F. and W. D. Criddle (Revision) "Determining Water Requirements in Irrigated Areas from Climatological and Irrigation Data," *United States Department of Agriculture*, SCS-TP 96 Washington, D.C. (1952), 48 pp.

Currie, B. W., "Precipitation, Prairie Provinces and Northwest Territories," *Saskatchewan Research Council,* Saskatoon (1953), 63 pp.

Laycock, A. H., "Precipitation and Streamflow in the Mountain and Foothill Region of the Saskatchewan River Basin," *Prairie Provinces Water Board,* Report 6, Regina, Sask. (1957), 48 pp.

Lowry, R. J. Jr., and A. F. Johnson, "Consumptive Use of Water for Agriculture," *Proc. Amer. Soc. Civ. Engr.,* Vol. 67 (1941), pp. 595-616.

Mohrmann, J. C. J. and J. Kessler, "Water Deficiencies in European Agriculture – A Climatological Survey," *International Institute for Land Reclamation and Improvement,* Publ. 5 Wageningen, Netherlands (1959), 60 pp.

Penman, H. L., "Natural Evaporation from Open Water, Bare Soil and Grass," *Proc. Royal Society of London*, Series A, Vol. 193 (1948), pp. 120-45.

Sanderson, M., "The Climate of Canada According to the New Thornthwaite Classification," *Scientific Agriculture*, Vol. 38 (1948), pp. 55-95.

Thornthwaite, C. W., "An Approach Toward a Rational Classification of Climate," *Geographical Review*, Vol. 38 (1948), pp. 55-94.

Thornthwaite, C. W. and J. R. Mather, "The Water Balance," *Publications in Climatology, Laboratory of Climatology,* Vol. VIII, no. 1, Centerton, N.J. (1955), 104 pp.

Thornthwaite, C. W. and J. R. Mather, "Instructions and Tables for Computing Potential Evapotranspiration and The Water Balance," *Publications in Climatology,* Vol. X, no. 3, Centerton, N.J. (1957), pp. 185-310.

7

Estimation of Latent Evaporation from Simple Weather Observations

W. Baier

G. W. Robertson

Introduction

Review of Literature

Potential evapotranspiration has been accepted as a useful climatic index as well as a tool in irrigation planning and management, but its direct measurement over large areas and over a period of time is impractical and simpler techniques for determining it are desirable. The numerous methods available for estimating potential evapotranspiration from one or several parameters or from instrument readings were recently discussed by Baier (1, 2). To standardize the measurement of evaporation as a climatic factor, Robertson (7) introduced a modified version of the black Bellani atmometer for the measurement of latent evaporation, which is closely related to potential evapotranspiration (4). Carder (3) compared different types of atmometer assemblies and found the black porous disk to be superior for the measurement of latent evaporation.

Comparison of various evaporimeters has shown the black porous disk atmometer to be responsive to solar radiation, vapor pressure deficit, and other meteorological factors affecting evaporation (9). At Summerland,

Reprinted from *Can. J. Plant Science,* Vol. 45 (1965), pp. 276-84.

B.C., Wilcox (13) investigated the effect of the four weather elements—solar radiation, temperature, wind, and humidity—on the rate of potential evapotranspiration from lysimeters and on evaporation from Bellani plate atmometers. Wilcox concluded that if weather records were used to estimate evapotranspiration, all four of the above-mentioned elements should be used, but better estimates of evapotranspiration can be obtained by using instrument evaporation records than by using any one weather record by itself.

Recently, Pelton (5) reported results of studies at Swift Current, Sask., to determine the influence of air temperature, radiation, wind, and vapor pressure on evaporation from the porous disk and the black Bellani plate atmometers, the U.S. Weather Bureau class A pan, and the Experimental Farm's buried tank. With the exception of the wind, highly significant simple correlation coefficients were obtained between each of these meteorological factors and evaporation from each of these instruments. Between 70 and 80% of the variation in evaporation could be accounted for by a linear combination of the four meteorological variables used.

Thornthwaite (12) proposed an empirical formula for calculating potential evapotranspiration involving only mean temperature and daylength. The formula was limited to these two factors simply because one was readily available in the climatological records of National Weather Services and the other was available in special tables (11). Penman's (6) formula on the other hand was basically more physically sound but required observations of vapor pressure, wind, and sunshine as well as temperature. Both of these formulae were first developed for use with monthly averages, but, because that of Penman has some physical basis, it was found to give reasonable estimates of daily evaporation under various climatic conditions.

Purpose of Investigation

In these and similar studies reported in literature, daily values of several meteorological parameters taken over a period of time were available for analysis. Unfortunately, meteorological records are often incomplete or there is sometimes doubt about their reliability. Direct measurements of latent evaporation are also still rare and provide only data over relatively short periods at a few stations.

It is felt that an empirical formula involving readily available climatological factors such as did Thornthwaite's but making use of them in a sounder physical sense could provide an estimate of evaporation approaching the accuracy of the Penman equation. Thus, instead of using mean temperature, it appears more logical to use maximum temperature and temperature range. Maximum temperature might be more closely related to the temperature of the evaporating surface while temperature range might be a manifestation of both solar energy and vapor pressure deficit, both of which are correlated positively with evaporation. Rather than use daylengths as a manifestation of the variation of solar energy with latitude, it might be better to use tabulated energy at the top of the atmosphere. (11) Since these three factors are readily available for all climatological stations, they are considered as first essentials in the empirical formula to be developed. Additional meteorological factors are then added for use as their availability warrants.

The purpose of the present study was to evaluate the influence of several meteorological parameters, either observed or obtainable from standard handbooks, on latent evaporation as measured with a black porous surface, either the black porous disk atmometer or the black Bellani plant atmometer. Special attention was paid to the fact that records of one or several meteorological elements, usually observed at ordinary weather stations, are sometimes incomplete or unreliable. Thus, the technique developed will be of special value whenever existing records of latent evaporation have to be completed or to be tested for their reliability.

Materials and Methods

The data used for analysis in this paper were collected in a country-wide crop-weather study in cooperation with the personnel at agricultural research establishments at Ottawa, Normandin, Swift Current, Lacombe, Beaverlodge, and Fort Simpson. Daily meteorological observations were taken from May to October each year over the period from 1953 to 1957. Because the computer program used in this study will handle a maximum of 999 cases, about 900 days with complete and reliable observations were selected at random from the material available for regression analysis and a similar number of different days was used for testing the results. Thus, the 150 random days of each station are from 5 months over the period from May to October of different years.

Simple and multiple linear correlation and regression calculations were employed throughout this study. Other more complex approaches such as logarithmic, quadratic, and cubic combinations of the independent variables were tested but did not significantly improve the estimation.

Daily records of latent evaporation as dependent variables were correlated with daily data of several meteorological and astronomical parameters as independent variables. These include:

1. *Latent evaporation* (E_L).—This entity was measured with black Bellani plate atmometers installed at 4-ft height above ground. In this paper latent evaporation is expressed in cc/day as originally recorded and adjusted for plate factors.

2. *Maximum temperature* (*max.*).—This was measured with mercury-filled maximum thermometers in a Stevenson screen 4 ft above the ground and expressed in °F.

3. *Temperature range* (*range*).—The difference in °F between daily maximum and daily minimum temperature both measured in a Stevenson screen is called daily temperature range.

4. *Wind.*—Total daily wind run in miles was recorded with totalizing cup anemometers or recording anemographs at least 5 ft above ground. For the purpose of this study the recordings were not reduced to a standard height.

5. *Duration of bright sunshine* (n).—Daily hours of bright sunshine were measured with a Campbell Stokes sunshine recorder.

6. *Vapor pressure deficit* ($e_w - e_s$).—Dew point temperature was measured twice daily with a sling psychrometer. The vapor pressure at mean air temperature (max. + min.)/2 and at mean daily dew point temperature was determined from saturation vapor pressure tables (11) and the difference expressed in mb was designated vapor pressure deficit.

7. *Solar energy at the top of the atmosphere* (Q_o).—Total solar radiation in cal cm^{-2} falling on a horizontal surface at the top of the atmosphere during 1 day was obtained from tables. (11)

8. *Day length* (N).—Duration of daylight, i.e. the interval between sunrise and sunset in hours, was also taken from standard tables. (11)

9. *Total sky and solar energy on a horizontal surface* (Q_s).—Total daily sky and solar energy received on a horizontal surface in cal cm^{-2} day^{-1} was calculated using a relationship involving duration of bright sunshine (n) and solar energy at the top of the atmosphere (Q_o) (or at the earth's surface in the absence of an atmosphere) (10). The form of the relationship is

$$Q_s = Q_o\,(a + b(n/N)).$$

One of the authors investigated this relationship some years ago using daily observations of total energy and bright sunshine at Ottawa and Edmonton for the growing seasons of 1952 and 1953. The values of a and b were determined by simple regression. The results of this analysis indicated very little difference between the coefficients from year to year or from station to station so that a single regression equation was determined for the two stations. This was

$$Q_s = Q_o(0.251 + 0.616(n/N)).$$

The coefficient of determination (r^2) for the analysis involving some 856 days was 0.85. The standard error of estimate for Q_s/Q_o was 0.0077.

This relationship was used for estimating total daily sky and solar energy (Q_s) for the present study.

Multiple regression equations are presented to estimate latent evaporation from different combinations of meteorological parameters available for processing.

Results

Simple Correlation

Daily data of meteorological and astronomical parameters affecting evaporation were correlated with each other to determine interrelationships (Table I). It is not surprising that all radiation terms were significantly correlated with each other and most of these also with temperature and vapor pressure deficit. But wind showed a correlation with only sunshine and daylength. Because there is no physical reason for a correlation between daylength and wind, it may be assumed that these two factors are related to a third common factor, for example, geographical location. It is important to note that solar energy (Q_s) is more closely correlated with energy at the top of the atmosphere (Q_o) than with daylength (N). Another feature is the close correlation between temperature range and solar energy (Q_s).

TABLE I. SIMPLE CORRELATION COEFFICIENTS[a] BETWEEN METEOROLOGICAL VARIABLES

	Q_o	*Range*	*Max.*	*Daylength*	Q_s	*Wind*	*Sunshine*	$e_w—e_s$
Q_o	1	0.05	0.20	0.73	0.44	0.13	0.22	0.13
Range		1	0.67	0.04	0.59	0.13	0.63	0.41
Max.			1	0.16	0.57	0.18	0.57	0.69
Daylength				1	0.30	0.24	0.24	0.27
Q_s					1	0.03	0.96	0.49
Wind						1	0.27	0.07
Sunshine							1	0.61
$e_w—e_s$								1

[a]For $P = 0.05$, $r = 0.20$; for $P = 0.01$, $r = 0.25$.

Each of the eight factors studied was correlated with latent evaporation (Table II). All correlation coefficients are statistically significant, but they range from 0.20 for the correlation with daylength to 0.71 for solar

TABLE II. SIGNIFICANCE OF SIMPLE CORRELATION BETWEEN LATENT EVAPORATION (y) AND METEOROLOGICAL VARIABLES (x)

Meteorological variables	r_{yx}[a]	*Significance between correlation coefficients*[b]
Q_s	0.71	
Sunshine	0.67	
$e_w—e_s$	0.65	
Max.	0.61	
Range	0.56	
Q_o	0.31	
Wind	0.26	
Daylength	0.20	

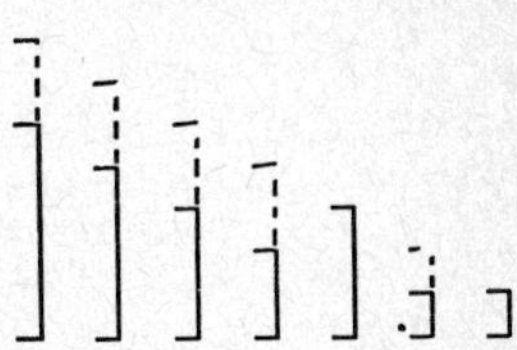

[a]For $P = 0.05$, $r = 0.20$; for $P = 0.01$, $r = 0.25$.
[b]Solid line indicates a significant difference at $P \leq 0.05$ and broken line indicates no significant difference between highest value and following values.

energy. Radiation terms such as solar energy and sunshine were both more closely correlated with latent evaporation than temperature terms such as maximum and range. Preliminary tests also showed that latent evaporation was more closely correlated with maximum temperature ($r = 0.60$) or with temperature range ($r = 0.52$) than with mean air temperature ($r = 0.46$).

The meteorological and astronomical variables to be used in the proposed multiple correlation and regression analysis were selected from these results. It may be seen from Table II that Q_o was correlated more closely with latent evaporation ($r = 0.31$) than was daylength ($r = 0.20$). Therefore, Q_o data were readily available from tables (11) included in the first formula. Sunshine was omitted, because solar energy on the earth's surface calculated from sunshine records resulted in a higher correlation with latent evaporation than sunshine alone. This is ascribed to the fact that total sky and solar energy in terms of cal cm^{-2} day^{-1} is physically more meaningful for the evaporation from a horizontal surface than the duration of sunshine expressed in hours per day.

Multiple Correlation

Multiple regression analysis was employed to evaluate the order of importance of the meteorological and astronomical factors involved and finally to develop equations for the calculation of latent evaporation.

The minimum data available from most climatological stations include observations of maximum and minimum temperatures while astronomical data are available from tables. These factors or their derivatives were included, therefore, as the basic part of all equations investigated. The multiple correlation coefficients increased in the order of importance: maximum, range, and Q_o (Table III). If only these three variables were involved, the multiple correlation coefficient (R) was 0.68. The inclusion of one of solar energy, vapor pressure deficit, or wind increased the R value to approximately 0.76. If any two of these factors were included, the R value increased to approximately 0.80, and with all six factors together R was 0.84. This means approximately 70% of the variations of latent evaporation can be explained by variations of the six factors involved.

Multiple regression coefficients a_0 to a_6 to be used in a linear multiple regression equation of the type

$$Y = a_0 + a_1x_1 + a_2x_2 + a_3x_3 + a_4x_4 + a_5x_5 + a_6x_6$$

were calculated (Table III).

This equation together with the coefficients facilitates the estimation of daily latent evaporation from any combination of up to six variables. The reliability of the estimates increases with the number of variables involved. The standard error of estimate for each method of estimation was also calculated (Table III). It may be seen that the standard errors decreased from 18.9 for method I with three variables involved to approximately 16.7 with either solar energy, vapor pressure deficit, or wind

TABLE III. MULTIPLE CORRELATION AND REGRESSION COEFFICIENTS OF METEOROLOGICAL VARIABLES ON LATENT EVAPORATION

				Regression data		*Test data*	
Method	*Meteorol. variables*	*Regression coeff.*	*Constant*	*R*	*S.E.* (cm^3)	*R*	*S.E.* (cm^3)
I	Max.	9.28×10^{-1}	−87.03	0.68	18.9	0.67	18.5
	Range	9.33×10^{-1}					
	Q_o	4.86×10^{-2}					
II	Max.	6.87×10^{-1}	−55.60	0.75	16.9	0.74	16.7
	Range	2.84×10^{-1}					
	Q_o	9.13×10^{-3}					
	Q_s	6.85×10^{-2}					
III	Max.	-2.28×10^{-2}	−42.28	0.76	16.7	0.81	14.6
	Range	1.09					
	Q_o	5.06×10^{-2}					
	e_w-e_s	2.99					
IV	Max.	1.13	−108.80	0.76	16.7	0.72	17.2
	Range	9.20×10^{-1}					
	Q_o	3.59×10^{-2}					
	Wind	1.31×10^{-1}					
V	Max.	-2.32×10^{-2}	−26.69	0.80	15.5	0.83	13.8
	Range	5.57×10^{-1}					
	Q_o	1.96×10^{-2}					
	Q_s	5.31×10^{-2}					
	e_w-e_s	2.41					
VI	Max.	8.97×10^{-1}	−78.68	0.80	15.4	0.80	15.0
	Range	3.40×10^{-1}					
	Q_o	1.66×10^{-3}					
	Q_s	6.13×10^{-2}					
	Wind	1.18×10^{-1}					
VII	Max.	3.50×10^{-1}	−69.30	0.81	14.9	0.82	14.1
	Range	1.04					
	Q_o	4.03×10^{-2}					
	e_w-e_s	2.31					
	Wind	1.01×10^{-1}					
VIII	Max.	3.37×10^{-1}	−53.39	0.84	14.2	0.86	12.9
	Range	5.31×10^{-1}					
	Q_o	1.07×10^{-2}					
	Q_s	5.12×10^{-2}					
	Wind	9.77×10^{-2}					
	e_w-e_s	1.77					

For $P = 0.01$ and two independent variates, $R = 0.30$.
For $P = 0.01$ and six independent variates, $R = 0.40$.

included. With five variables used, the standard error ranged from 14.9 to 15.5 and with all six variables it was 14.2.

To test the goodness of fit, latent evaporation was calculated by methods I to VIII using the meteorological data of about 900 test days other than those employed for the calculation of the regression equations (Table III). The correlation coefficients between calculated and observed daily latent evaporation (R) and the standard error of estimate (S.E.) for the test days are close to the values obtained for the data used for analysis. The significance of the correlation coefficients ($P = <0.001$) may be considered as a criterion for the accuracy of the method.

Sometimes estimates of latent evaporation accumulated over a number of days or even over a full month may be sufficient. This would simplify the data processing and at the same time probably increase the reliability of the estimates. Using the formulae above, estimates of latent evaporation were calculated for 1 day, 2 days, 4 days, 8 days, 16 days, and a full month. These estimates were then correlated with the observed latent evaporation for a corresponding period. For comparison, each set of data used for correlation consisted of 24 pairs of independent observations, each set referring to a different period.

In general, the correlation between observed and calculated values of latent evaporation improved slightly when the period of time used for calculation was increased from 1 day (methods II, III, and IV) or 2 days (methods V, VII, and VIII) up to 16 days (Table IV). The highest correlations were obtained when in method I the estimates were accumulated over 2 days, in method II over a full month, and in methods III to

TABLE IV. CORRELATION COEFFICIENTS BETWEEN DAILY AND ACCUMULATED VALUES OF LATENT EVAPORATION OBSERVED AND ESTIMATES CALCULATED BY METHODS I TO VIII

Method:	I	II	III	IV	V	VI	VII	VIII
Variables:	Max. Range Q_o	Max. Range Q_o Q_s	Max. Range Q_o $e_w - e_s$	Max. Range Q_o Wind	Max. Range Q_o Q_s $e_w - e_s$	Max. Range Q_o Q_s Wind	Max. Range Q_o $e_w - e_s$ Wind	Max. Range Q_o Q_s Wind $e_w - e_s$
No. of days accumulated								
1	0.65	0.76	0.85	0.69	0.90	0.82	0.88	0.90
2	0.82	0.82	0.87	0.77	0.86	0.80	0.84	0.85
4	0.79	0.79	0.88	0.76	0.88	0.77	0.86	0.85
8	0.73	0.81	0.88	0.78	0.90	0.84	0.88	0.89
16	0.79	0.82	0.93	0.86	0.93	0.89	0.93	0.94
30	0.80	0.82	0.91	0.82	0.91	0.85	0.90	0.91

For $P = 0.01$, $R = 0.52$.

VIII over 16 days. The correlation was lowest in methods I to IV for daily values, in methods V, VII, and VIII for values accumulated over 2 days, and in method VI accumulated over 4 days. Thus, with the exception of method I, the reliability of the estimates tended to improve, if the period of time used for calculation was extended from 1 or 2 days up to 16 days.

Discussion

The object of the present study was to develop a technique of estimating latent evaporation from a given set of meteorological data. Primarily, the single and combined influence on latent evaporation of several meteorological parameters which are ordinarily available at weather stations, had to be evaluated. It was found that on the average over all stations and years variations in solar energy alone accounted for 50% of the variations of E_L. This figure is in agreement with findings of Pelton (5) at Swift Current. But data on solar energy are not always available and a large amount of energy used for evaporation must still be attributed to factors other than direct radiation. Starting with observations of maximum temperature and temperature range and with solar energy at the top of the atmosphere, 46% of the variations of E_L could be explained. If records of any one of the additional variables, viz. solar energy, vapor pressure deficit, or wind were included, approximately 57% of the varieties of E_L were accounted for by variations in the meteorological parameters involved. With a suitable combination of five meteorological variables 64 to 66% of the variability in latent evaporation was accounted for and with six variables this value increased to 70%. This is the same value as given by Pelton (5) for a multiple correlation between disk atmometer evaporation and the four meteorological variables temperature, radiation, vapor pressure deficit, and wind at Swift Current. The coefficients presented here, however, refer to a wider variety of conditions across Canada.

This study indicates the feasibility of estimating daily E_L from astronomical data readily obtainable from tables and the available meteorological data. Regression coefficients for use in a linear, multiple regression equation involving up to six factors are presented. If only data of basic meteorological factors, such as maximum temperature and range, are available, the standard error of such estimates approaches 42% of the mean. But with the inclusion of additional meteorological factors, such as vapor pressure deficit, wind, and solar energy on the earth's surface, the standard error of the estimates decreases to 31% of the mean. The reliability of the estimates can be further improved, if the estimates are accumulated over a number of days.

Because the regressions were calculated from daily data recorded at six stations across Canada over several years, it is felt that the equations will provide reasonable estimates of latent evaporation under conditions in most parts of Canada.

References

Baier, W., "The interrelationship of meteorological factors, soil moisture and plant growth (a review)," *Agr. Meteorol. Tech. Bull.* 5, Agro-Meteorol. Sect., Can. Dept. Agr., Ottawa (1964).

Baier, W., "Studies on estimating potential evapotranspiration from empirical relationship," *S. Afr. J. Agr. Sci.*, Vol. 6 (1963), pp. 455-74.

Carder, A. C., "Atmometer assemblies, a comparison," *Can. J. Plant Sci.*, Vol. 40, no. 4 (1960), pp. 700-706.

Holmes, R. M. and Robertson, G. W., "Conversion of latent evaporation to potential evapotranspiration," *Can. J. Plant Sci.*, Vol. 38, no. 2 (1958), pp. 164-72.

Pelton, W. L., "Evaporation from atmometers and pans," *Can. J. Plant Sci.*, Vol. 44, no. 5 (1964), pp. 397-404.

Penman, H. L., "Natural evaporation from open water, bare soil and grass," *Proc. Roy. Soc. London*, Series A, Vol. 193 (1948), pp. 120-45.

Robertson, G. W., "The standardization of the measurement of evaporation as a climatic factor," Tech. Note No. 11, W.M.O., No. 42, T.P. 16 (1955).

Robertson, G. W., "Discussion of pans and atmometers," *Proc. Hydrol. Symp. No. 2, Evaporation.* (Ottawa, Queen's Printer, 1961), pp. 104-105.

Robertson, G. W. and Holmes, R. M., "A new concept of the measurement of evaporation for climatic purposes," *Intern. Assoc. Sci. Hydrol.*, I.U.G.G. Assemblée Générale de Toronto 1957 (Gentbrugge 1958) Tome III, pp. 399-406.

Smith, G. W., "Solar radiation and the duration of sunshine in Trinidad, West Indies," *Quart. J. Roy. Meteorol. Soc.*, Vol. 85, no. 366 (1959), pp. 421-23.

Smithsonian Institute, *Smithsonian meteorological tables*, 6th ed. Washington, D.C. (1951).

Thornthwaite, C. W., "An approach toward a rational classification of climate," *Geogr. Rev.*, Vol. 38, (1948), pp. 85-94.

Wilcox, J. C., "Effects of weather on evaporation from Bellani plates and evapotranspiration from lysimeters," *Can. J. Plant Sci.*, Vol. 43, no. 1 (1963), pp. 1-11.

8

Soil Moisture Regimes Under Summerfallow, Wheat and Red Fescue in the Upper Peace River Region

A. C. Carder

A. M. F. Hennig

Introduction

It has been determined that the upper Peace River region which extends eastward from the Rocky Mountains for 250 miles, in latitudes 54°-57° N has a climate that is fairly uniform within its margins (Carder, 1965). This climate is classed as boreal or subarctic (Trewartha, 1954). With an annual precipitation of 17.5 inches it appears to be almost semi-arid, but the distribution of precipitation is good with most falling in the summer months, and evaporation is low. Despite these redeeming features, field crops frequently suffer from drought and soil moisture is of considerable importance. Since the rivers generally course along deep ravines field irrigation is not practiced.

Because of the importance of moisture, this experiment was established to gain knowledge of the soil moisture balance under different cropping procedures. Its purpose was to determine: (*1*) the efficiency of summerfallow for the conservation and storage of water, (*2*) the amount of water used by a cereal and a grass crop typical of the region, (*3*) the water consumption of these crops during all stages of their development,

---

Reprinted from *Agricultural Meteorology,* Vol. 3 (1966), pp. 311-31.

(*4*) the soil depths whence this water is obtained, and (*5*) the source of this water, whether from soil reserves, rainfall or snowmelt.

Since the experiment was established on a soil common to the region and at a site free from any peculiar microclimatic influence, it is considered that the data obtained are quite representative.

Review of Literature

Summerfallow and moisture conservation

The study of the moisture regimes under summerfallow and field crops is not new. W. W. Burr, working in Nebraska, had by 1914 dispelled the myth that summerfallowing secured two years' rainfall for the production of one crop (Burr, 1914). He found that only under favorable conditions was it possible to store 33% of the rainfall of one season and that under unfavorable conditions only 10% was stored. He noted the great loss of water that occurs during the fallow year by evaporation, observing that: "A rain of one-half inch is of no value in storing water, unless it comes while the surface is still fairly moist from a preceding rain." This observation has been repeated by others (Bracken and Cardon, 1935; Kramer, 1949; Kelly, 1961). Mathews and Army (1960) in a study conducted over almost 20 years at 25 locations in the Great Plains (Montana to Texas) showed that the moisture stored by summerfallowing was a "pathetically small" 16.3% of the precipitation. They attributed most of the loss to evaporation from the soil.

Studies (Reimann et al., 1945; Kramer, 1949) have shown that much of the moisture lost by evaporation is from the top 6-8 inches of soil and that tillage sometimes substantially abets this loss (Veihmeyer, 1938). When the surface soil is very wet, evaporation is considerable and approaches that from a free-water surface (Tanner, 1957; Penman, 1963), but when the surface soil is dry, evaporation is slight unless the soil has contact with a saturated layer or a free-water surface (Veihmeyer, 1938; Veihmeyer and Brooks, 1954).

Water use by spring wheat

In this paper the water used by a crop means that transpired from the plant's tissues plus that evaporated from either the soil surface or the vegetal canopy (Cole and Mathews, 1923; Holtam and Musgrove, 1947). The amount of water used by spring wheat depends on many factors. Conditions unfavorable to growth such as drought decrease crop water use. This is partly because the limited vegetative matter produced when there is a dearth of water transpires less, and partly because the rate of transpiration is retarded by wilting and stomatal closure (Kramer, 1949; Harrold and Dreibelbis, 1951).

Perhaps the most precise experiment ever conducted to determine the amount of water used by a crop was that of Harrold and Dreibelbis (1951) in Ohio who used weighing monolith lysimeters. In four tests over two years they found that spring wheat used 13.2 inches of water from planting until harvest and yielded 35.2 bushels per acre of grain. Under more arid conditions such as in the northern Great Plains of the United States an average of the water used at a great many stations over many years showed that hard red spring wheat used 10.0 inches of water and yielded 15.3 bushels per acre of grain (Cole and Mathews, 1923). As will be pointed out, this water use is about the same as that at Beaverlodge, but the yield is much less.

The amount of water evapotranspired from crops even well supplied with moisture is never as great as that evaporated from a free-water surface. It is generally contended that under the same atmospheric conditions vegetation-covered land evapotranspires about two-thirds of the amount of water that is lost by evaporation from an extensive water surface (Neumann, 1953). A lush crop of corn, plentifully supplied with moisture, will for a brief few weeks during the later stages of its grand phase of growth evapotranspire 80% of the water lost from a free-water surface during the same time and under the same conditions (Denmead and Shaw, 1959).

In early spring when the grain crop is in the seedling stage much of the water loss is due to evaporation from the soil (England, 1963), but as the crop develops transpiration soon becomes uppermost (Kramer, 1949; Tanner, 1957). This is not only because of the greatly increased amount of leafy vegetation produced, but because the ground becomes shaded and protected from the wind. A complete vegetative cover will intercept almost all the radiant energy and will greatly reduce air flow near the ground. Even with a row crop, such as corn, the water loss from the soil in dry years may amount to only 30% of the total water evapotranspired (King, 1961).

Water use by grass

Harrold and Dreibelbis (1951) in their lysimeter measurements found that a mixed timothy-alfalfa meadow used 22.8 inches of water over its growing period and produced 2.5 tons per acre of hay. Two cuttings were taken each year and these usually resulted in marked reductions of water use. In Illinois, bluegrass pasture consumed 20.5 inches of water annually, March-November (Holtam and Musgrove, 1947). As will be shown, these figures are much above the amounts of water evapotranspired from the red fescue meadow at Beaverlodge.

The rate of evapotranspiration from a grass sward well supplied with moisture can be very rapid. In England from May to August, it has been found to amount to 80% of that from open water. Part of this difference is due to difference in reflectivity, for whereas, only about 5% of the incident short-wave radiation is reflected from water, some 20% is reflected from grass (Schofield, 1950). Root surface studies on different crops have indicated that the efficiency of bluegrass to extract water would be quite superior to certain other crops such as the cereals (Dittner, 1938).

Materials and Methods

Initial soil-moisture studies at Beaverlodge showed that the root penetration of cereal and grass crops does not normally exceed 3 ft. (Carder, 1954). The clayey and dense nature of the soil, general dearth of moisture in the deep subsoil, the cold temperatures of the subsoil, the short growing season, and the long summer days which force plant development all probably contribute to this condition. Accordingly, in the current experiment moisture determinations were not taken beyond a depth of 3 ft.

The test site was on one of the more prevalent and better agricultural soils of the region; i.e., a Landry silt loam, Black Solod developed on lacustro-till (Odynsky et al., 1961). The plot area sloped to the south at the rate of 1 ft. in 50. It was not overly exposed to cause loss of snow by blowing, yet contained no obstructions to encourage a build-up of drifts. The soil had fair to good surface drainage, but the subsoil was sufficiently dense to impede moisture penetration.

Comparisons were made of the soil moisture regimes under a wheat-summerfallow rotation and under a creeping red fescue ley. Moisture determinations were initiated in October 1956, just before freeze-up, and were terminated 5 years later. Precipitation records were obtained at the site through the course of the experiment.

The area was summerfallowed in 1956 to prepare it for planting to wheat and fescue in the spring of 1957. Six plots of grass sod and six of the wheat-fallow rotation were established in a randomized block design. Plots were 16 × 32 ft. separated by 4-ft. sodded pathways, while ranges were 18 ft. apart, separated by sodded roadways to permit turning of implements. There was an 18-ft. sodded margin around the test area and beyond on all sides were grain fields. It is considered that these latter substantially reduced the "oasis effect" inherent in experiments of this kind (Halstead and Covey, 1957; King, 1957; Tanner, 1957).

In the springs of 1957, 1959 and 1961, one-half of the plots were drilled to Saunders wheat at 1.5 bushels per acre along with ammonium phosphate (11-48-0) fertilizer at 40 pounds per acre. In the alternate years, i.e., 1958 and 1960, these plots were black fallowed. The other series of plots were sown to creeping red fescue, *Festuca rubra genuina*, L., in the spring of 1957 and retained in sod until the end of the experiment. Conventional farm machinery and methods of tillage were used throughout.

In August 1956, a stack of six fiberglass soil-moisture units was established at the center of each plot using the technique previously employed at Beaverlodge (Hennig and Carder, 1955), but modified in the light of Reinhart's (1953) work. The units were those described by Colman (1947) and were placed spirally in the wall of a 6-inch auger hole at depths of 3, 9, 12, 21, 27 and 33 inches. Three-inch layers of soil were removed with the auger and kept separately to ensure proper replacement. As the units were imbedded, the separate layers of soil were replaced, wetted to approximately field capacity and tamped to field density. The above depths were chosen because they best conformed to the soil horizon of the Landry series and because they lent themselves most suitably to the technique used to place the readings from them on a quantitative basis. The leads from the blocks of a single stack terminated at a common receptacle equipped with three-point sockets to accommodate the modified Bouyoucos bridge (G. W. Robertson, personal communication, 1954) used to obtain the resistance impulses. Two receptacles were attached to each terminal post placed between two test plots. On every plot a soil-moisture sampling grid was set-up around each 5-ft. square undisturbed stack area. Thus, it was possible to make gravimetric moisture determinations of soil cores from each 1-ft. square in a randomized sequence (Figure 1).

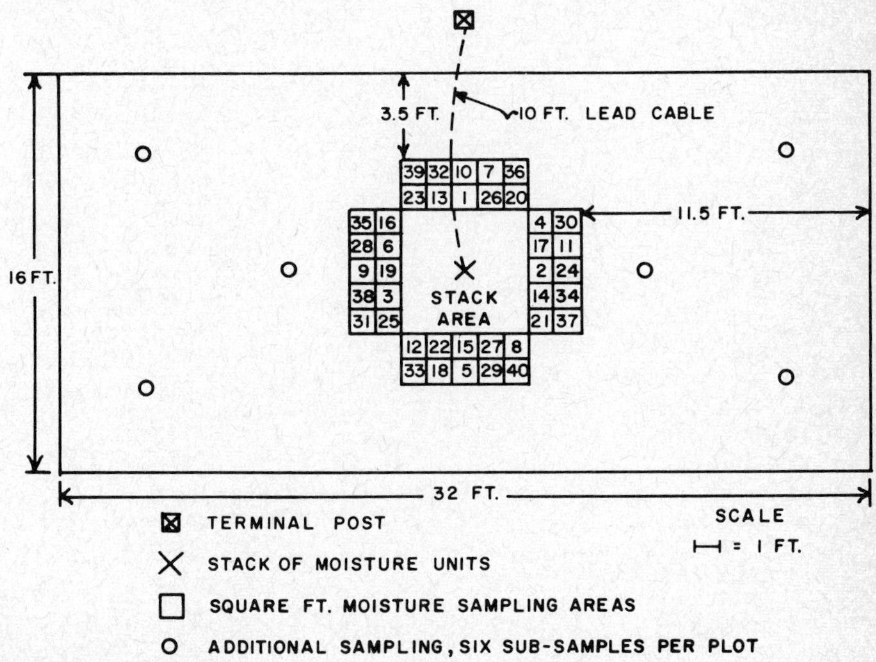

Figure 1. Plot detail.

Over the course of the experiment 40 sets of moisture samples were obtained from each plot with 30 taken to the full depth of 36 inches. The latter coincided with resumption of growth (fescue), time of seeding, emergence, four-leaf stage or initiation of shooting (wheat), heading, haying (fescue), milk stage (wheat), maturity (wheat), and just before freeze-up. Additional samples of the surface, 12-18 inches of soil, were taken as early as possible in the spring and after heavy rains. At least one complete set of samples was taken each year with a ¾ inch diameter soil sampling tube from each plot for volume-weight determinations. To facilitate removal of the 6-inch interval cores with the tube, the sampling hole was enlarged with a 3-inch auger to the previous sampling depth. This procedure was necessary because of the heavy clay soil and the occasional stone. Each year a composite of 48 samples was made up from each 6-inch soil layer under wheat, fallow and fescue sod for wilting point, field capacity and permeability determinations.

To convert gravimetric moisture determinations to inches available water, it was necessary to have the wilting point, field capacity and volume weight of each soil layer. Available soil moisture as defined by Baver (1948) is the moisture content between wilting point and field capacity. The quantity, inches-available-moisture, was obtained by multiplying percentage water by the fraction:

$$\frac{\text{volume weight} \times \text{depth in inches}}{100}$$

Wilting point percentages were obtained by the desiccator method of Lehane and Staple (1951), while normal field capacity percentages were determined by a method (J. A. Toogood, personal communication, 1955) similar to that evolved by Shaw (1927). Since T-tests of these parameters for the fallow-wheat rotation and for the fescue showed no significant differences, the means of the two series of analyses were used (Table I). Volume weights were obtained by oven drying and weighing cores of known volume. Here, T-tests showed differences between the two cropping sequences and, therefore, separate figures were used.

TABLE I. SOIL ANALYSES USED IN CONVERSION OF FIELD MOISTURE PERCENTAGES TO INCHES-AVAILABLE-WATER

Soil depth (inches)	*Wilting point fallow-wheat rotation, fescue ley (%)*	*Volume weight*		*Normal field capacity fallow-wheat rotation, fescue ley (%)*
		fallow-wheat rotation	*fescue ley*	
0- 6	14.86	0.91	0.99	36.66
6-12	15.67	1.11	1.23	32.52
12-18	18.79	1.09	1.26	41.36
18-24	17.15	1.34	1.42	38.24
24-30	15.47	1.51	1.59	33.92
30-36	15.50	1.43	1.43	34.41

Inches-available-water as determined gravimetrically for each soil depth under each cropping sequence were plotted on semi-logarithmic paper against readings taken at the same point and on the same date by use of the electrical resistance absorption blocks. Resistances were adjusted to a basic block coefficient and to a basic temperature of 60°F. and converted to ohms by use of the alignment chart prepared by Bethlahmy (1951) and adapted for use with the modified bridge (Hennig and Carder, 1955).

Some 14,600 resistance determinations were made, or 203 from each absorption block, over the 5 years of the experiment. The series of gravimetric checks amounted to 2,880, or about 40 per block. It was possible to work out the relationship between the electrical parameter, ohms, and inches-available-water as determined gravimetrically, because over the years of the test the soil was alternately wetted and dried by percolation and withdrawal of moisture. This action permitted the determination of a moisture spectrum from wilting point to field capacity. By charting on semi-logarithmic paper the absorption block readings in ohms against inches-available-water as determined gravimetrically, a curve was formed showing inches-available-water according to block response.

Such a curve was composed by transferring the dot scatter of ohms along the abscissa against inches-available-moisture along the ordinate from the semi-logarithmic arrangement to a linear arrangement, i.e., from semi-logarithmic paper to squared paper. The dot scatter on squared paper was then divided into about 40 equal parts along the abscissa and the average for each segment of dots determined. This series of averages in inches-available-water was then recorded numerically and smoothed by running average technique using intervals of five. The smoothed averages were then plotted on the linear graph and a freehand curve fitted to them. This curve was next superimposed on the scatter of dots on the semi-logarithmic paper. Twelve such curves were prepared: one for each of the six depths under the two cropping sequences. These curves were used to convert the remaining 11,720 block readings to inches-available-water.

A small percentage of the absorption blocks failed because of construction and installation faults, i.e., shorted or, more commonly, did not make good contact with the surrounding soil. To cull out spurious readings, because of these and other failings, each reading was plotted on a graph so that a continuous story of soil moisture was obtained for each block through the course of the experiment. Thus, six curves were obtained for each depth under each cropping sequence. In this way suspicious aberrations were spotted and discarded. The remaining determinations (about 14,000) were then used to delineate the moisture regimes under the two cropping sequences (Figures 2 and 3).

Moisture determinations by use of the blocks were taken about twice weekly from spring thaw to freeze-up. The phenology of the wheat and the fescue was recorded. Height measurements were taken. Wheat grain yields were determined, while the fescue was cut for hay and yields of forage obtained.

Inches of water evapotranspired from the crops or evaporated from the soil were determined by compiling a budget made possible by knowing the moisture reserves and precipitation amounts through the course of the experiment (Figures 4, 5, 6 and 7).

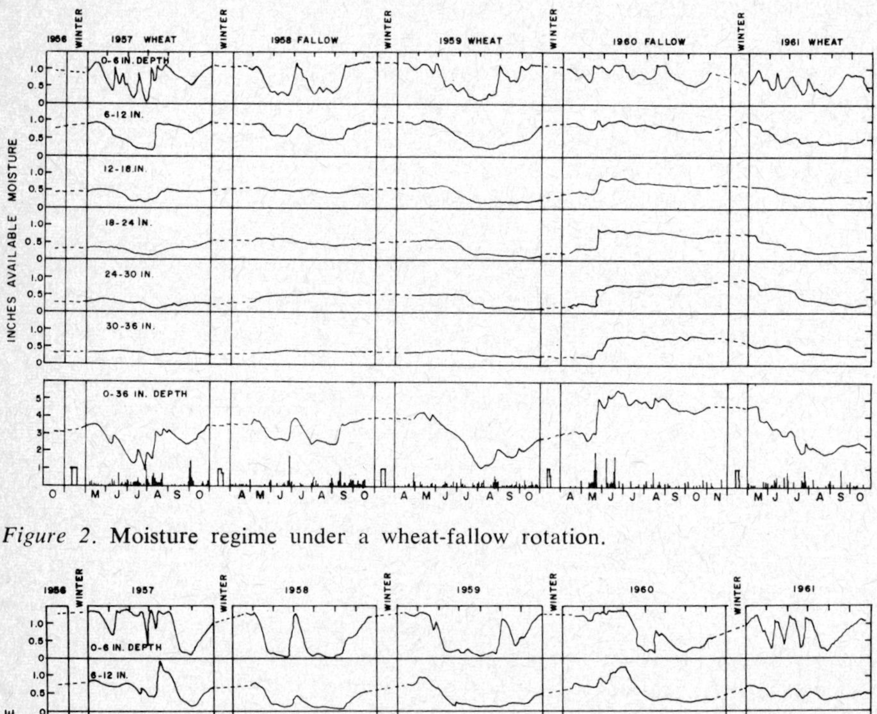

Figure 2. Moisture regime under a wheat-fallow rotation.

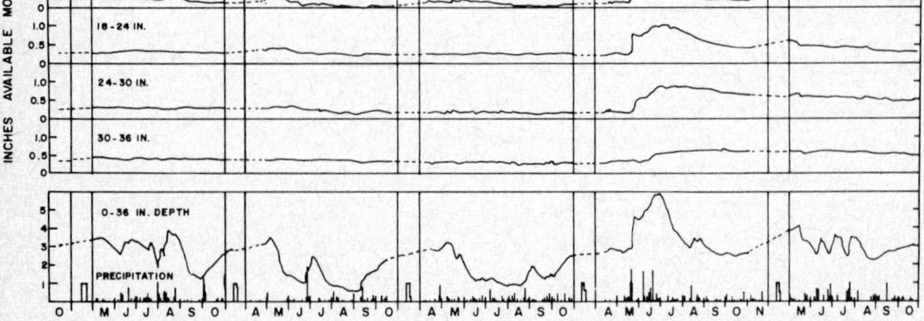

Figure 3. Moisture regime under a red fescue ley.

Results and Discussion

The moisture regime under summerfallow

The summerfallow procedure used was that standard in the region. In late May soon after the spring germination of weeds had occurred, the previous year's stubble was plowed under and the land left untouched until growth warranted further tillage. This was shallow and comprised a single passage of a tandem disc-harrow once or twice during the early summer. Later in the season weeds were controlled by shallow working with a cultivator. At no time were they permitted to advance beyond the seedling stage, thus securing an almost clean fallow condition.

Even under fallow considerable amounts of water were withdrawn from the soil during rainless periods in the summer (Figure 2). The summer of 1958 was dry except for one heavy rain at the end of June. Moderate rains fell in September. Losses were heavy from the top 6 inches of soil. Some loss occurred down to 1 ft. and possibly trace amounts to 18 inches. From spring thaw to freeze-up almost 9 inches of water were lost by evaporation in 1958 (Figure 4). In 1960, when the land was again fallowed, exceptionally heavy rains occurred through May and June with very dry weather subsequently, except for one moderate rain in August. In that year moisture was again withdrawn by evaporation from a depth of 18 inches, although most was lost at shallower depths. Over 13 inches of water

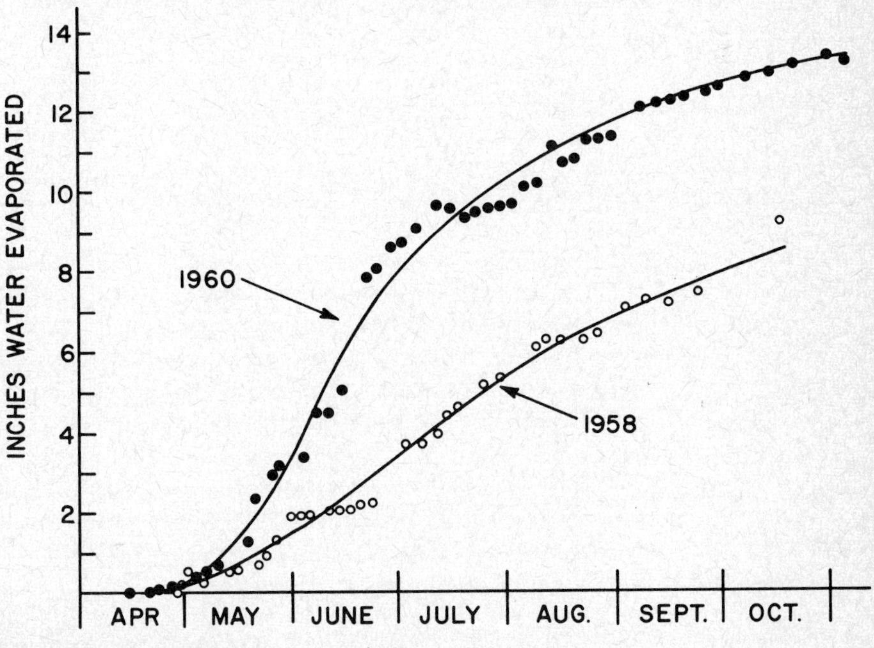

Figure 4. Moisture loss from summer fallow.

were lost to evaporation in 1960 (Figure 4), much of it during the time of the heavy May and June rains, presumably because of the high rate of evaporation which occurred from the surface soil between rains. This surface evaporation was abetted by the low permeability of the soil to water.

The above results indicate that heavy losses of moisture by evaporation do occur on well-worked fallow and that these vary from about 8 inches in comparatively dry years to at least 13 inches in summers with a pronounced wet period. Although the experimental field has a 2% slope it is not thought that substantial amounts of moisture were lost by surface run-off. Rain-intensity records showed that at no time during the course of the experiment did rain fall at a rate above 0.25 inches per hour. Heavy rains were of long duration and the soil was able to absorb them as they fell. At the height of the rain in late June 1958, 1.60 inches fell in 10 h. The 3.60-inch rain of May 1960, extended over 3 days. It is believed that no important moisture loss from lateral seepage occurred since the slope of the test site decreased from upper to lower level so that, if anything, more lateral moisture flow would be expected to occur at the upper than at the lower margin. However, Figure 2 indicates that during the excessive rains of May and June 1960, a small amount of moisture did percolate below the rooting zone of the cereal crop and most certainly did below the 3-ft. depth selected as maximum for this study. Greater loss than this was prevented by the stiff clay subsoil which deep borings have shown to extend to at least 8 ft., or far below the normal penetration of crop roots, except those of certain legumes.

An aspect of summerfallowing is the amount of water conserved. For convenience of study, the 20-month summerfallow period was divided into several subperiods (Table II). During the total period at Beaverlodge, 26.2 inches of precipitation fell, of which only 6.3% was conserved—an average for two fallow periods[1]. The breakdown shows that much of this was stored early in the fallow period. For instance, of the precipitation that fell during the short time between harvest anad freeze-up 13% was held in the soil. Over the first winter of the fallow period 5% of the precipitation was stored, while over the second winter a slight loss took place. During the fallow year (spring break-up or seeding time to freeze-up) only 8% of the moisture was conserved because of high evaporation (Figure 4).

It is interesting to compare the Beaverlodge findings with data obtained elsewhere (Table II). Studies have shown that from Texas in the southern Great Plains area to southern Saskatchewan there is an increase from 12 to 21% in the moisture conserved by summerfallowing (Cole and Mathews, 1923; Thysell, 1938; Doughty et al., 1949; Staple and Lehane,

---

[1] Data from a more recent experiment at three sites in the Beaverlodge area confirm this finding.

1952; Mathews and Army, 1960; Hass and Willis, 1962). Apparently, this trend for greater conservation of moisture toward the north does not hold for Beaverlodge on the extreme northern edge of the Great Plains area. The reason for this is thought to be due to the impermeable nature of the subsoils of the upper Peace River region and the consequent extensive freezing in winter of the shallower depths, an action curtailing infiltration of snowmelt.

As at Beaverlodge, the data for the summerfallow period at Swift Current, southern Saskatchewan, and at Mandan, North Dakota, have been divided into subperiods. The data of these confirm the observation at Beaverlodge that much of the moisture conserved by summerfallowing is obtained early in the fallow period. In these instances, however, the percentage amounts are much greater.

The moisture regime under wheat and some phenological aspects

Spring wheat was grown in the wet year of 1957, the dry year of 1959, and the rather normal year of 1961 (Figure 2). In 1959, and more particularly in 1961, wheat withdrew small quantities of moisture from as deep as 3 ft., or to depth measurements were made. In the rainy summer of 1957 it withdrew very little moisture from below the 18-inch level, while in 1959 and 1961 it took considerable amounts from as deep as 2½ ft. As it developed wheat withdrew moisture from successively deeper depths.

On an average of 3 years, wheat used 10.4 inches of water from planting till ripe and yielded 44.1 bushels per acre of grain. The average loss from fallow during this period in the 2 years it was practiced was 8.5 inches. From spring break-up until winter freeze-up the average moisture loss where wheat was grown was 13.2 inches (Figure 5). This amount compares to an 11.2-inch average loss from fallow for the same period. On this basis the moisture saved by fallowing is 2.0 inches (note also Table II).

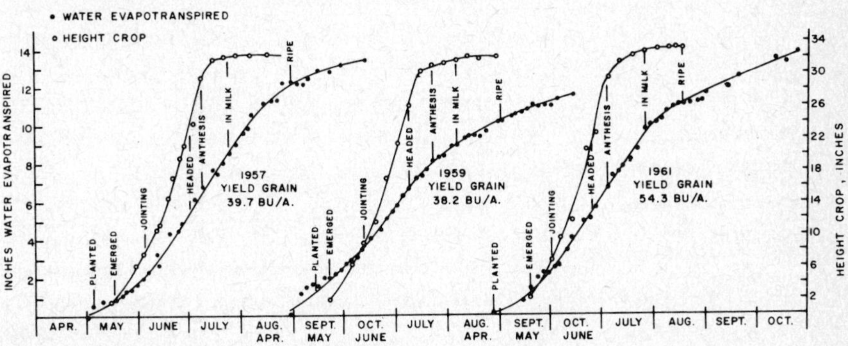

Figure 5. Moisture use, growth and phasic development of spring wheat.

Only 5% of the winter precipitation was conserved by the wheat stubble (Figure 2 and Table II). However, there was actually a small loss following fallow. Since the winter precipitation at Beaverlodge approaches 6 inches, or ⅓ of the annual precipitation, these relationships are most unfortunate.

The rate curve of water uptake by wheat is sigmoid in shape, as is the increase in height of the wheat (Figure 5). At about jointing (4-leaf stage) there was a quickening of the crop's water requirement along with acceleration of growth in height. If moisture was plentiful, as it was in 1957, this rapid consumption of water was maintained until after the milk stage following which, there was a gradual decrease. Curtailment in height was rather abrupt and occurred about the time the crop came into flower. There was practically no increase in height after the kernel was in the milk stage.

It has been shown that the photoperiodic effect in the upper Peace River region strongly compensates for the low heat budget prevailing there (Carder, 1957). This same effect is believed to assist the Beaverlodge crops to economize water. To throw light on this point the moisture uses by spring wheat at Beaverlodge, at Swift Current in southern Saskatchewan 450 miles south, and at Dickinson in North Dakota 750 miles south, were compared. Seasons were selected at each location during which the wheat crop did not at any time suffer from lack of moisture. The Beaverlodge crop was grown in a year (1961) considered about normal in precipitation, but the rains were well distributed and the soil in the spring was unusually well supplied with moisture (Figure 2). The year selected for Swift Current was one of high available moisture at seeding time and good distribution of adequate rainfall through the season (Doughty et al., 1949). At Dickinson the year was one considered exceptionally favorable to wheat (Cole and Mathews, 1923). Whereas at Dickinson 18.1 inches of water were used to produce 35.7 bushels per acre, 15.9 were required for 50.0 bushels per acre at Swift Current and 11.1 inches for 54.3 bushels per acre at Beaverlodge (Figure 8).

Not all this economy of moisture per unit grain weight at Beaverlodge can be ascribed to the photoperiodic effect. For instance, at Dickinson heat surplus to the maximum performance of the crop may be used to evapotranspire water, a fact which curtails the moisture-use-efficiency of wheat grown there. However, the heat requirements of hard spring wheat do not indicate that the heat budget in the year selected at Dickinson or that at Swift Current were unduly high (Blair, 1915; Wilsie, 1962) so that to conclude that much of the water economy at Beaverlodge is due to favorable photoperiod seems justified.

TABLE II. MOISTURE CONSERVED BY SUMMERFALLOW AT DIFFERENT POINTS IN WESTERN CANADA AND U.S.

Location	*No. of fallow periods*	*Period total*[1]		*Stubble*				*Fallow*			
				harvest–freeze-up		*freeze-up–seeding time*		*seeding time–freeze-up*		*freeze-up–seeding*	
		prec. (inches)	*cons. (%)*	*prec. (inches)*	*cons. (%)*	*prec. (inches)*	*cons. (%)*	*prec. (inches)*	*cons. (%)*	*prec. (inches)*	*cons. (%)*
Beaverlodge, Alta.	2	26.2	6.3	3.7	13	5.2	5	12.1	8	5.2	—1
Swift Current, Sask.	49	18.7	21.4	2.2	36	4.3	33	7.8	13	4.4	16
Mandan, N.D.	38	22.7	19.2	3.6	25	3.5	46	12.1	15	3.5	3
Dickinson, N.D.	20	24.3	18.6								
U.S. Northern Plains	112	22.7	18.9								
U.S. Central Plains	22	26.6	17.2								
U.S. Southern Plains	7	25.8	12.4								

[1] Prec. = precipitation.
Cons. = conservation or storage efficiency.

The moisture regime under red fescue

This grass is a perennial with slowly creeping habit. Depending on density of the original stand, it tends to form a continuous sward within 2 or 3 years after planting. It is an important grass in the upper Peace River region where it is much used for pasture and as a seed crop.

In this experiment the fescue was sown without a companion crop in late April, 1957, and the stand, assisted by frequent rains, was well established by the fall of that year. During the first part of the season it extended its root system and after early July used considerable quantities of water (Figure 6). The grass did not seem capable of withdrawing moisture from much below the 1-ft. depth although in the very wet year of 1960 a rank growth presumably withdrew moisture from as deep as 2 ft. (Figure 3). Established fescue took up moisture much more rapidly than wheat, particularly in the spring before wheat had a fully developed root system. Because fescue produced a large mass of fibrous roots in the shallow layers of the soil the top foot of soil under it was capable of holding more moisture than that under wheat or fallow (Figures 2 and 3). Moreover, the grass sod collected and absorbed more winter precipitation than did the fallow or stubble fields. With fescue about 10% was saved.

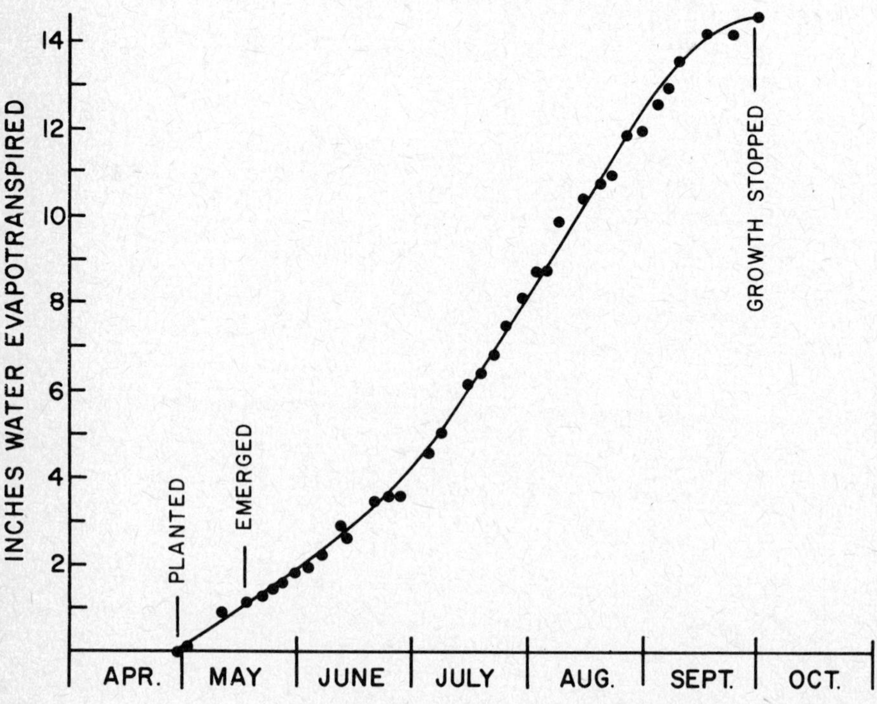

Figure 6. Moisture use by an establishing stand of red fescue, 1957.

Whereas from opening to close of the growing season 12.9 inches of water were used by established fescue, 11.2 inches were lost from fallow (Figures 4 and 7). On this basis the moisture saved by fallowing amounts to only 1.7 inches. From the opening to the close of the season, wheat and established fescue used about the same amount of moisture, 13.2 and 12.9 inches, respectively (Figures 5 and 7).

Established fescue grew rapidly following resumption of growth in the spring, then slowed slightly until the appearance of the panicle after which there was another surge in height (Figure 7). Unless stimulated by extremely moist conditions such as took place in the early summer of 1960, growth in height rapidly decreased after flowering. Cutting the fescue for hay had little sustained effect on its rate of water uptake. The fescue was not cut in 1957, its year of establishment. It was cut twice in 1958 and yielded 0.90 and 0.30 tons per acre of cured hay. A single cutting only was made in the subsequent 3 years with yields of 0.57, 0.48 and 0.37 tons per acre. It is considered that fertilizer application would have increased hay yields, but would not have affected the moisture uptake of

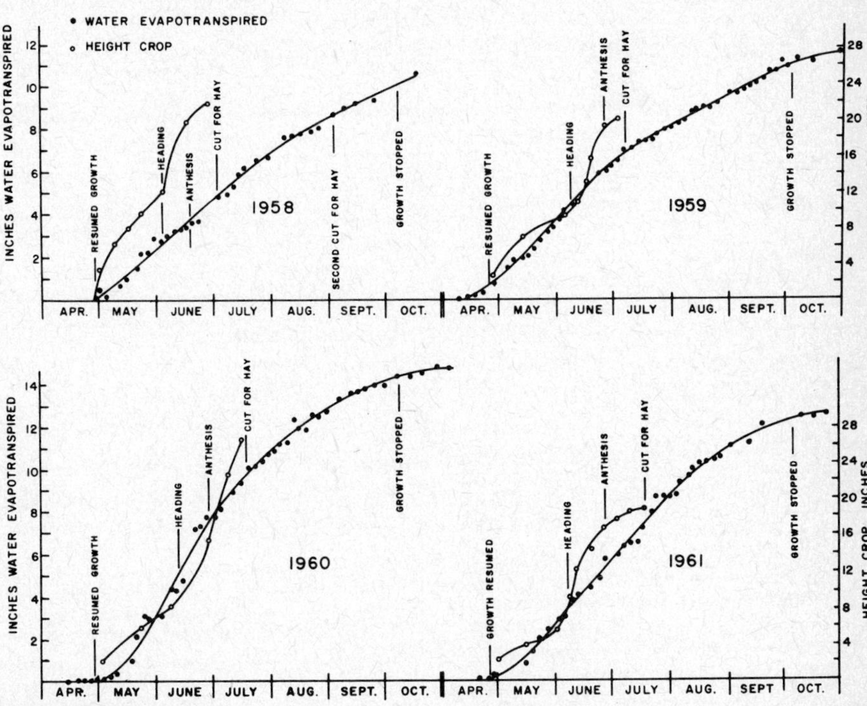

Figure 7. Moisture use, growth and phasic development of an established stand of red fescue.

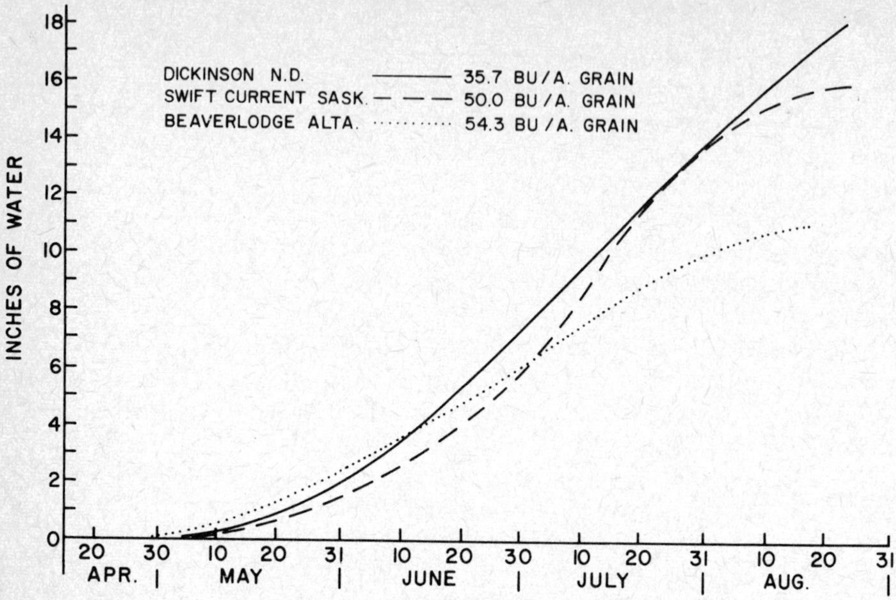

Figure 8. Moisture use efficiencies of spring wheat on summer fallow.

the grass, since existing stands consumed practically all available supplies within reach of their roots (Figure 3). Although fertilizer would not have increased greatly the amount of moisture used, it would have caused it to be used more efficiently (Ferguson, 1963; Warder et al., 1963).

Summary

By use of the absorption block electrical resistance method an almost daily record of the soil-moisture content in fallowed land and under crops of spring wheat and red fescue was obtained over a 5-year period. Changes in soil moisture were collated with the growth and phenological stages of the crops.

During the fallow year (spring break-up to winter freeze-up) 11.2 inches of water were lost by evaporation. This loss of moisture from fallow compared to 13.2 and 12.9 inches evapotranspired over the same period from fields where crops of spring wheat and established stands of fescue were grown.

A feature of the study was the small amount of precipitation that was stored in the soil during the fallow period, which extends from the harvest of the crop preceding the fallow to the planting of the next crop some 20 months later. At Beaverlodge only 6% of the precipitation falling during this period was stored as compared with 12—21% elsewhere.

When the ground was bare of residue as after the fallow year none of the winter precipitation was saved, while with wheat stubble only 5% was saved. Since the winter precipitation at Beaverlodge is about 6 inches or ⅓ of the annual, such loss is serious. Fescue sward proved superior to wheat stubble for with fescue 10% of the winter precipitation was stored.

At the latitude of Beaverlodge the photoperiod effected a water economy. Thus, in selected seasons when water was not limiting it was found that only 11.1 inches of water were needed at Beaverlodge to produce 54 bushels per acre of wheat, while 15.9 inches were needed at Swift Current, 450 miles south, for 50 bushels per acre, and 18.1 inches were needed at Dickinson, North Dakota, 750 miles south for 36 bushels per acre.

Wheat with its extensive root system withdrew moisture from at least a depth of 3 ft., while fescue did not withdraw it from much beyond 1½ ft. Established fescue withdrew water more rapidly and thoroughly than wheat, particularly in the spring before wheat had a fully developed root system. If moisture was plentiful, wheat used it rapidly at all stages from jointing (4th leaf) to near maturity. Under the same conditions fescue was fairly constant in water uptake from soon after start of spring growth to reduction of growth in the fall. Hay harvest had little effect on water consumption.

References

Baver, L. D., *Soil Physics* (New York, Wiley & Sons, 1948), 398 pp.

Bethlahmy, N., "An alignment chart for use with the fibreglass soil-moisture instrument," *Soil Sci.*, Vol. 71 (1951), pp. 377-80.

Blair, T. A., "Temperature and spring wheat in the Dakotas," *U.S. Weather Rev.*, Vol. 43 (1915), pp. 24-26.

Bracken, A. F. and Cardon, P. B., "Relation of precipitation to moisture storage and crop yield," *J. Am. Soc. Agron.*, Vol. 27 (1935), pp. 8-20.

Burr, W. W., "The storage and use of soil moisture," *Nebraska Univ., Agr. Expt. Sta. Bull*, Vol. 5 (1914), 88 pp.

Carder, A. C. "*The Influence of certain Climatic Factors on the Growth and Development of some Crop Plants*," unpublished doctoral thesis, University of Wisconsin (1954).

Carder, A. C., "Growth and development of some field crops as influenced by climatic phenomena at two diverse latitudes," *Can. J. Plant Sci.*, Vol. 37 (1957), pp. 392-406.

Carder, A. C., "Climate of the upper Peace River Region," *Can. Dept. Agri. Publ.*, Vol. 1224 (1965), 20 pp.

Cole, J. S. and Mathews, O. R., "Use of water by spring wheat on the Great Plains," *U.S. Dept. Agr. Bull.*, Vol. 1004 (1923), 33 pp.

Colman, E. A., *Manual of Instructions for Use of the Fiberglass Soil-Moisture Instrument*, (Richmond Calif.), Berkeley Scientific Co. (1947), 20 pp.

Denmead, O. T. and Shaw, R. H., "Evapotranspiration in relation to the development of the corn crop," *Agr. J.*, Vol. 51 (1959), pp. 725-26.

Dittner, J. G., "A quantitative study of the subterranean members of three field grasses," *Am. J. Botany*, Vol. 25 (1938), pp. 654-57.

Doughty, J. L., Staple, W. J., Lehane, J. J., Warder, F. G. and Bisal, F., "Soil moisture, wind erosion and fertility of some Canadian prairie soils," *Can. Dept. Agr. Publ.*, Vol. 819 (1949), 78 pp.

England, C. B., "Water use by several crops in a weighing lysimeter," *Agr. J.*, Vol. 55 (1963), pp. 239-42.

Ferguson, W. S., "Effect of intensity of cropping on the efficiency of water use," *Can. J. Soil Sci.*, Vol. 43 (1963), pp. 156-65.

Halstead, M. H. and Covey, W., "Some meteorological aspects of evapotranspiration," *Soil Sci. Soc. Am. Proc.*, Vol. 21 (1957), pp. 461-64.

Harrold, L. L. and Dreibelbis, F. R., "Agricultural hydrology as evaluated by monolith lysimeters," *Tech. Bull.* (1950), no. 1050, 149 pp.

Hass, H. T. and Willis, W. O., "Moisture storage and use by dryland spring wheat cropping systems," *Soil Sci. Soc. Am. Proc.*, Vol. 26 (1962), pp. 506-509.

Hennig, A.M.F. and Carder, A. C., "The electrical resistance method as a means for quantitative determination of seasonal moisture trends under a growing crop," Unpubl. MS (1955).

Holtam, H. N. and Musgrove, G. W., "Soil water and its disposal under corn and bluegrass," *U.S. Dept. Agr. Soil Conserv. Serv., Res. Rept.*, Vol. 68, (1947), 36 pp.

Kelly, O. J., "Soil evaporation is a major water loss," *U.S. Dept. Agr. Soil Conserv.*, Vol. 64, (1961), 3 pp.

King, K. M., "Effects of soil, plant and meteorological factors on evapotranspiration," *Roy. Meteorol. Soc., Can. Branch*, Vol. 8 (1957), pp. 1-10.

King, K. M., "Evaporation from land surfaces," Evaporation, *Proc. Hydrol. Symp.* Vol. 2 (1961), pp. 55-73.

Kramer, P. J., *Plant and Soil Water Relationships*, (New York, McGraw-Hill, 1949), 347 pp.

Lehane, J. J. and Stape, W. J., "Desiccator method for determining permanent wilting percentage of soils," *Soil Sci.*, Vol. 72 (1951), pp. 429-33.

Mathews, O. R. and Army, T. J., "Moisture storage on fallowed wheatland in the Great Plains," *Soil Sci. Soc. Am. Proc.*, Vol. 24 (1960), pp. 414-18.

Neumann, J., "On a relationship between evaporation and evapotranspiration," *Bull. Am. Meteorol. Soc.*, Vol. 34 (1953), pp. 454-57.

Odynsky, W., Lindsay, J. D., Reeder, S. W. and Wynnyk, A., "Soil survey of the Beaverlodge and Blueberry Mountain Sheets," *Alberta Soil Survey, Rept.*, Vol. 20 (1961), 123 pp.

Penman, H. L. "Vegetation and hydrology," *Commonwealth Bur. Soils, Harpenden, Tech. Commun.*, Vol. 53 (1963), 124 pp.

Reimann, E. G., Van Doren, C. A. and Stauffer, R. S., "Soil-moisture relationships during crop production," *Soil Sci. Soc. Am. Proc.*, Vol. 10 (1945), pp. 41-65.

Reinhart, K. G., "Installation and field calibration of fiberglass soil-moisture units" in Soil-moisture measurement with the fiberglass instrument, *U.S. Dept. Agr., Forest Serv., Southern Expt. Sta. Occasional Paper*, Vol. 128 (1953), pp. 40-48.

Schofield, R. K., "Soil moisture and evaporation," *Trans. Intern. Congr. Soil Sci.*, 4th Amsterdam (1950-51), pp. 20-28.

Shaw, C. F., "The normal moisture capacity of soils," *Soil Sci.*, Vol. 23 (1927), pp. 303-17.

Staple, W. J. and Lehane, J. J., "The conservation of soil moisture in southern Saskatchewan," *Sci. Agri.*, Vol. 32 (1952), pp. 36-47.

Tanner, C. B., "Factors affecting evaporation from plants and soils," *News Notes*, Vol. 12 (1957), pp. 221-27.

Thysell, J. C., "Conservation and use of soil moisture at Mandan, N.D.," *U.S. Dept. Agri., Tech. Bull.*, no. 617 (1938), 40 pp.

Trewartha, G. T., *An Introduction to Climate* (New York, McGraw-Hill, 1954), 402 pp.

Veihmeyer, F. J., "Evaporation from soils and transpiration," *Trans. Am. Geophys. Union*, Vol. 19 (1938), pp. 612-19.

Veihmeyer, F. J. and Brooks, F. A., "Measurements of cumulative evaporation from bare soil," *Trans. Am. Geophys. Union*, Vol. 35 (1954), pp. 601-607.
Warder, F. G., Lehane, J. J., Hinman, W. C. and Staple, W. J., "The effect of fertilizer on growth, nutrient uptake and moisture use of wheat on two soils in southwestern Saskatchewan," *Can. J. Soil Sci.*, Vol. 43 (1963), pp. 107-16.
Wilsie, C. P., *Crop Adaptation and Distribution* (San Francisco, Freeman, 1962), 448 pp.

9

Precipitation Variations in a Small Forested Watershed

D. Storr

Introduction

Interest in watershed management in Canada has lagged behind that in the United States but has been increasing in Alberta since about 1960. As part of a program set up to evaluate and improve land management on the east slopes of the Rocky Mountains for water yield. Marmot Creek Basin was instrumented to study the water balanace in a typical sub-alpine spruce-fir forested watershed. Instrumentation began in 1962 and was mostly completed in 1964 although minor changes are still being made. This program has been described in detail by Jeffrey (1964, 1965). This paper is a preliminary report on the methods and first results of the precipitation measurement program, so the conclusions are somewhat tentative and may be subject to some future revision.

Marmot Creek Basin is located about 50 miles west of Calgary, Alberta, in the narrow steep-walled Kananaskis Valley between the Fisher Range and the Continental Divide. Elevation in the basin ranges from 5,200 ft. MSL to 9,200 ft. MSL with tree-line at approximately 7,500 ft. The general aspect of the basin is easterly. The topography is steeply sloping, averaging 39 percent for the basin as a whole (Figure 1). This limits vehicle access to the lower portion of the basin and limits reading the upper precipitation gauges to once a week in the brief summer and about once or twice during the 8-10 month snow season.

Reprinted from Proceedings, Western Snow Conference, Boise, Idaho (April, 1967).

Precipitation Gauge Network and Observing Procedure

Because of the steep terrain and the high proportion of convective type precipitation, a very dense network of raingauges has been installed. In this area of about 3½ square miles, there are now 30 sites at which summer rainfall is measured weekly, while 10 Sacramento storage gauges

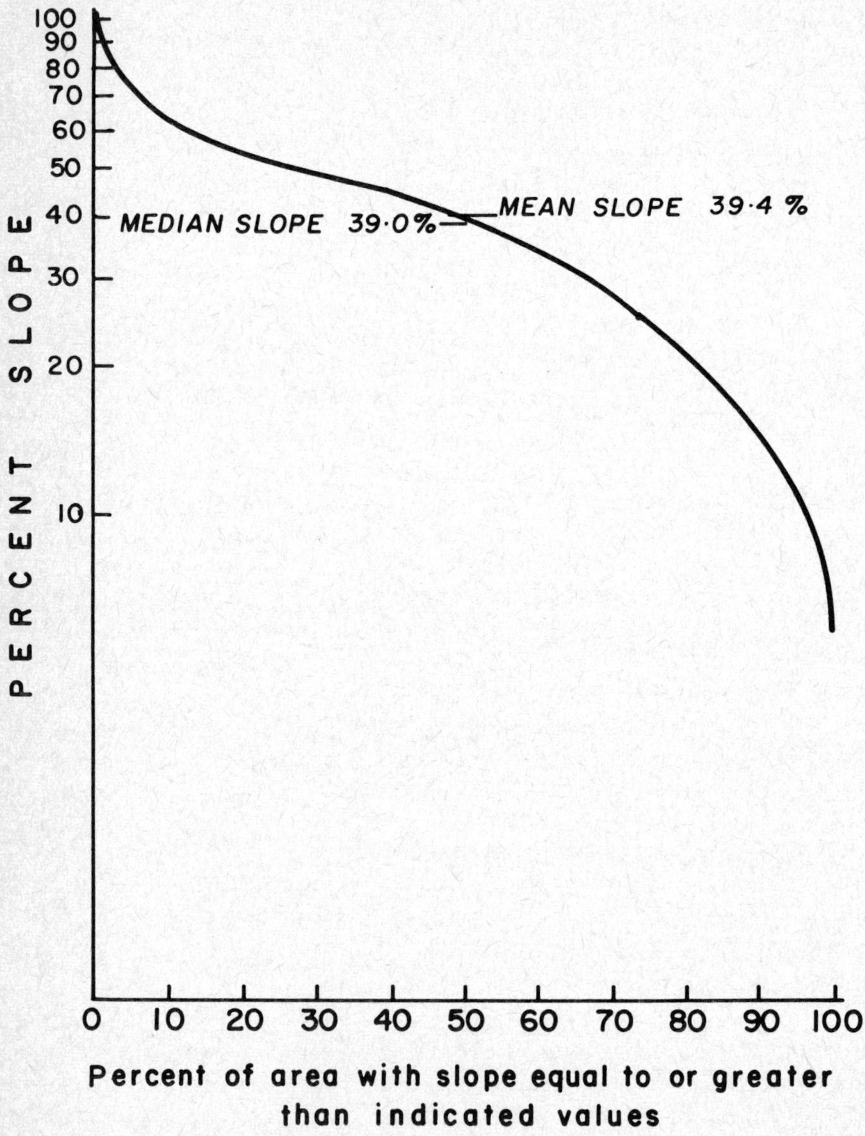

Figure 1. Marmot Creek basin land slope distribution.

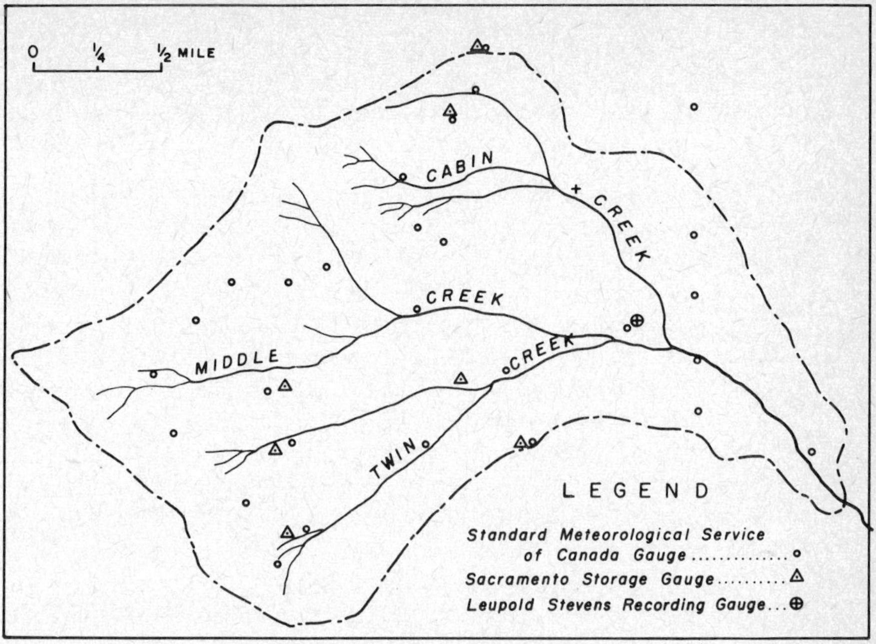

Figure 2. Marmot Creek basin precipitation network.

and one Leupold-Stevens recording gauge measure seasonal and annual precipitation (Figure 2). The raingauge used is the standard Meteorological Service of Canada gauge, with an orifice of 10 square inches standing 12 inches above ground. Its capacity of 4.5 inches has proved adequate so far. A supplementary network of 10 small orifice gauges with a capacity of 18 inches guards against complete loss of data in case a weekly visit has to be missed in very wet weather. The standard MSC gauge measures rainfall to one one-hundredth of an inch while the Sacramento and the Leupold-Stevens gauges measure both rain and snow to the nearest third and tenth of an inch respectively. The MSC gauge is virtually useless for measuring snowfall. The gauges are exposed with the orifice horizontal. An exhaustive test (Storr, 1966) over a two-year period proved that gauges with the orifice parallel to the slope consistently catch less than those with the orifice horizontal in this valley on the lee side of the Rockies.

In addition to the gauge network 20 snow courses have been laid out in the basin to sample the snow pack under various elevations, aspects and types of tree cover. They are read monthly from January to April, then weekly until the snow is gone in May or June .

The two main difficulties in the observing program arise first from the steep terrain which limits the frequency of gauge readings, and secondly from the frequent occasions in the Spring and Fall when rain falls in the lower levels and snow at higher elevations. This produces a situation where precise measurements are obtained in the lower areas and less precise measurements in the area where the heaviest precipitation falls.

Summer Variations in Rainfall

Because the raingauge network is much denser than the Sacramento network and the gauges can be read much more frequently, the rainfall pattern over the basin can be delineated more precisely than that of snow. Charts of the total summer rainfall have been drawn for the three years 1964, 1965, and 1966 (Figures 3, 4, and 5) and although the rainfall values naturally vary from year to year, the pattern is remarkably consistent. Even though a high proportion of the rain is caused by convective activity, over a summer season this variability smooths out as several studies have reported (McKay 1964, Linsley and Kohler 1951, Rainbird 1965). There is a progressive increase of amount with elevation, amounting on the average to about 1.5 inches per thousand feet (Figure 6) but there is considerable scatter from this value, indicating the importance of such other factors as aspect and slope. This "best fit" line is obviously only a rough guide and the next step in refining the rainfall map will be to use the graphical correlation technique as Spreen (1947) has done for western Colorado to relate rainfall totals to all three factors. It is interesting to

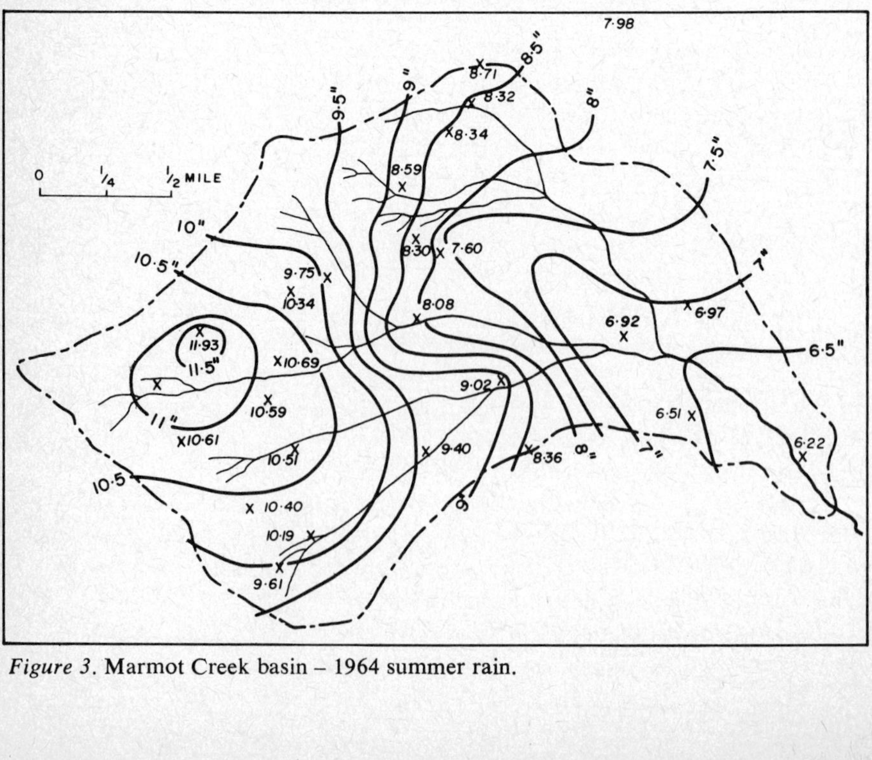

Figure 3. Marmot Creek basin – 1964 summer rain.

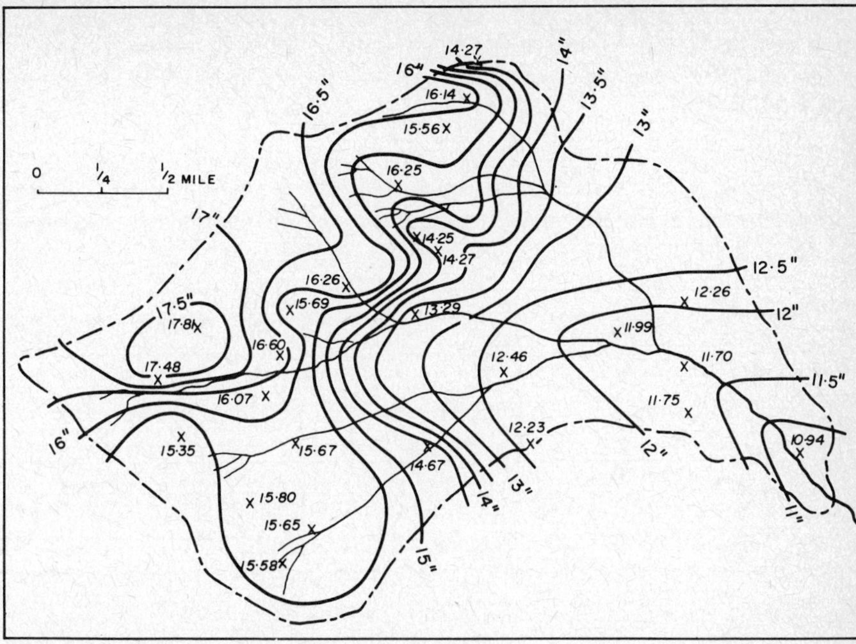

Figure 4. Marmot Creek basin – 1965 summer rain.

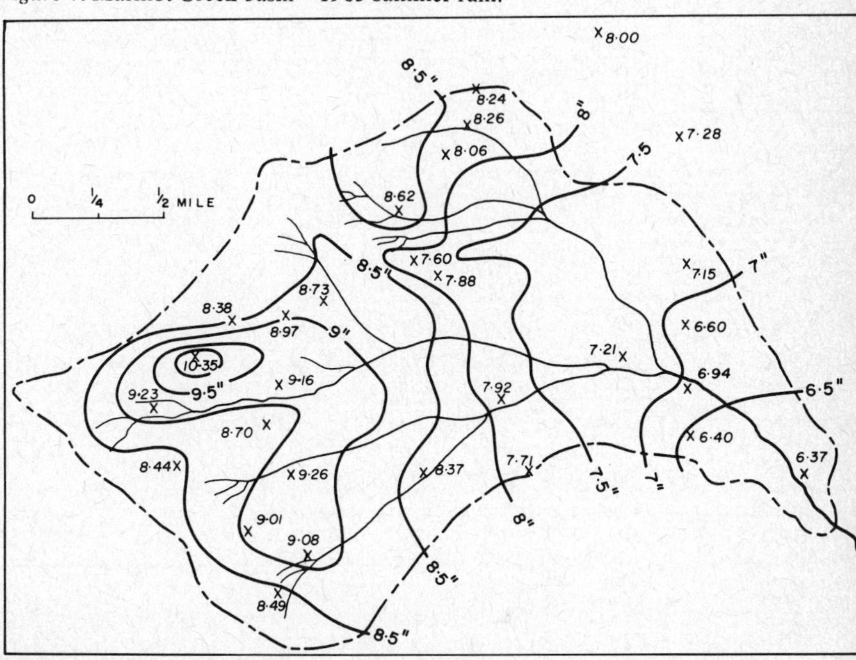

Figure 5. Marmot Creek basin – 1966 summer rain.

note that the curve levels off by about 7,600 MSL with a possible decrease near 8,000 MSL. This requires further study.

Annual Variation in Total Precipitation

Assuming that the annual precipitation pattern will be similar to the consistent summer rainfall pattern, the rainfall pattern was used as a guide in the construction of the annual precipitation map. The annual Sacramento data were plotted and the summer pattern applied to them with only one discrepancy found. Figure 7 shows the average annual precipitation for the three water-years October 1963-September 1966 and it can be seen that the highest precipitation area is in the headwaters of Twin Creek instead of the headwaters of Middle Creek as on the rainfall charts. This difference is easily explained. The majority of heavy summer rains come with winds from the west or southwest, so slopes facing this direction will receive the largest amounts. This accounts for the heavy falls on the south-facing slopes in the headwaters of Middle Creek. In contrast, many of the heaviest snowfalls are caused by outbreaks of Arctic air bringing northerly winds to the basin, and dropping most of the falls on north-facing slopes such as those in the Twin Creek headwaters. Also some "blow-over" of snow from the next basin to the west has been noted visually in the Twin Creek headwaters and may contribute to the high precipitation area there. The previously mentioned assumption therefore seems justified.

Over the past three water years 25-30% of the annual precipitation has occurred as rain with the remaining 70-75% as snow or a mixture of

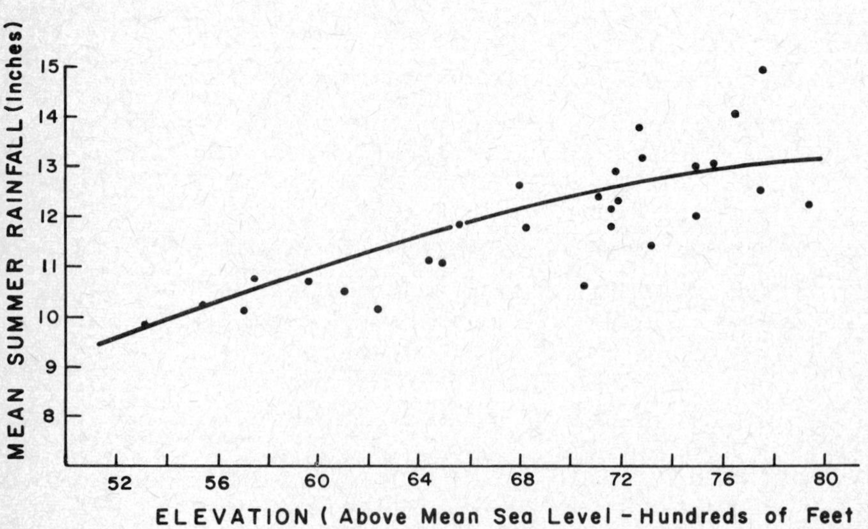

Figure 6. Rainfall-elevation relationship.

rain and snow. This indicates the importance of snow to the water budget of the area and also the importance of studies in the fields of snow melt, snow accumulation, snow evaporation, etc., in the hydrologic investigation.

Accuracy of Measurement

The obvious question that must arise concerns the accuracy of the precipitation measurements. Most recent studies (McKay 1964, Allis et al., 1963) conclude that "true areal precipitation remains unknown" but the degree of accuracy required depends on the purpose of the investigation (Hershfield 1965, Bruce and Clark 1966). An accurate estimate of the total volume of precipitation falling on an area is not as difficult as estimating the spatial and temporal distribution. Obviously the degree of accuracy of estimation depends on two factors; the accuracy of the devices measuring the point precipitation, and secondly the density of the gauge network in sampling the area adequately.

It is not the purpose of this paper to compare the relative performance of raingauges, but since the standard Meteorological Service gauge stands only 12 inches above ground it is felt to be at least as accurate as the standard U.S.W.B. gauge. Curry and Mann (1965) found that the Sacramento storage gauge undercatches rain by 15-20% but Bruce (1961) found no significant difference between the Sacramento and the MSC

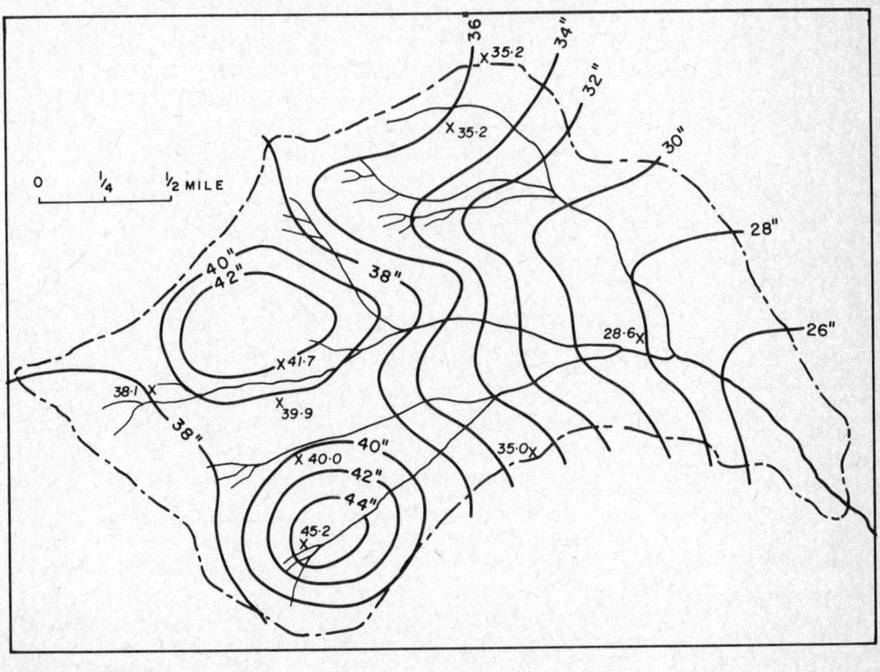

Figure 7. Marmot Creek basin—3 year average water year precipitation, 1963-1966.

standard gauge. Some preliminary incomplete data for 1966 tends to support Curry's conclusion, but until more data are available the annual precipitation map will not be adjusted.

Gauge exposure in the basin is as uniform as possible, but with some gauges in clearings in the forest and some possibly over-exposed in the alpine area, there will be some difference in catch efficiency. It is a matter of continuing study to improve gauge exposure.

Concerning the network density, it was laid out to sample as adequately as possible, within the limits of access and available clearings, all ranges of elevation, slope and aspect as recommended by many investigators (Rainbird 1965, Neff 1965, Corbett 1965). The network density of 8.3 per square mile is the most dense in Canada, as far as can be ascertained, and is similar to that at Coshocton, Ohio. This is not to imply that the network is perfect, in fact there are a few areas of uncertainty remaining and three more standard raingauges will be installed this year. These however are mainly to improve definition of storm rainfall patterns and would affect seasonal or annual precipitation by less than 1 inch.

An indirect method of assessing the accuracy of the precipitation measurement is by comparison with surface run-off figures. The basic water budget equation:

$P = Q + ET + \Delta gw + \Delta sm$ where P is precipitation,
Q is surface run-off,
ET is evapotranspiration,
Δgw is change in groundwater storage,
Δsm is change in soil moisture storage,

may be modified in the following manner for this purpose. Soil moisture storage in this basin is relatively minor because of the shallow mantle over most of the area, and because of its high percolation rate as shown by the rapid response of groundwater wells to precipitation (Stevenson*, personal communication). Also the groundwater wells and the surface run-off are on the "draw-down" phase each fall and tend to be comparable each year. Therefore, if the annual water-year (ending September 30) figures for P, Q, and ET are used, the terms Δgw and Δsm will be negligible in comparison and may be ignored. Any small error created by this step may be further minimized by taking average annual water year figures. Now, $P = Q + ET$.

A comparison of the indicated evapotranspiration loss of 15.6 inches for the basin with the estimated annual lake evaporation of 22 inches (Bruce and Weisman, 1967) suggests three possibilities:

1. Actual evapotranspiration is less than potential evapotranspiration as estimated from the lake evaporation by 6.4 inches.

2. The estimate of lake evaporation may be too high.

---

*R. D. Stevenson, Groundwater Geologist, Alberta Research Council.

TABLE I. AVERAGE PRECIPITATION, RUN-OFF AND INDICATED EVAPOTRANSPIRATION AT MARMOT CREEK

	Precipitation (inches)	*Run-off (inches)*	*Indicated Evapotranspiration (inches)*
Total Marmot Creek Basin	35.26	19.63	15.63
Twin Creek Sub-basin	38.70	24.09	14.61
Cabin Creek Sub-basin	32.10	15.58	16.52
Middle Creek Sub-basin	38.28	23.44	14.84

3. The measurement of precipitation by storage gauges may be too low.

It is suspected that all three factors may be involved and each will be the subject of further study.

The sub-basin figures however are consistent. It can be seen that Cabin Creek has a higher evapotranspiration loss than the other two sub-basins, but this was expected since it has a much higher proportion of south-facing slopes and therefore receives more radiant energy and will lose more water as a result.

It is therefore felt that although the annual precipitation figures may have to be adjusted for an undercatch by the storage gauges as previously mentioned, the adjustment will not be great, and the basic pattern is correct.

Intensity Variations

Precipitation intensity is measured at four locations in the basin with tipping bucket gauges, and eventually some conclusions may be drawn about variations in intensity with elevation, slope, aspect, etc. However, for a number of reasons (mainly mechanical failure) no significant amount of data has been collected from the two upper stations for analysis.

Applications and Discussion

This study has shown that, within limits, it is possible to measure with present instruments the precipitation pattern even in a complex mountainous watershed. The question of temporal variations is a matter of continuing study.

Improvements in instrumentation are needed to overcome the problems of measuring mixed rain and snow, and of frequency of observation.

The study also points up the importance of careful selection of gauge sites both for point measurement of precipitation and for network planning.

The wide variation in precipitation amount over such a short distance confirms the well known fact that researchers and hydrologists should not expect to measure the precipitation in mountainous watersheds using a single gauge without accepting the risk of a large degree of error. Experience in this and two other watersheds in Alberta would indicate that the more rugged the terrain the greater will be the precipitation variation.

Although the precipitation figures are tentative and subject to revision as noted, they do not indicate any major roadblocks to the attainment of a reasonably accurate annual water budget for the basin. Other studies are being undertaken to assess evapotranspiration by the energy budget and Penman methods, and to enable the assignment of values for groundwater and soil moisture storage changes for shorter periods and smaller areas. When the water budget has been delineated, experiments in various forms of manipulation of vegetation will be carried out to determine the optimum means of improving or sustaining water yield, regime and quality.

References

Allis, J. A., Harris, B. and Sharp, A. L., "A Comparison of Performance of Five Raingauge Installations," *Journal of Geophysical Research,* Vol. 68 (1963), pp. 4723-29.

Bruce, J. P., "A Note on Precipitation Measurements by the U.S.W.B. Storage Precipitation Gauge and Meteorological Service of Canada Standard Measurement Methods," Canada Department of Transport Technical Circular Series Cir – 3520 Tec – 366 (1961).

Bruce, J. P. and Clark, R. H., *Introduction to Hydrometeorology* (London, Pergamon Press, 1966), p. 107.

Bruce, J. P. and Weisman, B., "Provisional Evaporation Maps of Canada," Canada Dept. of Transport, Technical Circular Series (in press).

Corbett, E. S., "Measurement and Estimation of Precipitation on Experimental Watersheds," *Proceedings of International Symposium on Forest Hydrology*, Pennsylvania State University (1965).

Curry, G. E. and Mann, A. S., "Estimating Precipitation on a Remote Headwater Area of Western Alberta," Proc. 33rd Annual Western Snow Conference (1965).

Hershfield, D. M., "On the Spacing of Raingauges," Proc. WHO-IASH Symposium on Design of Hydrometeorological Networks (1965).

Jeffrey, W. W., "Watersheds in the Rocky Mountains, Alberta, Canada," Publication No. 66 "Symposium of Budapest" of the I.A.S.H. (1964).

Linsley, R. K. and Kohler, M.A. "Variations in Storm Rainfalls over Small Areas," *Trans A.G.U.*, Vol. 32 (April, 1951), pp. 245-50.

McKay, G. A., "Meteorological Measurements for Watershed Research," *Proc. of Hydrology Symposium* No. 4, National Research Council of Canada (1964).

Neff, E. L., "Principles of Precipitation Network Design for Intensive Hydrologic Investigations," *Proc. WHO-IASH Symposium on Design of Hydrometeorological Networks* (1965).

Rainbird, A. F., "Precipitation – Basic Principles of Network Design," *Proc. WHO-IASH Symposium on Design of Hydrometeorological Networks* (1965).

Spreen, W. C., "A Determination of the Effect of Topography upon Precipitation," *Trans. A.G.U.,* Vol. 28, no. 2 (April, 1947).

Storr, D., "A Raingauge Comparison in a Mountainous Area," Canada Dept. of Transport, Technical Circular Series, Cir. 4452 Tec. 617 (1966).

10

Some Climatological Characteristics of Hailfall in Central Alberta

P. W. Summers

A. H. Paul

Introduction

The Alberta Hail Studies project (ALHAS) is a joint study operated by the Research Council of Alberta in conjunction with the National Research Council of Canada, the Canadian Meteorological Service and McGill University. The primary aim is to determine the causes and behaviour of hailstorms in Alberta, and devise models of hailstone growth as a basis for developing hail suppression techniques. The project has grown steadily in size since its inception in 1957 and is located on the Canadian Forces Base near Red Deer. Current project operations are as follows.

Radar records are obtained with a 10cm set equipped with a 22-foot diameter parabolic dish antenna and a variable polarisation feed. The radar data are presented in both PPI and HARPI* displays using "grey scale" intensity steps. Stereo cloud photography from the ends of a 3 mile base line is used to record the visual appearance of the clouds and provide quantitative data on rates of growth, etc. A radiosonde system

*HARPI, Height Azimuth Range Position Indicator display recently developed by the Stormy Weather Group at McGill University.

Previously published as Scientific Report, McGill University Stormy Weather Group MW-57 (December, 1957) and Proceedings, Fifth Conference on Severe Local Storms, St. Louis, Mo. by *American Meteorological Society* (October 1967), pp. 313-24.

is operated on a routine daily basis during the hail season, and in addition serial ascents are made on active days. Mobile PIBAL crews obtain wind measurements in the vicinity of storms. Ten high-resolution automatic rain gauges and networks of rain and hail sample collectors are operated by selected farmers. A mobile sampling unit is directed into storms to collect samples and measure rain and hailfall rates.

The basic hail reporting system consists of about 25,000 farmers living within the 22,000 sq. mile project area outlined in Figure 1. Every spring these farmers are issued with a supply of business-reply postcards which they are requested to fill in and return after every storm. Estimates put the response of farmers at between 20 and 30 per cent. Whilst this response is good enough for climatological studies, a higher reporting density is desirable for special case studies and correlation with radar and other data. A system of surveys was initiated in 1962, and has been steadily improved to the point where 500 to 600 reports can now be solicited by telephone within two days of a storm. Reporting densities as high as one report per 2 sq. miles can be obtained under favourable conditions.

All of the hail reports collected by these two methods during the last ten years have now been placed on IBM punched cards. Some 30,000 such reports are available for analysis, and in this paper a first analysis of some of the more important parameters is presented.

Hail Frequency

The hail season in Alberta extends from May to September. The average number of days on which hail was reported from somewhere within the ALHAS project area during the ten-year period was 66.5 per hail season. The greatest number of hail days was 80 in each of the years 1957, 1962 and 1963. The quietest year was 1961 with only 37 hail days.

TABLE I. SUMMARY BY MONTHS OF VARIOUS HAIL PARAMETERS

Hail Parameter	*May*	*June*	*July*	*Aug.*	*Sept.*	*Total or average*
Average no. of hail days	5.3	20.3	20.3	15.7	4.9	66.5
Probability of a hail day	17%	68%	66%	51%	16%	48%
Aver. no. major hail days	0.2	5.4	5.4	2.3	0.2	13.5
Prob. of a major hail day	1%	18%	17%	7%	1%	10%
Total no. mailed-in reports	484	8968	9672	4725	378	24227
Mean hail onset time (M.S.T.)	1527	1607	1655	1612	1517	
Mean hailfall duration (min.)	11	9	11	11	8	10
% report with hail $\geqslant$ 1″ diam.	8%	11%	22%	23%	1%	18%
% report with soft hail (1965-66)	52%	59%	35%	42%	59%	44%

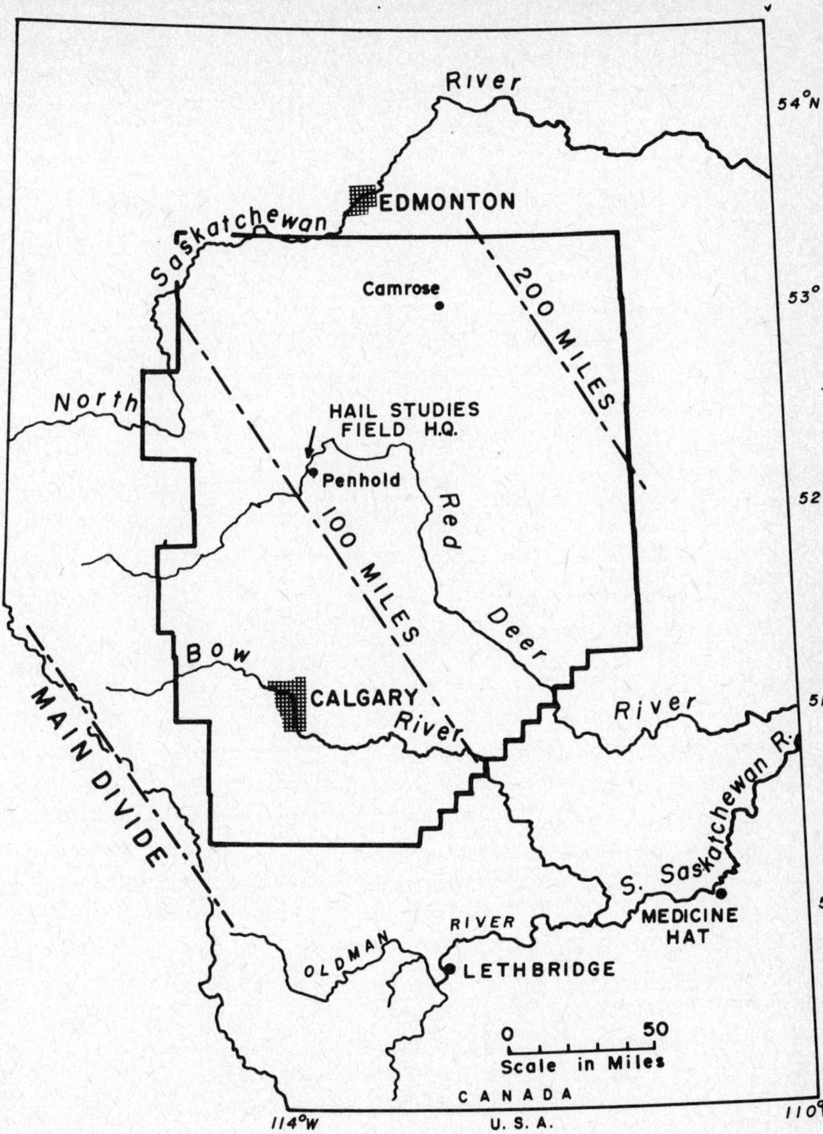

Figure 1. Map of southern Alberta showing hail studies farmer reporting area.

Table I shows the month by month breakdown. The season rapidly reaches its peak in June, when the probability of a hail day is at a maximum of 68%. The probability gradually falls off through July and August, with the season effectively over by mid-September. This is illustrated more clearly in Figure 2. A major hail day is defined, on the basis of areal extent

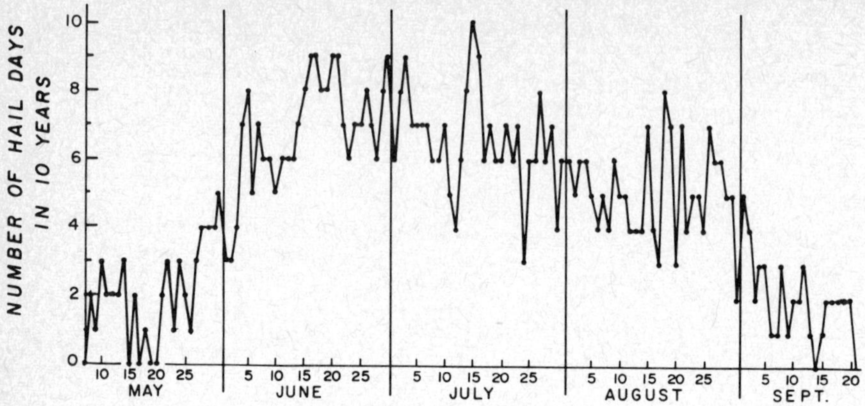

Figure 2. Seasonal variation of hail days in ALHAS study area, 1957-1966.

of activity, as a day when more than fifty reports are mailed in, and is most likely to occur in June also. However although storms are much less frequent in August, those that do occur have the highest probability of producing large hail ($\geqslant$ 1″ diam.).

Singularities. Figure 2 shows a tendency for hail activity to peak on certain dates. In particular it has hailed on 15 July in each of the ten years (and also again in 1967). Ten years is not a long enough record for detecting singularities with any confidence. The Alberta Hail Insurance Board (AHIB) has maintained records of the number of claims for damage on each calendar date for the last 22 years for the whole province of Alberta. These statistics are shown in Figure 3. Four dates immediately stand out as having more damage reported than the few days immediately preceding or following. These are 14 July, 28 July, 13 August and 12 September. No singularities show in June since little insurance is written until the crops have germinated. The most significant peaks in the AHIB records, for the number of times claims have been made on a specific date, are 15 July and 28 July, both with claims in 19 out of 22 years. Thus the most pronounced peaks in Alberta hail activity occur on 14-15 July and 28 July, with lesser peaks on 13 August and 12 September. The 14-15 July, 13 August and 12 September peaks coincide well with those observed in Switzerland, but none of the Alberta dates agree with those found in Illinois (Changnon 1962).

Persistence. There is a strong tendency for clustering of hail days and no-hail days into spells, and major hail days in particular often occur adjacent to each other. This effect is illustrated in two ways. Firstly, Table II shows the probability of certain events given what happened the previous day. The probability of a no-hail day is highest (52%) following a no-hail day, and is very low (16%) following a major hail day. Likewise

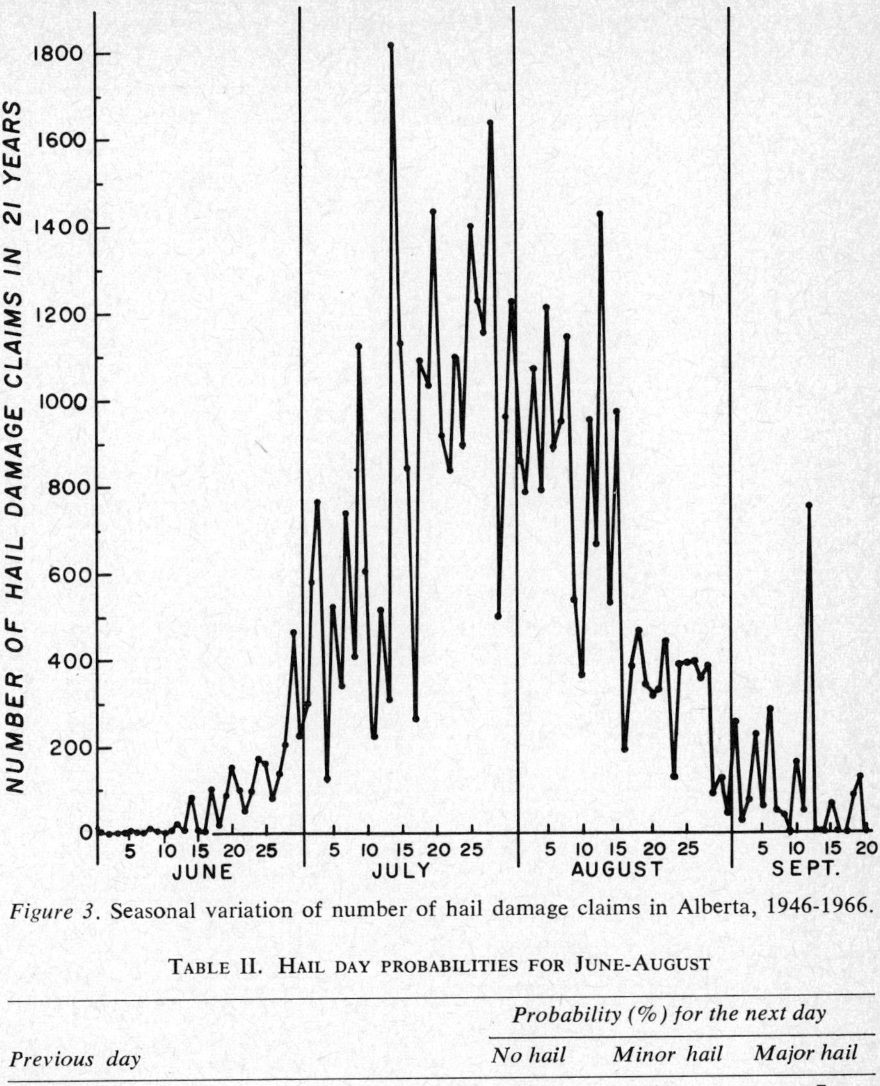

Figure 3. Seasonal variation of number of hail damage claims in Alberta, 1946-1966.

TABLE II. HAIL DAY PROBABILITIES FOR JUNE-AUGUST

	Probability (%) for the next day		
Previous day	*No hail*	*Minor hail*	*Major hail*
No hail	52	41	7
Minor hail	33	50	17
Major hail	16	58	26
Overall expected probability	39	47	14

the probability of a major hail day after a no-hail day is extremely low (7%), and is almost four times as likely (26%) after a major hail day. This shows a strong serial correlation of events on consecutive days. Another way of looking at persistence is to compare the frequency of occurrence of hail day runs of various lengths with those expected on the basis of a purely random distribution of hail days with no persistence. If it is assumed that there is no persistence, then the probability p of a hail day, and the probability q $(=1 - p)$ of a no-hail day are independent of what has gone before. Thus the probability that a day begins a run of hail days is qp. The chance that the following day is a hail day is p, and so the probability of a run of at least two days is qp^2. Similarly the chance of at least three consecutive hail days is qp^3, and so on. If the total number of days in the test period is N, the expected numbers of runs of at least one, at least two, at least three... hail days are Nqp, Nqp^2, Nqp^3 . . . . Thus the expected number of runs of exactly one day is $Nqp - Nqp^2$, exactly two days is $Nqp^2 - Nqp^3$ and so on. Using the period 15 May to 15 Sept., then for all ten years Table III was prepared comparing the expected (E) number of runs, based on no persistence, with the actual (A) observed number of runs of n days. The χ^2 test can be applied to the data in Table III. Since this test should only be used when the expected frequency of an event is >5, then all the data for $n > 7$ are combined into one class. Thus $\chi^2 = \Sigma(A - E)^2/E = 105.1$. With six degrees of freedom this value gives a probability of much less than 0.001 that the differences are due to chance. Table IV was prepared for major hail days only, giving $\chi^2 = 31.4$. With one degree of freedom, this gives a probability of much less than 0.001 that the observed distribution occurred by chance. It can be seen clearly that there are 2½ times as many runs of 2 or more major hail days as expected, again showing a strong tendency for persistence.

Hail frequency index. A card file is maintained of all the farmers who have ever sent in a hail report. This file then gives the number of actual observers in each township. The total number of hail cards received during

TABLE III. COMPARISON OF THE EXPECTED AND ACTUAL NUMBER OF RUNS OF n HAIL DAYS

Value of n	1	2	3	4	5	6	7	8	9	10	11	12	13
Runs of exactly n days, assuming no persistence (E)	147	73	37	20	11	5.8	3.1	1.8	1.0	0.6	0.3	0.2	0.1
Observed runs of n days (A)	69	44	25	19	11	9	11	4	4	1	3	1	1

TABLE IV. COMPARISON OF THE EXPECTED AND ACTUAL NUMBER OF RUNS OF MAJOR HAIL DAYS

Value of n	1	2	3	4
Runs of exactly n days assuming no persistence (E)	98	10	1	0.1
Observed runs of n days (A)	69	23	3	1

the ten years from each township divided by the number of file cards thus gives a measure of the hail frequency, corrected for population and reporter density. The hail frequency index obtained in this way is an underestimate of the true frequency by an unknown factor, but is still a useful indicator for comparing frequencies in various parts of the project area. Townships were combined into larger areas to give a better sample size, and the map in Figure 4 prepared from 122 data points. The most obvious feature of

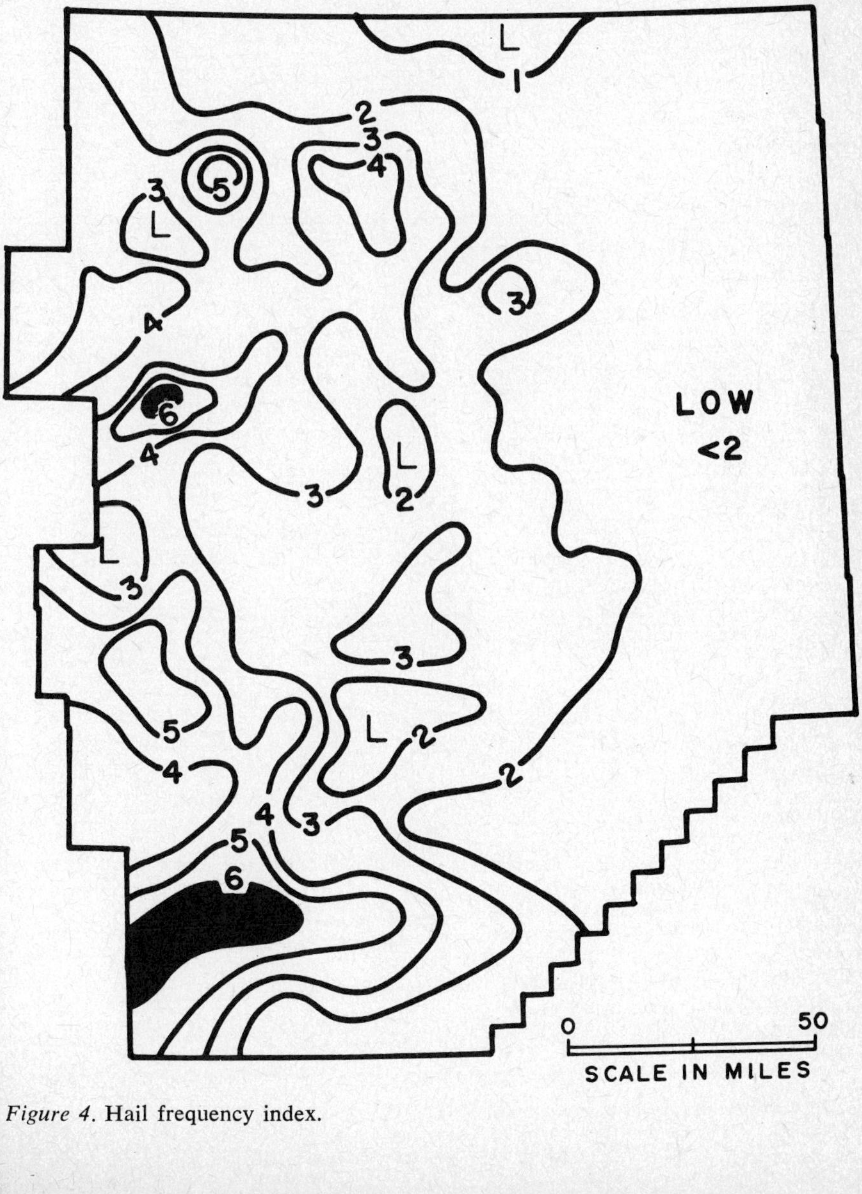

Figure 4. Hail frequency index.

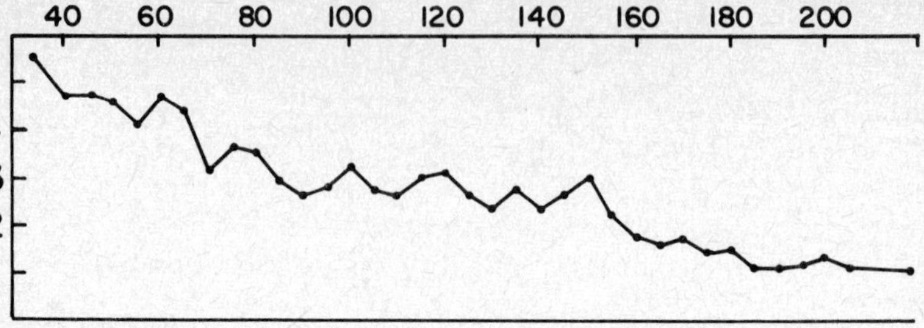

(a) Hail frequency index.

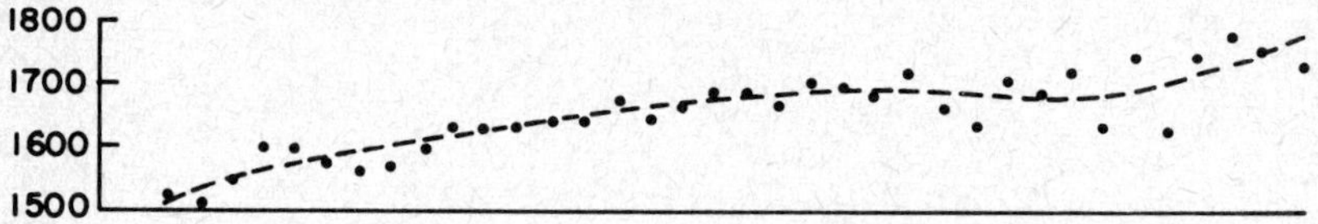

(b) Mean hail onset time.

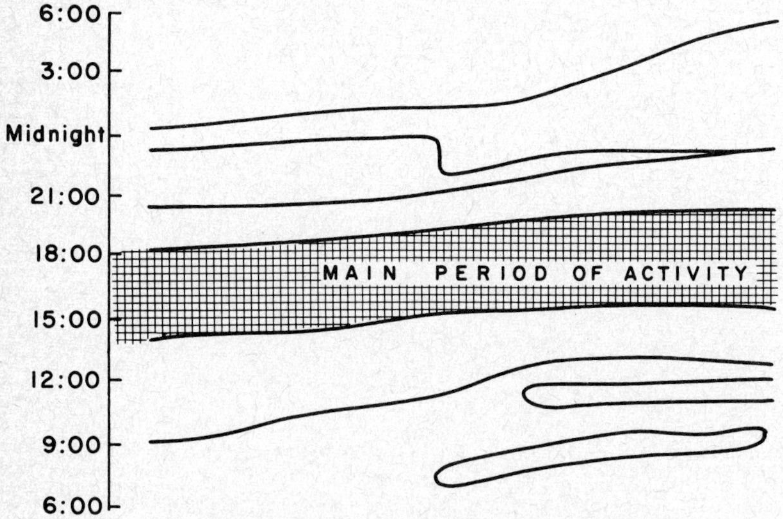

(c) Schematic representation of frequency distribution of hail onset time for each strip parallel to the mountains.

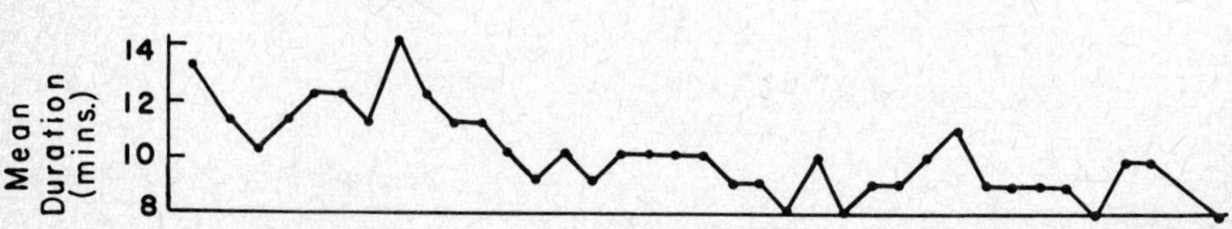

(d) Mean hailfall duration.

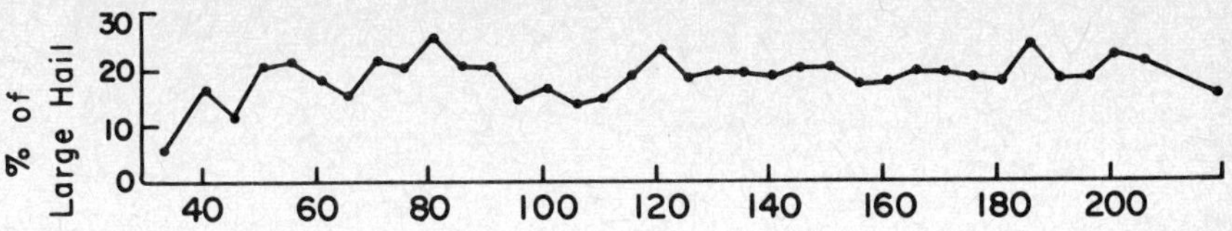

(e) Percentage of large hail.
Figure 5. Variation of parameters with distance from the mountains.

Figure 4 is the sharp decrease in the hail frequency from west to east. But even in the west there is a wide variation, with several pockets of high frequency showing up.

Variation with distance from mountains. The computer was programmed to sort the townships into strips running parallel to the main continental divide of the Rocky Mountains. The mean distance from the divide of each strip turned out to be a multiple of 5 miles. Obviously sample sizes in the extreme southwest and northeast of the project area are very small and so strips were grouped together. The mean value of many parameters could then be calculated at 5 mile intervals over a range 30 to 220 miles from the mountains. Figure 5a shows a steady decline in the hail frequency index with increasing distance from the divide. Thirty miles from the divide the index is 5.6, dropping to half this value at 105 miles, and one fifth at 220 miles. Since most Alberta hailstorms are generated in the foothills immediately east of the main divide, this graph can be interpreted as follows. A hailstorm which starts in the foothills has a 50% chance of reaching a distance of 105 miles from the divide and only a 20% chance of persisting for a distance of 220 miles. In actual fact some storms (perhaps 10%) develop further east away from the foothills, and so the probabilities of foothills-generated storms reaching the above distances are even less.

Onset Time of Hail

The farmers are requested to report the hail onset time to the nearest minute if they can. In actual practice there is a strong bias towards reporting times to the nearest 5 minutes, quarter hours and whole hours (Douglas 1961). In order to minimize this bias, onset times were sorted into quarter hour class intervals with the divisions at 7½, 22½, 37½ and 52½ minutes past the hour. These were then further combined in this paper into one hour classes with the divisions at 22½ minutes past the hour.

Seasonal Variation. The mean onset time of hail for each month is given in Table I. A marked seasonal trend shows up, with the time of onset becoming later from May to July and then earlier again through August to September. Frequency histograms for onset times by month are shown in Figure 6. In May and June the main period of hail activity

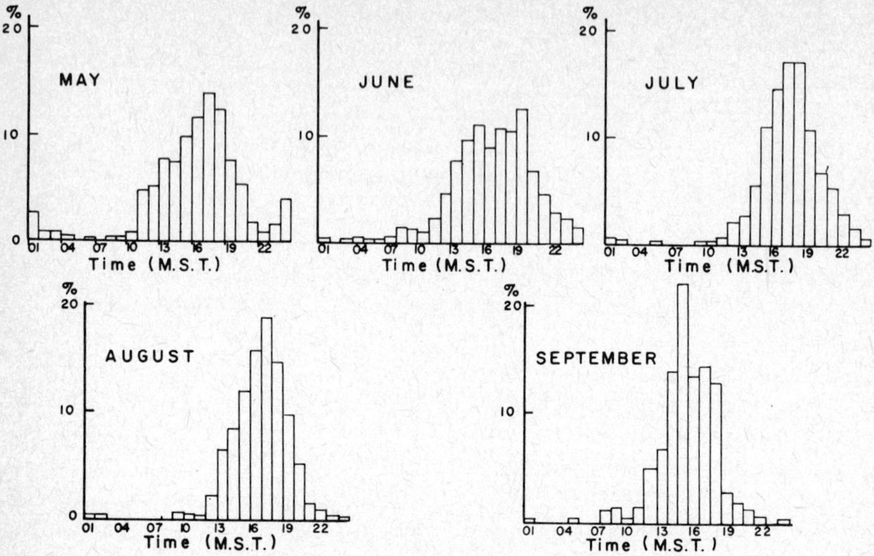

Figure 6. Frequency histograms for onset time of hail by months.

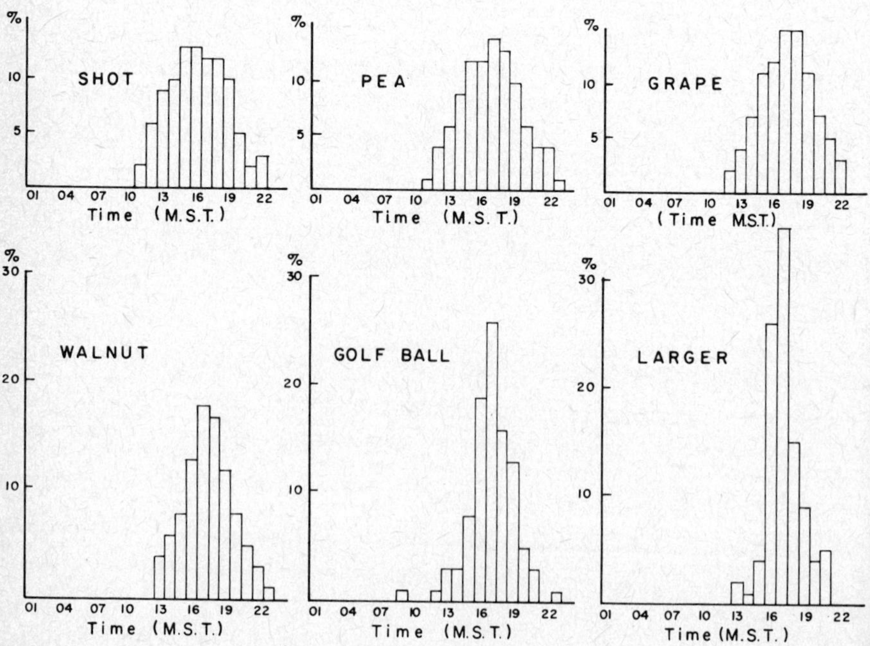

Figure 7. Frequency histograms for onset time of hail for each maximum size category.

is quite long, extending from 1100 MST to 2000 MST with a rather flat late-afternoon maximum. As the hail season progresses, the period of activity becomes shorter with a more pronounced peak becoming earlier in the afternoon. Night-time reports form a relatively high percentage of the total in May, when a secondary peak shows up at midnight. In the other months night-time activity is infrequent, at least in comparison with other times.

Relation to maximum hail size. For each of the six maximum size categories (see *Maximum Hail Size*), hail onset time frequency histograms are shown in Figure 7. The percentages were rounded off to the nearest whole per cent and thus give an inadequate representation of night-time hail, since at many hours the frequency is still finite but less than 0.5%. The smallest hail is experienced over the widest range of onset times, with the largest stones concentrated into a very narrow range in the later afternoon. This information can be presented in another way, as in Figure 8. The percentage frequency of hail onset times between 1523 and 1822 MST for each maximum hail size increases steadily from only 37% for shot size hail to 76% for larger than golfball size hail. Strong updrafts are required for the growth of large hailstones, and so the greatest frequency of large hail coincides with the period of maximum convective activity due to surface heating.

Variation with distance from mountains. The mean hail onset time for the whole season versus distance from the divide is shown in Figure 5b.

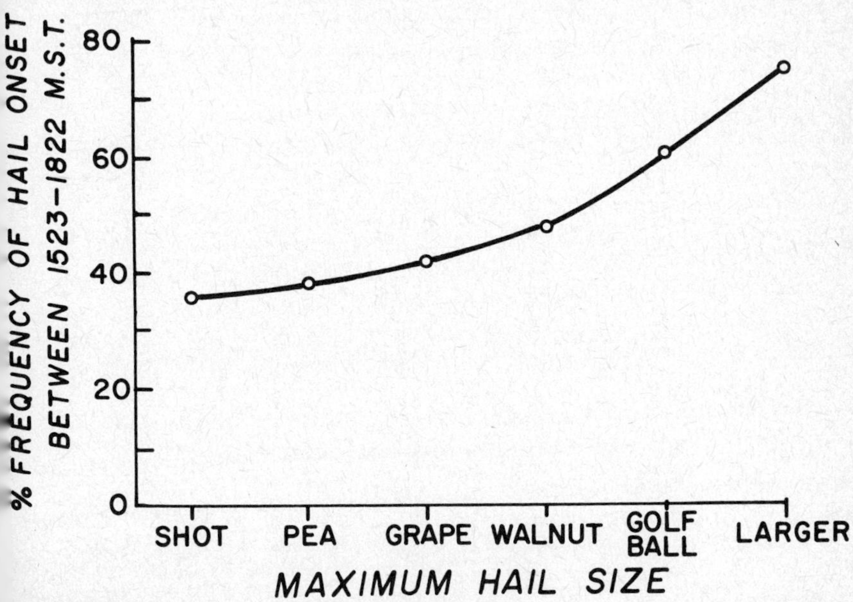

Figure 8. Hail onset time vs. maximum hail size.

Hail onset time is earliest (1500 MST) at 30 miles from the divide, and becomes progressively later up to a distance of 150 miles. At greater distances the times show a wide scatter. The reason for this can be seen more clearly in Figure 5c, which is a schematic representation of the onset time frequency distributions for each 5-mile strip parallel to the mountains. The main band of activity (which represents 60 to 75 per cent of the total hailfalls) is centred on 1600 MST close to the mountains, and becomes progressively later eastwards. This hail activity is due to the main daytime cycle of surface heating inducing convective activity. The slope of this main period indicates a time delay of 2 hours in 180 miles. Obviously this cannot be due solely to the translation of storms since their mean speed is 20-30 mph. There must thus be some propagation also, with new storms developing further east ahead of already existing ones. In fact the cold surface outflow of decaying storms has often been observed to set off new activity several miles further east.

In addition to the main period of activity, several interesting secondary periods show up. The earliest activity commences at 0730 MST at a distance of 115 miles from the divide. Another band of activity breaks out over the whole distance range of 140-220 miles at 1030-1100 MST. These morning bands of activity may be due to a mountain wave phenomenon releasing potential instability at preferred distances downwind. After the main period of activity in the foothills, another brief period of activity commences just before midnight, and almost simultaneously right across the area. This activity could possibly be due to a katabatic wind effect coming off the cooler eastern slopes of the mountains, undercutting the warmer air on the plains and releasing instability. An alternative explanation could be the spill-over of showers that developed further west in the mountains during the afternoon and evening. It is worth noting that Beckwith (1957) detected a secondary peaking of thunderstorm development in the mountains west of Denver, Col., in the evening. The data in Figure 5c should be further investigated to see whether the secondary bands are general in the project area or confined to certain localities.

Hailfall Duration

A strong bias towards multiples of 5 minutes is evident in the hailfall durations reported by farmers. In order to minimize this bias the data were therefore stratified into 6-minute class intervals as shown in Table V. The mean hailfall duration at a point for all observations made over the 10-year

TABLE V. HAILFALL DURATION

Duration (min.)	⩽6	7-12	13-18	19-24	⩾25
Frequency (%)	47	26	13	7	7

period was 10 minutes. Nearly half of the reported durations were $\leqslant 6$ minutes, and only 14% were in excess of 18 minutes. The duration at a point is dependent on three factors, a) the size of the hail-producing unit, b) its speed of motion and c) whether the centre or edge of the hail area passes over the observer. Assuming a circular hail-producing area, moving at 30 mph, and an equal probability of the observer being at any distance off the centre of the track, then a mean duration of 10 minutes gives a mean diameter of such an active hail area as 6.4 miles. However, many of the reported hailfalls in Alberta are intermittent, with two or more bursts occurring during one storm. This suggests the existence of several smaller hail-producing cells within the active area (Douglas 1965).

Seasonal variation. The mean duration for each month is shown in Table I. Durations are longest in July and August when storms are most severe, and shorter in June and September. May also has longer durations, but this is probably due to the fact that the farmers are not requested to start reporting until 1 June, and the only reports are for severe early storms which are reported by keen observers using the previous year's cards.

Spatial variation. The map in Figure 9 shows the spatial variation of hailfall duration based on the 122 data points. The pattern is very irregular with several pockets of long and short durations evident. The most striking feature is the region east of Calgary with exceptionally long durations. A contiguous area of 250 square miles has a mean duration of 21½ minutes, with 15% of the recorded hailfalls lasting longer than 25 minutes. Perhaps this is some effect due to the city of Calgary causing a regeneration effect and longer persistence of storms immediately downwind.

Variation with distance from mountains. This is shown in Figure 5d. There is some evidence for longer durations (11-14 minutes) close to the mountains, decreasing gradually to 8-10 minutes in the northeast corner of the project area. If it is assumed that hail-producing units remain roughly the same size, then this suggests that they move more slowly during the early stages of development in the foothills, and then accelerate as they reach maturity. There is some radar evidence for acceleration of storms as they move east.

Maximum Hail Size

The farmers are asked to check off one of six boxes for both the largest and the most common sizes of hail observed in each storm. The choice is between six sizes associated with familiar objects, namely:

shot (1-4 mm.); pea (4-12); grape (12-20); walnut (20-32); golfball (32-50); larger ($\geqslant 50$).

The numbers in parentheses are the assumed size ranges in mm. Williams and Douglas (1963) suggested that a greater percentage of small hail reports are received by surveys compared with the mail-in response.

However, a study of all the survey data for the years 1962 to 1966 revealed that the ratio of walnut and larger hail to all hail reported was exactly the same as for mailed-in cards during the same period. Therefore in this analysis all data on maximum hail sizes will be utilized regardless of the method of collection. The most frequently reported maximum hail size during the period 1957-66 was pea, with 18.2% of the total reports being walnut or larger (see Table VI). A large variation occurs from year to year. The least amount of large hail fell in 1958 when only 6.8% of the

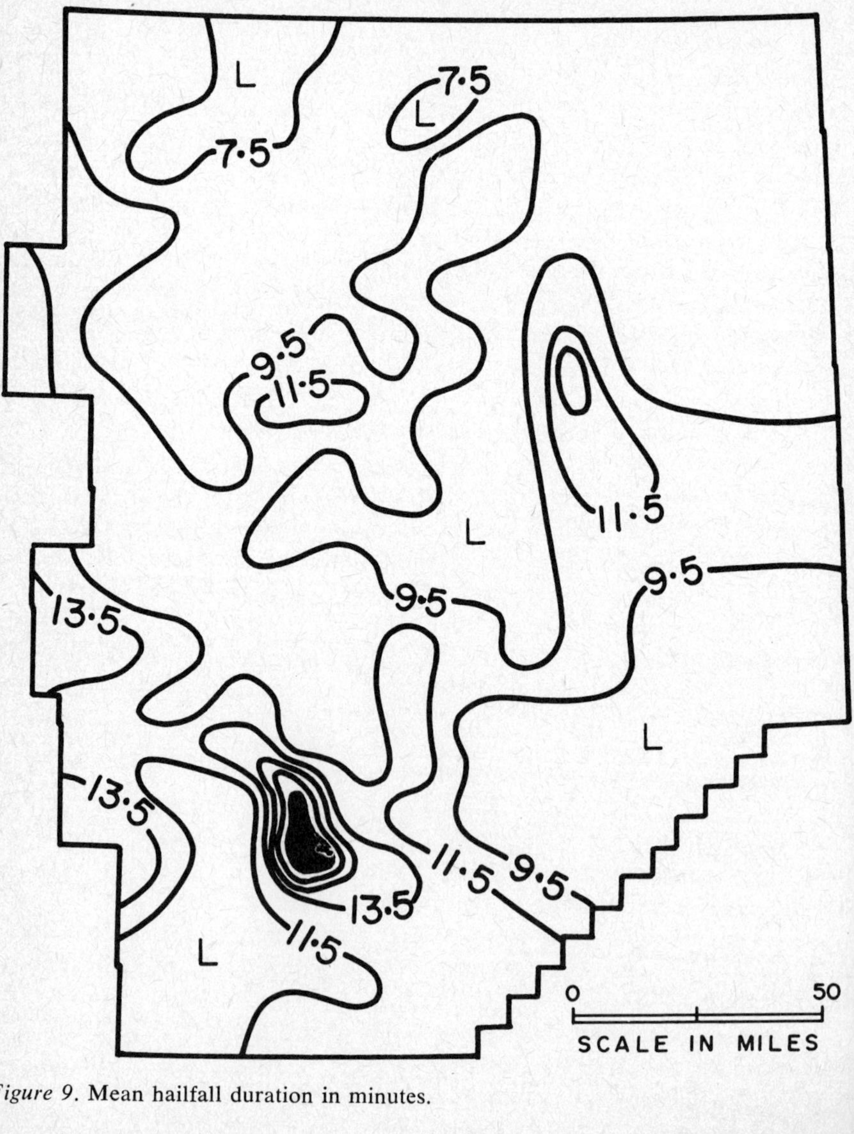

Figure 9. Mean hailfall duration in minutes.

TABLE VI. PERCENTAGE FREQUENCY OF MAXIMUM HAIL SIZE

	Frequency (%) of maximum hail size					
Year	*shot*	*pea*	*grape*	*walnut*	*golfball*	*larger*
1957-66	9.4	38.7	33.7	12.2	4.7	1.3
1958	12.2	52.7	28.2	5.4	1.4	0.0
1960	6.3	28.6	33.4	15.3	11.1	5.3

reports mentioned walnut and larger hail. In contrast, the modal maximum hail size in 1960 was grape instead of pea, and 31.7% of the reports indicated large hail.

Spatial variation. The map in Figure 10 shows a very irregular pattern for the percentage frequency of large hail ($\geqslant$ 1″ diameter). Several small pockets occur where the frequency is greater than 30%. It is interesting to note that one of these pockets coincides with the area of maximum hailfall duration just east of Calgary. These two factors combined suggest that this area was the location of the most severe hailstorms in the project area during the period 1957-66.

Variation with distance from mountains. In this case there is no overall trend discernible but rather a series of peaks and trough (see Figure 5e). The occurrence of large hail is infrequent close to the mountains in the storm breeding area, but increases to a maximum frequency of 26% at a distance of 80 miles away. This is followed by a sharp decrease in frequency from 90 to 110 miles away. Another peak occurs at 120 miles and a less well defined maximum at 180 to 200 miles from the mountains. This suggests that the first wave of hail activity which develops in the foothills reaches a maximum intensity 80 miles from the mountains and then the storms weaken, with further regeneration or new development taking place between 110 and 120 miles away. It is interesting to note that Figure 5c shows that 120 miles from the mountains is an area for development of hailstorms in the early morning and late evening.

Soft Hail

During the last few years an increasing number of farmers in Alberta has been reporting soft or slushy hail. In 1965 a question regarding this property was therefore added to the report card. The question simply asked whether or not any of the hail observed was soft or slushy, and if so what was the estimated maximum size of the soft hail. The 1965-66 hail seasons produced 6021 answers to this question, and in 2672 (44%) of the reported hailfalls some soft hail fell.

Seasonal variation. The seasonal trend is shown in Table I. The greatest percentage frequency of soft hail is at the beginning and end of the season, and reaches a minimum in July. In 97% of the cases it is the largest, or one size less than the largest, hail falling at a point that is soft. In Figure 11 a systematic seasonal change is evident when the percentage frequency of soft hail is related to the maximum size of hail. In May there is a strong tendency for the soft hail to occur in association with large hail. In June there is little variation with size. In July and August most of the large hail tends to be hard. September reverts to a regime similar to May.

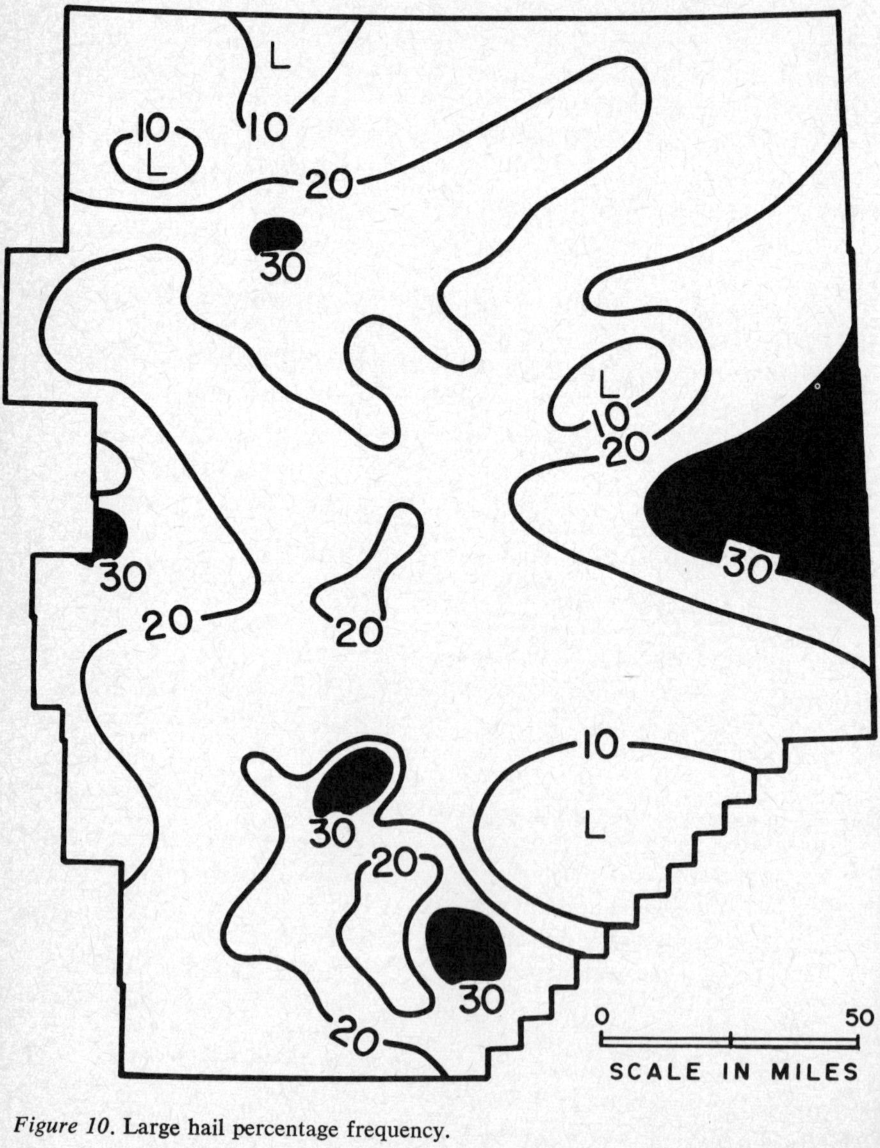

Figure 10. Large hail percentage frequency.

There is a wide variation in the percentage of soft hail from storm to storm. By the end of the 1967 season about 10,000 answers re. soft hail will be available for a more detailed analysis of the spatial and temporal variations.

During the period 1 June to 10 August 1967, a team from the National Centre for Atmospheric Research at Boulder, Col., visited Alberta with a truck equipped with a calorimeter for measuring the water content of freshly fallen hailstones. On 14 July in a storm at Winfield (60 miles northwest of Penhold) several stones with diameters between 1.25 and 2.0 cm. were tested and found to have water contents running between 11% and 16%. There is some doubt about the accuracy of data from samples tested on other days, but only in one case was no liquid water found in the

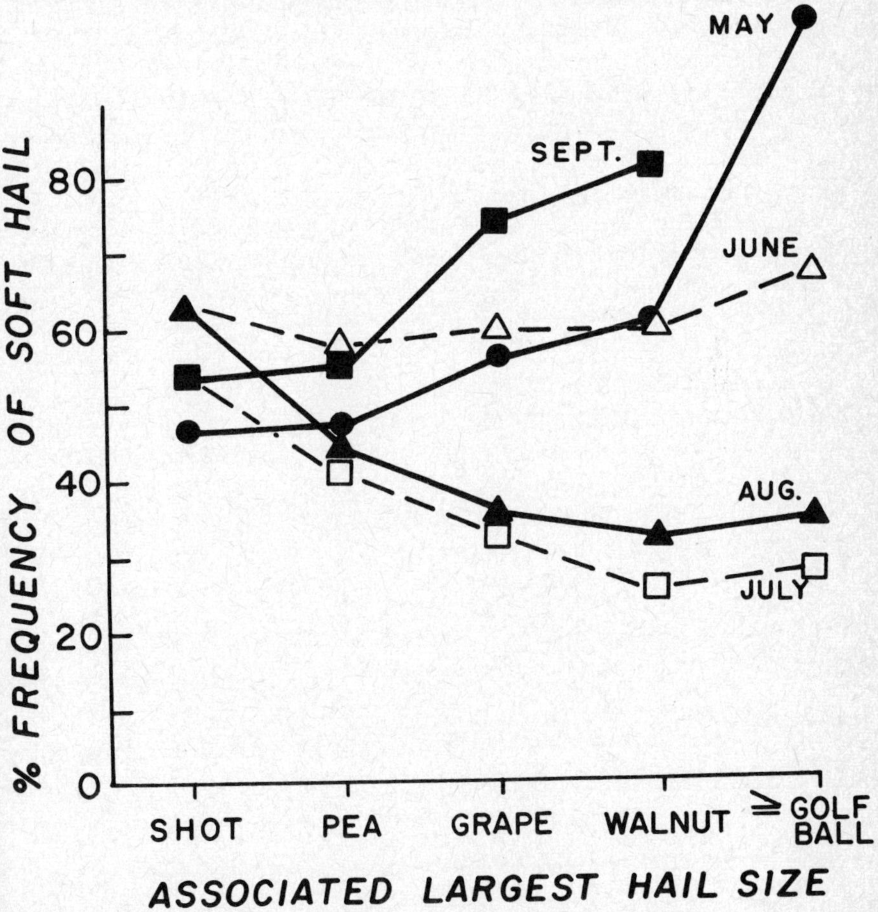

Figure 11. Relation between percentage of soft hail and associated maximum hail size by months.

stones. These data therefore serve to support the farmers' reports and suggest that Alberta hail is very often soft or spongy, and can have very high liquid water contents on arrival at the ground.

Discussion

Some of the more important hail parameters reported by Alberta farmers are analyzed in this paper. Similar data are available from only a few other places around the world, but such comparisons as can be made suggest that Alberta hailstorms behave much like those observed in northeastern Colorado. Colorado State University has recently set up a hail reporting network using a card format almost identical to the Alberta card, and this will allow direct comparisons of many parameters to be made.

The spatial variation of the parameters analysed is very irregular, but when they are related to the distance from the continental divide of the Rocky Mountains certain trends emerge. Hailstorms appear to be at their maximum intensity at a distance of 70-80 miles from the divide, on the basis of long durations and a high percentage frequency of large hail. The region 110-120 miles downwind is a favourable location for the generation of early morning and late evening hailstorms, as well as for the intensification of already existing storms. Another region of high percentage frequency of large hail exists 185-200 miles from the mountains. Thus there is some evidence of phenomena occurring at distances from the mountains which are multiples of 35-40 miles; these may be related to some downwind wave effect.

Other information provided by the farmers, such as the rain-hail sequence and amount of hail falling, is being analysed. Together with the climatological data presented here, this provides extremely valuable background data for the design and evaluation of hail suppression experiments. In addition this climatological information can be fed into the hailstorm models being developed for Alberta. Correlations between various parameters predicted by these models can also be checked out against the actual observations.

Finally, indices characterising the hailfall on any given day are being developed. These will be placed on upper-air data summary cards for the last ten years. Correlations can then be made between the hail activity indices and synoptic parameters such as height of freezing level, height of tropopause, jet streams, etc.

References

Beckwith, W. B., "Characteristics of Denver Hailstorms," *Bull. Amer. Met. Soc.*, Vol. 38 (1957), pp. 20-30.

Changnon, S. A., "Singularities in Severe Weather Events in Illinois," CHIAA Res. Rep. no. 13, *Crop-Hail Insurance Actuarial Association*, Chicago (22 October, 1962), 19 pp.

Douglas, R. H., "Intermittency in Western Canadian Hailfall," *Proc. Int. Conf. Cloud Physics* (24 May – 1 June 1965), Tokyo and Sapporo, Japan.

Williams, G. N. and Douglas, R. H. 1963, "Continuity of Hail Production in Alberta Storms," Sci. Rep. MW-36, Stormy Weather Group, McGill University, Montreal (August, 1963).

11

A Study of the Causes of Hail

R. W. Longley

C. W. Thompson

1. Introduction

Although hail has attracted the attention of mankind for centuries, primitive people had little success in finding a cause for the phenomenon. The early explanations of meteorologists recognized that it was associated with the more severe thunderstorms and that the stones were maintained in the atmosphere long enough to grow to hail size only because of strong updrafts. The past decade has seen much concentrated study of hail, prompted by the extreme damage that a severe hailstorm can cause. Much of this work has been directed to the microstructure of the storm and the method by which the stones grew. (See Carte, 1963; Douglas, 1963; Douglas and Hitschfeld, 1959; List, 1960; Ludlam, 1958; Williams and Douglas, 1963.)

Some scientists have instead directed their research to the structure of the weather situations which produce hail. In one of the first studies on the synoptic situations leading to hail, Beckwith (1956) showed that hail in the Denver area was usually associated with west-southwest winds at the 500-mb level. Ramaswamy (1956) related convective activity, in his study of storms in India, to southwest winds aloft. He chose his reference level at 300 mb. Schleusener (1962) also studied the relationship between upper winds and the occurrence of hail in the lee of the Rockies. Hail,

Reprinted from *J. of Applied Meteorology*, Vol. 4, no. 1 (1965), pp. 69-82.

according to his study, came when a wind maximum moved southward across the region, as would occur when a jet associated with an approaching trough moved into the area.

Dessens (1960) believed that the strong winds aloft with lighter winds near the surface provided an important factor in the production of a destructive hailstorm, but Ratner (1961) questioned the conclusion. Dessens did not discuss the wind direction in considering the effect of strong winds, although it is probable that most of the winds with which he dealt had a westerly component. Das (1962), in a theoretical study on hailstone growth, concluded that shear is not necessary to have a hailstorm, but that the chances are better when shear is present.

Another factor in the development of hailstorms is the instability present in the atmosphere. Since hail comes from cumulonimbus clouds with strong updrafts, the relationship should appear to be clear-cut. Fawbush and Miller (1953) used a measure of this instability as a clue to the size of the maximum hailstones. Beckwith (1956, 1960) also demonstrated that instability was necessary to the formation of hail. Moisture, providing a source for hailstones, has been studied in relation to hail. Huff (1961) used surface dew points as a significant measure of moisture. Appleman (1958) and Pappas (1962) emphasized the moisture within the cloud as the instability developed.

The presence of fronts, usually cold but at times warm, has been shown to be correlated with hail. Beckwith (1956) showed this for the Denver area. Huff (1961) found that hail in Illinois was correlated with frontal activity. Frisby (1962), in her study, found that almost all the hail in South Dakota was associated with fronts and frontal cyclones although few came with air mass instability.

Summarizing, we find that the different studies have shown that the occurrence of hail can be correlated with the following meteorological variables:

a) Instability at least to the 500-mb level;
b) Southwest winds aloft;
c) Strong winds at high altitudes;
d) Moisture in the lower atmosphere;
e) Frontal activity, usually, but not always with cold fronts.

Although an association has been shown, the studies have failed in most cases to show the complete picture. For example, Beckwith's study (1956) showed that, in general, the hail came with 500-mb winds from the southwest quadrant. But his data also showed that hail did at times come with winds aloft from other sectors and that on many occasions with southwest winds aloft there was no hail in the Denver area. Pappas (1962) did consider the more complete picture. Yet even after he analyzed 70 storms and, from his study, developed a "predictor" on which to base his forecast, he found that the "predictor" failed him in 12 out of 70 instances. He would have forecast hail on 7 days when no hail was reported, and would have failed to forecast the hail that fell on 5 other days.

2. Previous Studies of Alberta Hail

Hail has caused serious damage to Alberta crops, with the center of maximum damage about 50 miles northeast of Calgary. Powell[1] showed that many sections (i.e., one mile square) receive hail 9 or more times in a decade, and some farmers report having been hailed out seven years out of ten. Since the time of maximum frequency is between June 20 and July 31, after the crops are well started, there is little chance for complete recovery after hail flattens a crop.

In order to study the causes and characteristics of hail, the Research Council of Alberta, the Meteorological Branch, Department of Transport, and McGill University united to organize the Alberta Hail Studies Project. Information has been obtained by means of a radar set located at Penhold, by reports from farmers and others of hail observations, and by other means. [See Figure 1 for location of places mentioned in the report.]

Using data from these sources, Douglas, Hitschfeld, and others have reported on a number of aspects of the microstructure of the hail cloud. Many of these reports are contained in the publications of the Stormy Weather Research Group, McGill University. Thompson[2] began his series of reports on the synoptic situations leading to hail in Alberta before the initiation of the Alberta Hail Research Group. During the past three years Longley and Thompson[3] have combined in some studies. Many other reports on Alberta hail are found in the series of circulars printed by the Meteorological Branch, Department of Transport.

3. Classification of Days

In order to relate the synoptic weather map with hail, it is necessary to identify a "day with hail," meaning a day when hail was widespread and severe. In the southern half of Alberta, convective showers are frequent during the summer. If we consider only the North Saskatchewan River

---

1 Grace L. Powell, "The relationship of physiography to the hail distribution pattern in central and southern Alberta," *Unpub. M.Sc. thesis*, Univ. of Alberta, Edmonton (1961).

2 C. E. Thompson, "The problem of hail forecasting in Alberta," *Tech. Air Series Cir. 2804*, Tech. 236 (Toronto, Dept. of Transport, Meteor. Branch, 1956). Other technical circulars, by various authors, dealing with hail are nos. 276, 317, 388, 443 and 493.

3 R. W. Longley and C. E. Thompson, "A synoptic study of the occurrence of hail in central Alberta during the summer of 1959," *Tech. Cir. Series Cir. 3521*, Tech. 367 (Toronto, Dept. of Transport, Meteor. Branch, 1961). Similar studies for 1960 and 1961 are contained in Cir. 3567, Tech. 381 and Cir. 3774, Tech. 442 respectively.

Basin in Alberta, extending east-west through Edmonton, measurable rain fell at some observing station on 85 per cent of the days included in this study. If one includes the area of Alberta south to the United States boundary, but east of the Rocky Mountains, the percentage increases to 92. A number of these convective showers produce hail, many just briefly, during their life cycle with the result that light hail is relatively frequent. During June 1963, when reporting to the Alberta Hail Studies Project was excellent, hail was reported on 21 days.

To investigate the causes of severe hail storms, the days were classified into: days with major hail; days with minor hail; and days without hail.

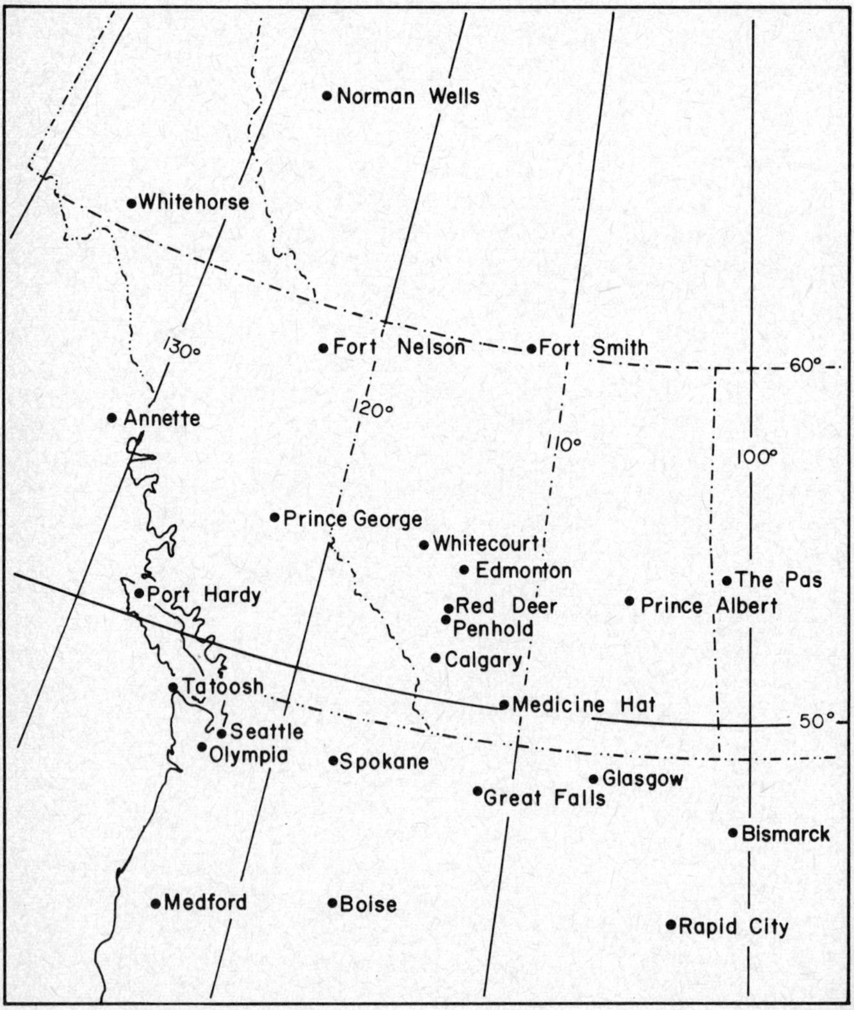

Figure 1. Location of stations.

The dividing lines were made by subjective criteria based upon the number of reports of hail received by the Alberta Hail Studies Project, reports of insurance claims, and reports from other sources. The number of days in each class is given in Table I.

TABLE I. DISTRIBUTION OF MAJOR, MINOR, AND NON-HAIL DAYS

	Major hail				*Minor hail*			
	June	*July*	*Aug.*		*June*	*July*	*Aug.*	
1959	10	5	9		5	6	7	
1960	5	10	2		8	3	11	
1961	4	4	0		4	5	4	
1962	8	8	3		8	9	14	
1963	5	9	5		4	6	3	
Totals	32	36	19	87	29	29	39	97

	No hail			
	June	*July*	*Aug.*	
1959	15	20	15	
1960	17	18	18	
1961	22	22	27	
1962	14	14	14	
1963	21	16	23	
Totals	89	90	97	276

F. H. Ludlam, in a personal communication, suggested that the classification could be done equally well and more objectively by using as a basis the size of the largest stones reported. Since maximum size is correlated with the heights of the cloud tops (see Donaldson, Chmela and Shackford, 1960), this method of classification would be to link classes with instability. The method would shift a small number of days from one class to another, but it is doubtful if the conclusions would be greatly different.

4. Hail and Warm Weather

In considering hail, one immediately meets its annual variation and geographical distribution. In Alberta, large stones fall only during the summer. Occasionally severe hail comes in late May or early September, but the major concentration is in the intervening summer months. A clue to this annual variation comes by considering the temperatures above Calgary, obtained for most days by interpolation. During the five years, major hail fell on only four days when the 700-mb temperature was as low as –2C. Yet there were 15 days with this temperature below –1C on which minor hail fell. The minimum value of the temperature for minor hail days was –6C.

In order to have more uniform data, the analysis was based upon those days when the Calgary 700-mb temperature was –1C or higher. With all five years included, the analysis is based upon 83 days with major hail, 83 days with minor hail, and 257 days without hail. When, as in Section 7, the analysis used data for 1959-1962 only, the number of days were, respectively, 65, 70 and 199.

Mayençon (1961), in a study of hail in Algeria, distinguished between small cold-season hail (November-March) and warm-season hail (April-October) when the storms were more spotty but more severe. He did not establish any clear dividing line, but certainly temperatures aloft were significant.

5. Moisture

Hail can form when there is sufficient moisture from which to obtain the required liquid. Appleman (1958) has related the occurrence of hail to the height of the lifting condensation level and its relation to the freezing level. If the moisture is sufficient so that the cloud forms between 1 and 2 km below the freezing level, hail may form within the cloud. Too thick a cloud below the freezing level inhibits the formation of hail.

Some consideration was given to atmospheric moisture and its relation to the incidence of hail in Alberta. But this was limited because of the lack of data. As described in Section 6, some afternoon upper-air observations were made in 1961 and 1962 over Calgary. From these, mean mixing ratios were computed for 850 mb for the three classifications. The mean dew points were: for days of major hail, 4.5C; for days of minor hail, 5.3C; and for days without hail, 3.8C. Although there was some difference, the data failed to distinguish clearly the type of weather on the basis of upper-air moisture.

Although the upper-air data were limited, surface moisture is measured regularly and reported by dew points. These were examined in several ways without finding any satisfactory relationship. For instance, the maximum dew points reported in the vicinity of Medicine Hat were examined for the years 1959 and 1960. For these years, the mean maximum dew points were: for days with major or minor hail, 55F; and for days without hail, 54F. Lack of moisture must at times inhibit hail, but as yet no satisfactory measure to clarify the relationship for Alberta has been examined.

6. Instability

It has long been recognized that unstable air is associated with hail. The analysis of the relationship in the Alberta situation has been difficult.

Upper-air ascents are taken regularly at Edmonton. Because the station is north of the major areas of hail, the ascent was frequently in a colder air mass and so could not be used to determine the stability of the air in which the hail is formed. Great Falls, Montana, is the closest radiosonde station to the south, and so the reports from there have been used to try to determine the relationship between severe hail and instability.

The instability at Great Falls for 1700 MST was measured by means of the Showalter stability index, and also by the temperature drop from 700 mb to 300 mb. Table II gives the distribution of the stability index for the five summers classified according to the weather in Alberta.

TABLE II. STABILITY INDEX AT GREAT FALLS AT 1700 MST, SUMMERS, 1959-1963, CLASSIFIED ACCORDING TO THE WEATHER IN SOUTHERN ALBERTA

	Major hail	Minor hail	No hail	*p* for major hail	*p* for major and minor hail
−5	1			1.00	1.00
−4	2	2		0.50	1.00
−3	3	1	1	0.60	0.80
−2	4	4	5	0.31	0.62
−1	7	7	9	0.30	0.61
0	15	2	22	0.38	0.44
1	22	16	37	0.29	0.51
2	12	13	34	0.20	0.42
3	9	12	45	0.14	0.32
4	3	7	52	0.05	0.16
5	2	4	23	0.07	0.21
6	3	8	11	0.14	0.50
7		3	10	0.00	0.23
8		1	4	0.00	0.20
9		1		0.00	1.00
10		2	1	0.00	0.67
11					
12					
13			1	0.00	0.00
Total	83	83	255		

The table shows that there was some correlation, with hail more probable on days of unstable air at Great Falls. This is also shown in Figure 2. Here are plotted the quotients: number of occurrences of hail/number of observations, for occurrences of major and minor hail and for major hail only, for each value of the stability index.

By statistical methods, one may determine the trend lines of Figure 2 to be:

for major hail, $p=0.307-0.0455x$,

for major and minor hail, $p=0.489-0.0394x$

where x is the stability index, and p the probability. In the computation, each point is given a weight equal to the number of observations with the given stability index.

The coefficient of correlation, $r=-0.88$, for major hail, has little relationship to the original data, but does indicate that the points on the diagram do not deviate far from the trend line. Thus it would appear that sufficient data have been obtained to establish a reliable relationship. The correlation for major and minor hail is only -0.50, a value confirmed by the points on the diagram. It would appear that the relationship is not so well determined.

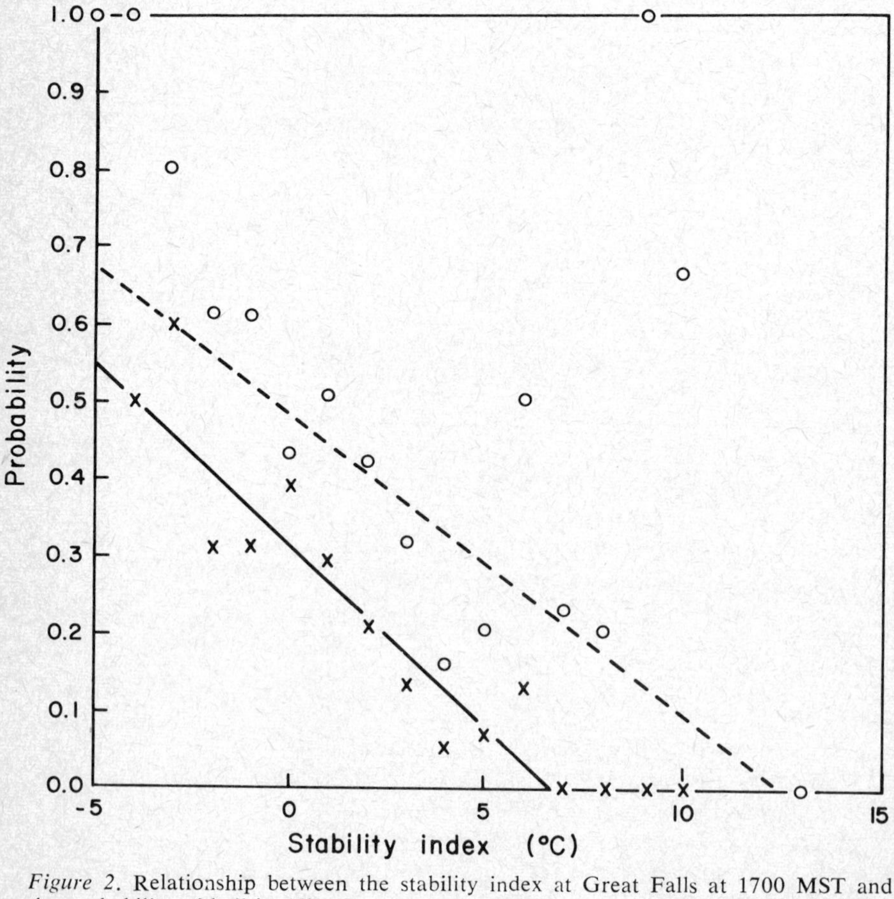

Figure 2. Relationship between the stability index at Great Falls at 1700 MST and the probability of hail in Alberta.

From the analysis, one may determine, for instance, the value of the stability index when the probability is 0.3. This for major hail is 0C, and for major plus minor hail 5C. It is also of interest to know by how much the probability changes over the normal range of the stability index. The mean index is 2.4C, and the standard deviation 2.5C. Considering the range 2.4$\pm$ 2.5C, which includes approximately two thirds of all observations, the value of p changes by 0.23 for major hail and 0.20 for minor hail. The probability does then change significantly in the normal range of the stability index. Nevertheless one cannot project with confidence the line to determine the indices corresponding to $p=1.00$ and $p=0.00$.

The statistics on probability are summarized in Table III. The stability index considers the instability only to the 500-mb level. Yet hail is not probable unless cloud tops reach 25,000 ft. (Douglas and Hitschfeld, 1959). A rough measure of the instability to higher levels is given by the drop in temperature from 700 to 300 mb. These were determined again for Great Falls for 1700 MST for the individual days for the five summers.

The data showed some relationship between the difference and the weather in Alberta. Median values were: for major hail days, 48C; for minor hail days, 47C; and for days without hail, 46C. The data were treated in the same manner as those for the stability index. The results are given in Table III. These confirm a relationship between the temperature drop and the probability of hail. But the lower values for the change of p as the temperature difference moves through its normal range of values indicates that it is not as good for identifying days with hail.

Upper air values from Great Falls were taken, even though it is 300 miles from the center of major activity, because it has given the best picture of the warm air in which the hail was formed. In July 1961, regular afternoon ascents began at Calgary. With the data from two months of 1961, and three of 1962, some comparisons may be made between the values at Great Falls and those near the center of activity.

During the two summers, there were 139 days which were warm enough for hail (Calgary 700-mb temperature over —2C) and on which observations were taken at both Calgary and Great Falls.

Computations showed that, as expected, the mean Calgary temperatures were colder than the mean Great Falls temperatures, the difference decreasing from near 5C at the surface to about 3C at the top of the troposphere. Higher relative humidities at the 850-mb and 700-mb levels more than compensated for the colder temperatures, so that the air at Calgary was more moist than that at Great Falls. At 850 mb, the mean mixing ratios were: for Calgary, 6.5 g kg^{-1}, and for Great Falls, 5.5 g kg^{-1}. Figure 3 gives mean temperature curves for the two stations, classified according to weather.

TABLE III. PROBABILITY (p) OF HAIL IN ALBERTA RELATED TO VARIOUS METEOROLOGICAL VARIABLES

*Parameter No.**	*Trend line*	*Value for* $p = 0.3$	*Coeff. of correlation*	*Change in p for change of* 2σ *in x*
	For major hail			
1	$0.307 - 0.0455x$	0°	− 0.88	0.23
2	$-1.151 + 0.0296x$	49°	0.78	0.15
3	$0.135 + 0.00541x$	30	0.85	0.15
4	$0.046 + 0.0201x$	13°	0.79	0.19
5	$0.198 - 0.0172x$	− 6°	− 0.54	0.14
6	$0.078 + 0.0134x$	17°	0.78	0.20
7	$0.003 + 0.0185x$	16°	0.76	0.24
8	$-0.251 - 0.0264x$	− 21°	− 0.77	0.20
9	$0.174 + 0.00278x$	45	0.82	0.23
10	$0.193 + 0.00225x$	47	0.79	0.17
	For major plus minor hail			
1	$0.489 - 0.0394x$	5°	− 0.50	0.20
2	$-0.676 + 0.0236x$	41°	0.55	0.14
3	$0.309 + 0.00748x$	0	0.89	0.21
4	$0.180 + 0.0285x$	4°	0.89	0.27
5	$0.396 - 0.0250x$	4°	− 0.74	0.20
6	$0.232 + 0.0183x$	4°	0.75	0.27
7	$0.148 + 0.0234x$	7°	0.81	0.31
8	$-0.180 - 0.0338x$	− 14°	− 0.80	0.26
9	$0.371 + 0.00275x$	− 26	0.80	0.22
10	$0.389 + 0.00315x$	− 28	0.90	0.24

*Parameter No.:

1. Great Falls stability index, 1700 MST (°C)
2. Great Falls temperature difference, 700 to 300 mb, 1700 MST (°C)
3. Flow index at 850 mb, 1700 MST (10's of ft)
3. Temperature difference, Great Falls-Edmonton, 850 mb, 1700 MST (°C)
5. Edmonton 850-mb, 24-hr temperature change, 1700 MST (°C)
6. Maximum temperature at Medicine Hat less that at Edmonton (°F)
7. Temperature difference, Great Falls-Seattle, 850 mb, 1700 MST (°C)
8. Temperatures at 500 mb over Prince George at 0500 MST (°C)
9. Vorticity over British Columbia at 0500 MST (10's of ft)
10. Vorticity over Alberta, 1700 MST (10's of ft)

Analysis showed that stability at Calgary at well as at Great Falls was correlated with Alberta hail. By means of the stability index, the Calgary ascent was more useful in identifying a day with any hail than the Great Falls ascent, but the identification of a day with major hail was slightly better using the Great Falls report. Lack of data prohibited a complete analysis such as was done with the Great Falls stability index.

One then concludes that instability is a factor in causing hail, but one cannot identify the stability which will produce hail. Major hail fell on 4 August 1962, when the stability index was +2 at Calgary and +4 at

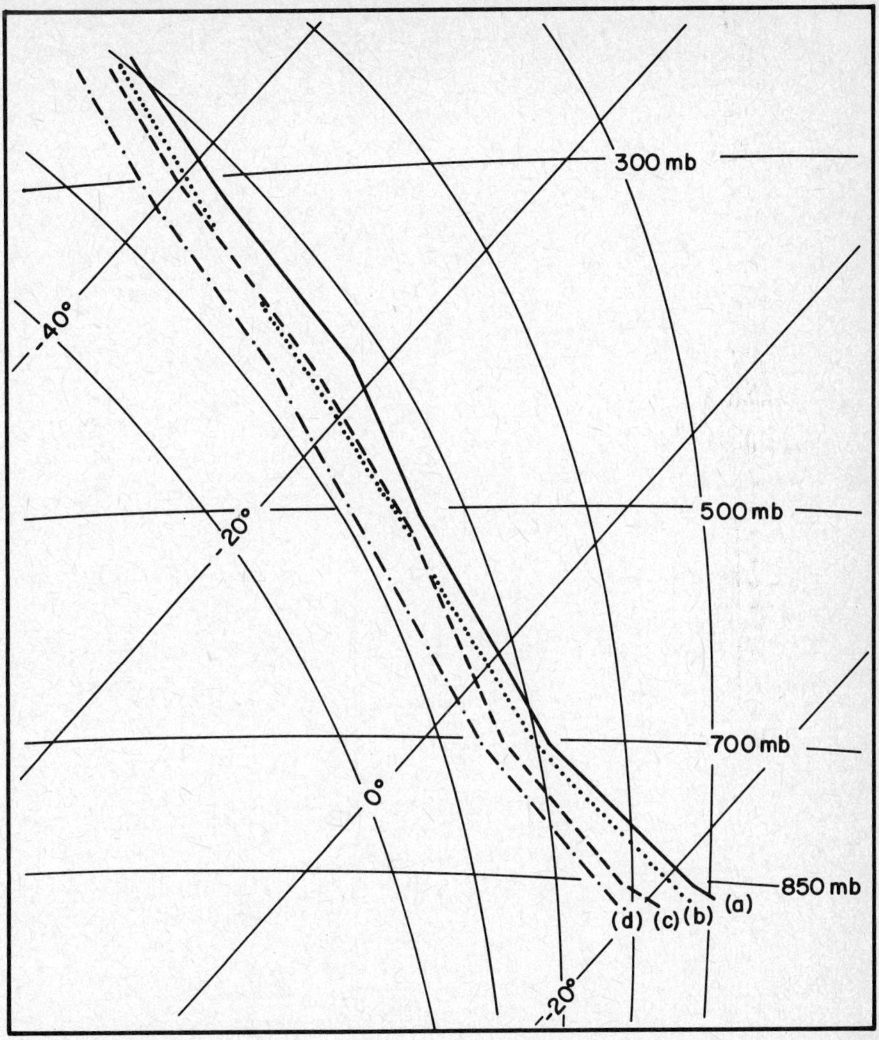

Figure 3. Mean temperature curves. (a) Great Falls for days without hail. (b) Great Falls for days with major hail. (c) Calgary for days without hail. (d) Calgary for days with major hail.

Great Falls. Also on eleven out of the 82 days without hail, both ascents showed zero or negative values of the index. Other factors than stability must be considered in differentiating days with hail from those without hail.

7. Mean Map and Alberta Hail

Since some factor other than stability is important in causing hail, an examination of the weather maps should be of value. The work of Frisby

(1962), Dessens (1960), Reed (1960), and others showed that certain features of the weather distribution were frequently present when hail fell.

Forecasters in Edmonton associated certain features of the synoptic chart with the incidence of hail. To clarify these relationships, composite maps have been obtained for 850 mb and 500 mb for 0500 MST (1200 Z) and 1700 MST (0000 Z of the following day). Mean heights, temperatures, and, at 850 mb, dew points, were obtained for each of 18 radiosonde stations for each of the three classes; major hail days, minor hail

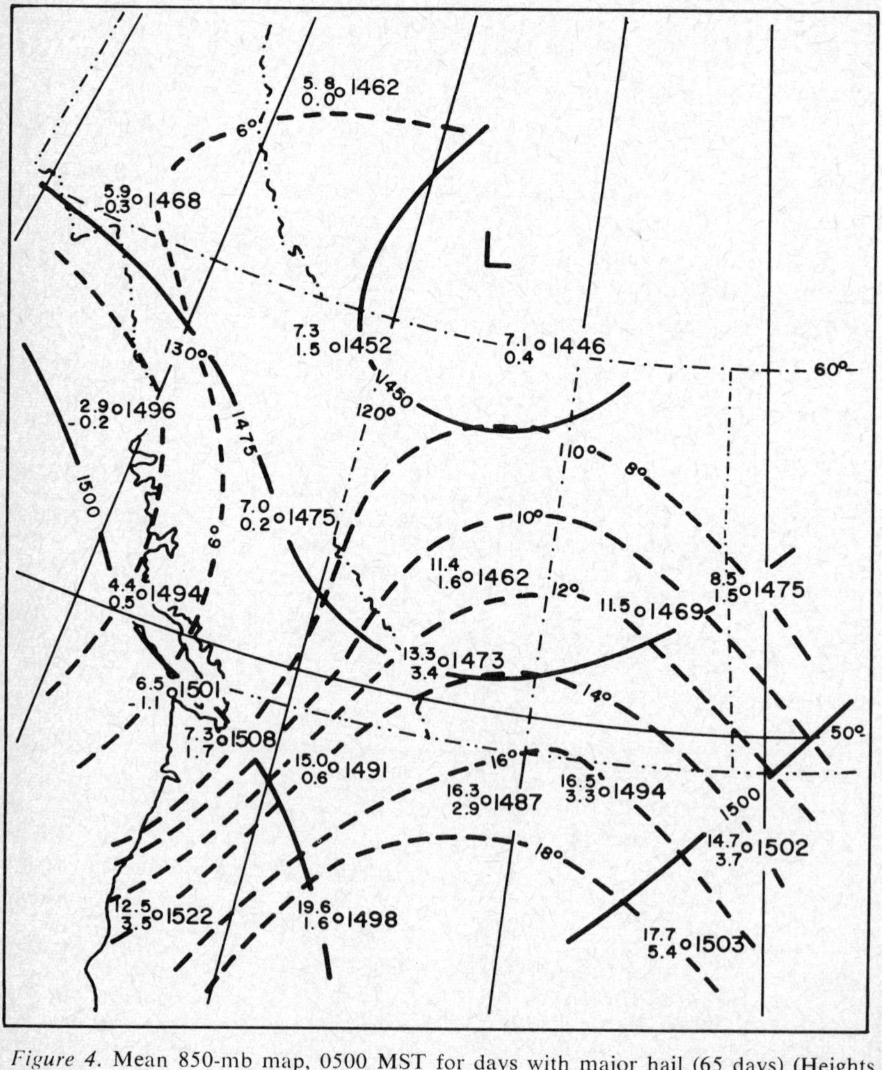

Figure 4. Mean 850-mb map, 0500 MST for days with major hail (65 days) (Heights in metres).

days, and days without hail. Computations were also made for Calgary, and Prince Albert, Saskatchewan, the individual values being obtained by interpolation. In this analysis data from 1959-1962 only were used. Figures 4-8 give five of these composite maps for 850 mb.

The 850-mb maps show significant differences. The most significant are the differences in the trough over Alberta and Saskatchewan, and the tongue of warm air extending northward from Montana.

Figure 8 for 1700 MST for days with no hail, shows a broad weak

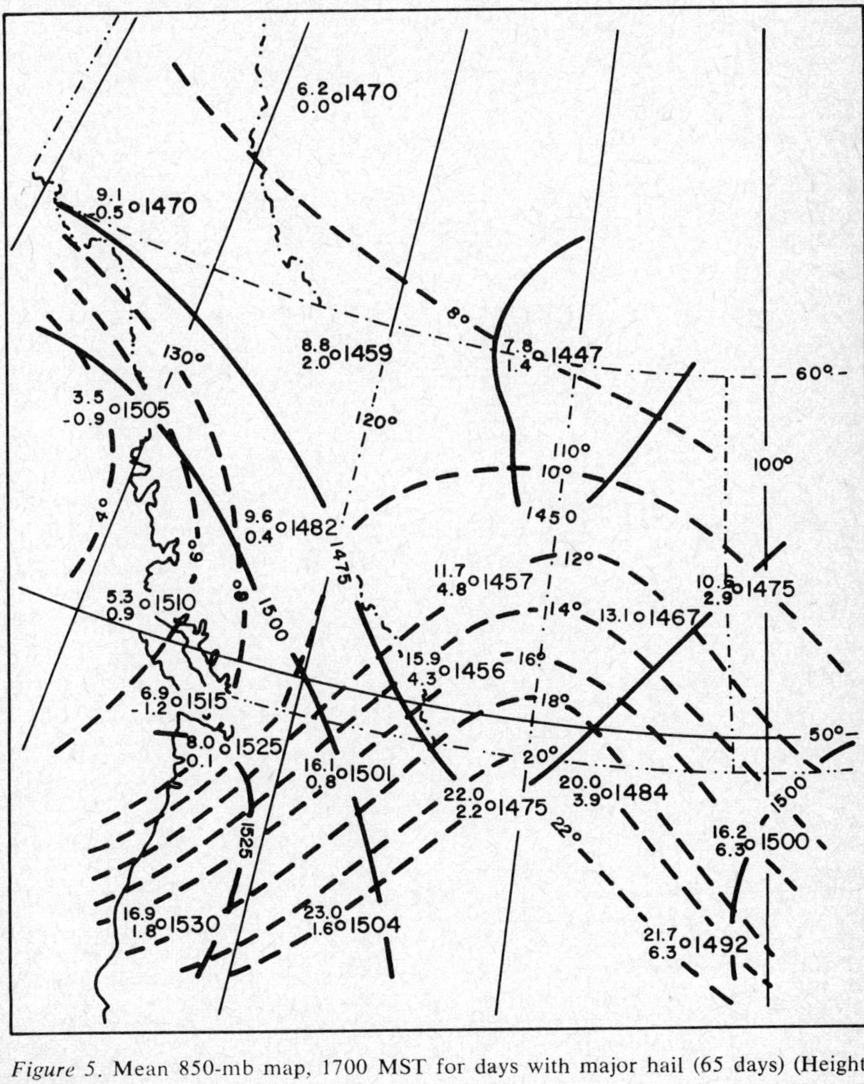

Figure 5. Mean 850-mb map, 1700 MST for days with major hail (65 days) (Heights in metres).

trough over the two prairie provinces, a trough which changed little during the day. In contrast, Figures 4 and 5 for days with major hail show a deep trough over Alberta. Comparison between the two figures shows that this trough deepened during the day, with the heights at Edmonton and Calgary dropping 5 and 17 m respectively in contrast to a drop of 2 m only at Prince Albert and a rise of 7 m at Prince George. Associated with this deepening trough was a flow of colder air from the northwest into

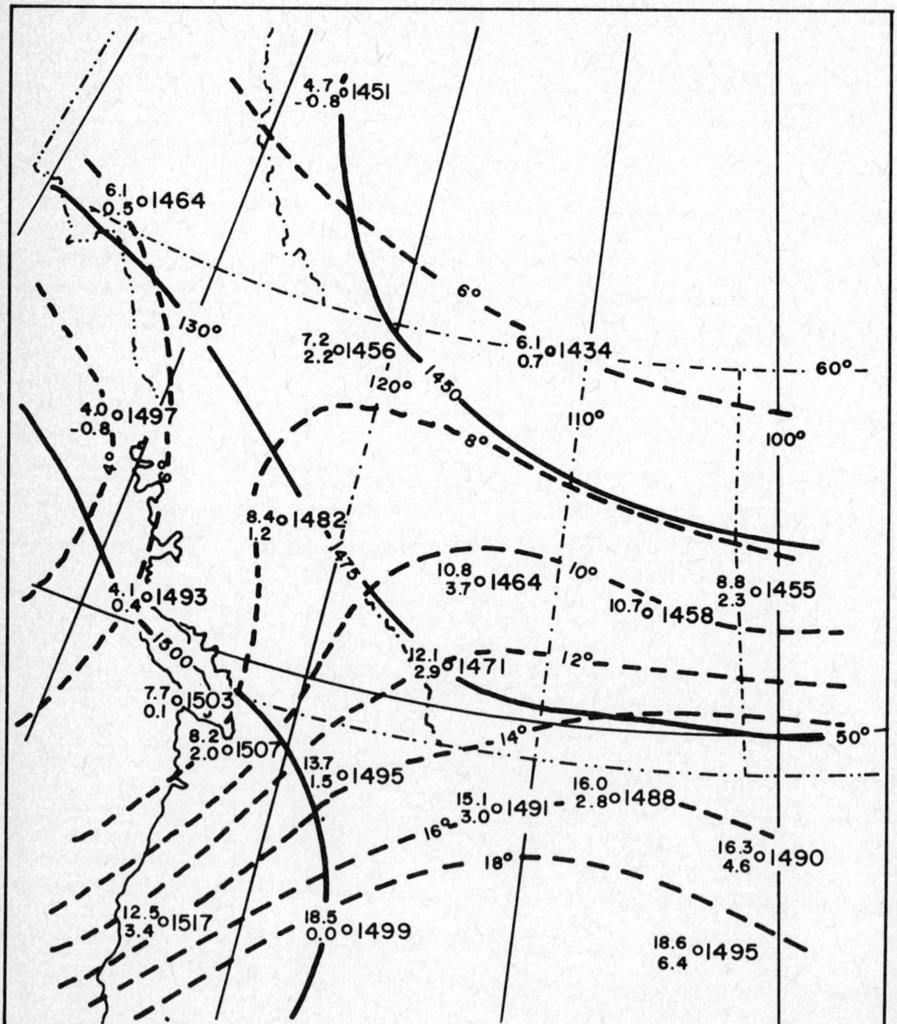

Figure 6. Mean 850-mb map, 0500 MST for days with minor hail (70 days) (Heights in metres).

Western Alberta. Because of this flow, the Edmonton 850-mb temperature rose only 0.3C during the 12 hours between maps, in contrast to a rise at Calgary of 2.6C and at Great Falls of 5.3C.

The second contrast is related to the temperature distributions. On days of major hail, the warmest air lay along the Alberta-Saskatchewan boundary, northward from Great Falls, with Prince George 3.5C colder than Prince Albert at 1700 MST. On days of minor hail (Figure 7), the

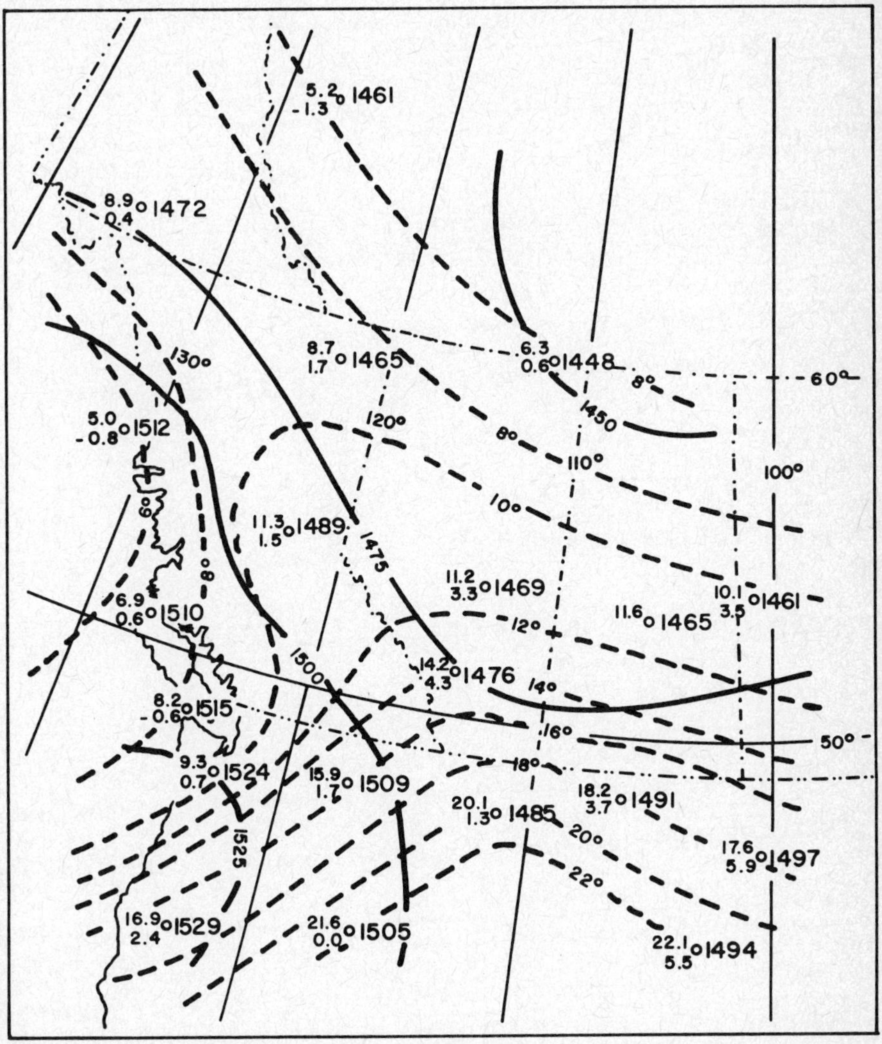

Figure 7. Mean 850-mb map, 1700 MST for days with minor hail (79 days) (Heights in metres).

change of temperature at 1700 MST from Prince George to Edmonton to Prince Albert was slight. On days without hail, the tongue of warm air extended from Edmonton northwestward to Fort Nelson, B.C.

The deepening trough over Alberta, the strengthening wind to the west, and the tongue of warm air to the east, all indicated on the mean map for 1700 MST for major hail days, suggest the presence of a low-pressure area over southern Alberta, with a cold front to the west. As Beckwith (1956, 1960), Frisby (1962), and Huff (1961) have shown,

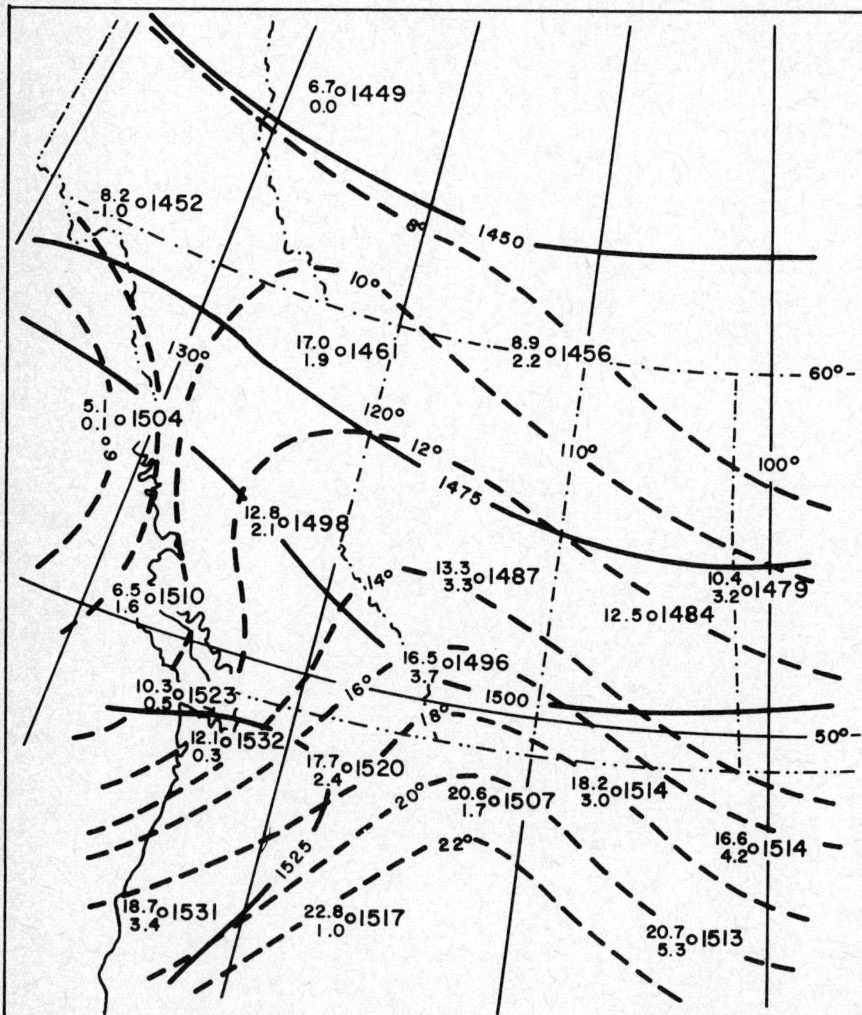

Figure 8. Mean 850-mb map, 1700 MST for days without hail (199 days) (Heights in metres).

frontal activity is associated with the development of hail. They left unanswered the question about the possibility of a cold front moving through the area and yet no hail falling.

In order to assess the presence and strength of the cold front and associated phenomena, four measures were used. The first was the strength of the northwesterly flow to the west of an Edmonton-Great Falls line. This was computed by taking the heights of the 850-mb surface at Prince George and Spokane and subtracting those at Edmonton and Great Falls. It is recognized that this is a theoretical value, for mountain masses in the area penetrate the 850-mb surface. For the years of study the values have been computed in tens of feet. The second measure was the temperature difference between Great Falls and Edmonton at 850 mb. The third measure was the drop in temperature at 850 mb over Edmonton in the preceding 24 hours. A fourth measure which is related to the presence and strength of a cold front over Alberta during the afternoon was the difference in maximum temperature between Medicine Hat and Edmonton.

One advantage of all these measures is that they can be obtained objectively. There is a disadvantage since the measures do not allow for a variation in the location of a cold front in Alberta. For instance, if the cold front associated with the hail lay near the U.S. boundary, with the hail falling in the extreme southern part of Alberta, Medicine Hat might lie in the cold air and the strength of the cold front would not be measured by the difference in temperature between this station and Edmonton.

When the values of each of the variables were related to the incidence of hail, it was found that for each of the measures, major hail is more probable with higher values. The data were treated as described for the stability index. The statistics are given in Table III. The trend lines support the hypothesis that hail is related to these measures of a cold front moving into the area. But the original distributions (not shown) and the small changes in probability over the normal range of values of flow index and of the temperature fall at Edmonton show that there is no clear cut division between cold fronts associated with hail and those which are not.

Although the Montana area was much warmer on days with major hail than on non-hail days, the changes along the Pacific coast were in the opposite direction. Seattle-Olympia was 4C warmer on days with no hail than on days with major hail. As a result, the temperature gradient from Great Falls to the coast was great on those days when hail fell in Alberta. This temperature gradient gave a thermal wind which tended to produce southwest winds aloft.

This change in temperature gradient among the three mean maps suggested that there was a correlation between the east-west temperature gradient and the incidence of hail in Alberta. The measure taken was the difference in the 850-mb temperatures at Great Falls and Seattle (changing to Olympia when the site of the radiosonde observations was changed). The data on this temperature difference were also analyzed, and the results included in Table III. Changes in this temperature gradient apparently change the probability of hail significantly.

8. Upper Level Flow

Differences in the flow at upper levels associated with hail in Alberta

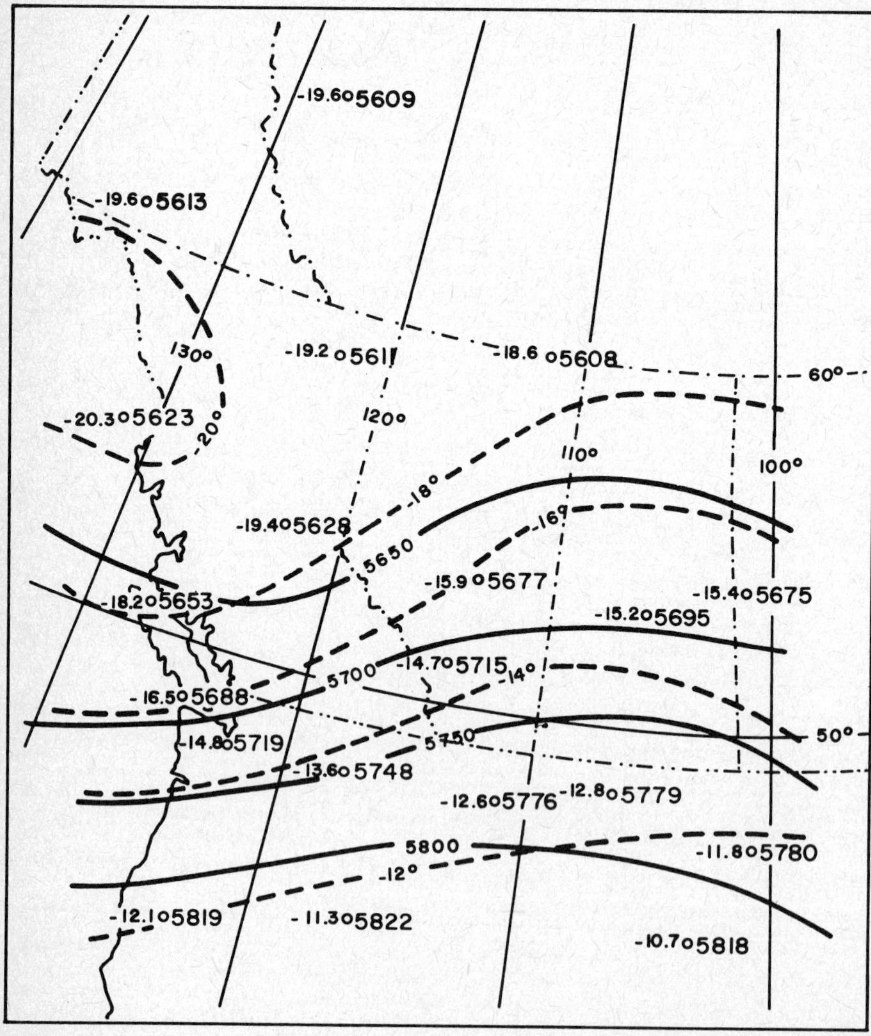

Figure 9. Mean 500-mb map, 0500 MST for days with major hail (65 days) (Heights in metres).

are seen by comparing the maps (Figures 9-13) showing the mean flow at 500 mb for major hail days, minor hail days, and days without hail. Differences are found in the temperature patterns and the wind flow.

In the temperature field, the variations among the maps at the northern station of Norman Wells and the southern stations of Rapid City and Boise are small. The major variations are found in the intermediate areas. With much colder temperatures in Alberta and British Columbia on the map for major hail, the temperature gradient is much greater than on the map

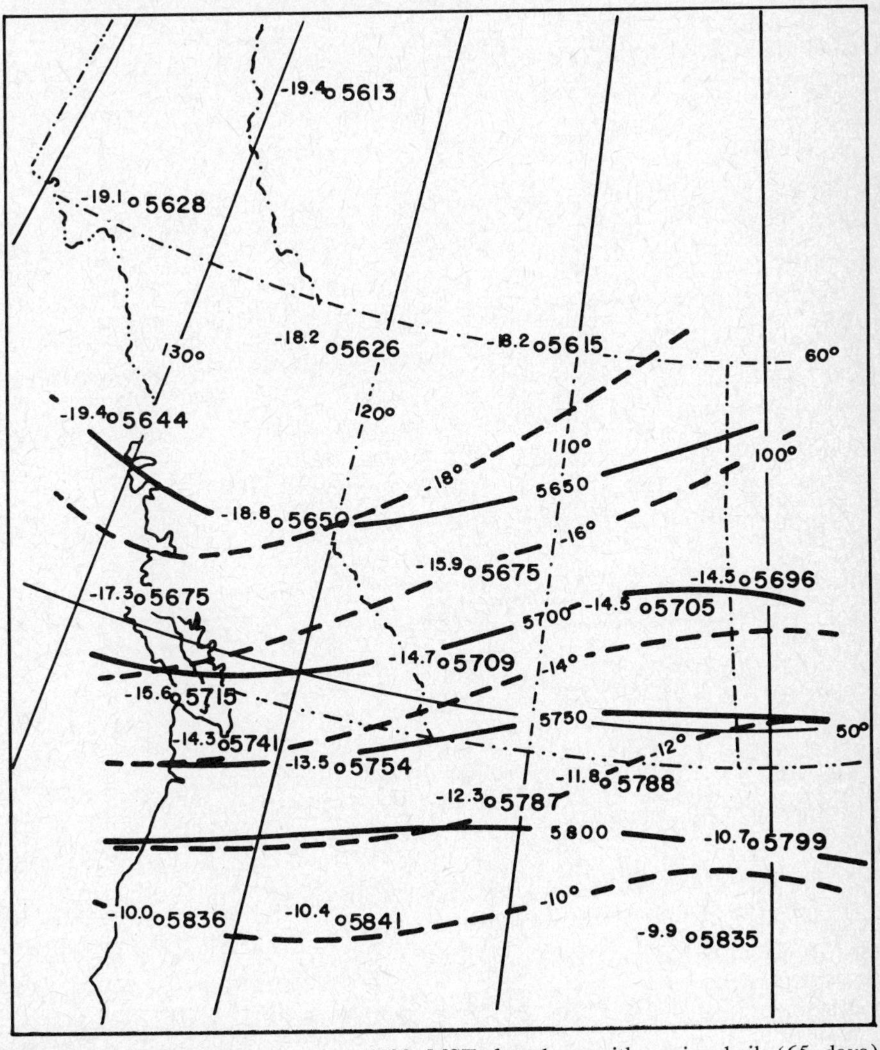

Figure 10. Mean 500-mb map, 1700 MST for days with major hail (65 days) (Heights in metres).

for no hail. The distance from the —12.5C isotherm to the —17.5C isotherm at 0500 MST along the Rocky Mountains is: on the map for major hail (Figure 9), 800 km; for minor hail (Figure 11), 1000 km; for no hail (Figure 13), 1600 km.

This temperature contrast is also reflected in the value of the temperature at Prince George at 500 mb. A study of the temperatures for individual days showed again what is seen on the mean maps, that there is a correlation between hail in Alberta and Prince George 500-mb tempera-

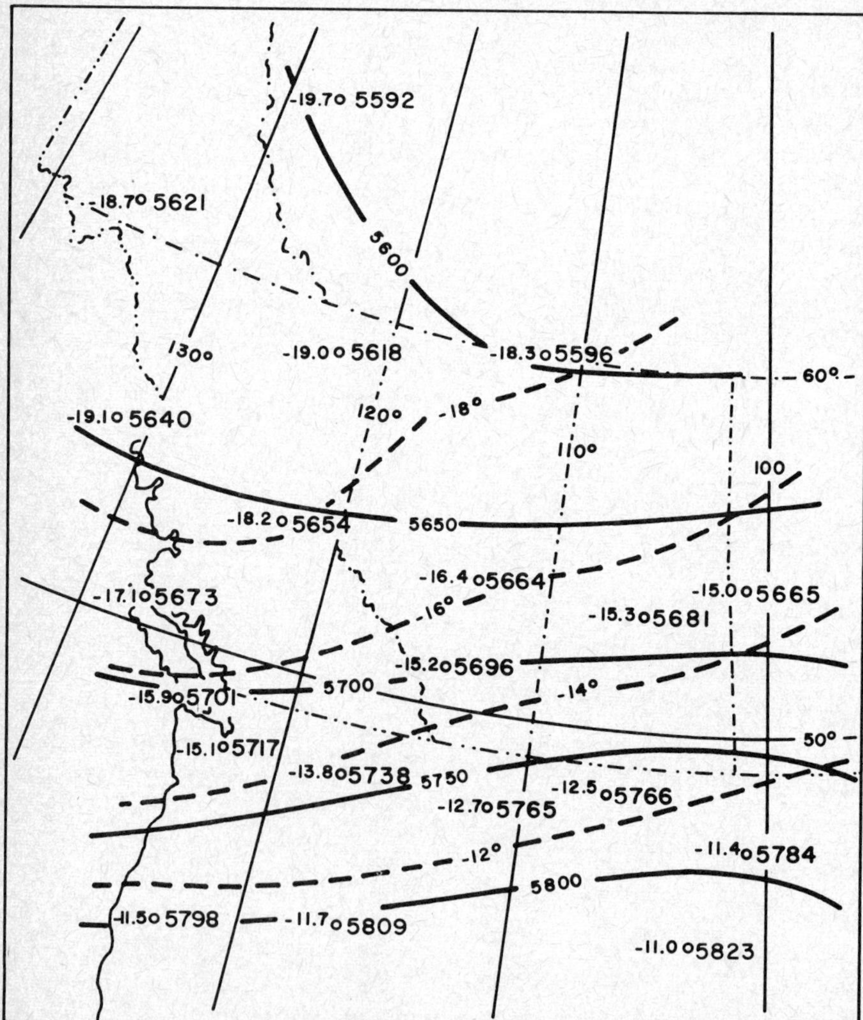

Figure 11. Mean 500-mb map, 0500 MST for days with minor hail (70 days) (Heights in metres).

tures, particularly those at 0500 MST. Results of computations, given in Table III,—show that the probability of hail is clearly a function of these temperatures.

9. Upper Winds

The temperature fields between 850 mb and 500 mb gave rise to different wind fields at the 500-mb level. For major hail days, the thermal wind produced on the average a strong southwest wind at 500 mb, a wind

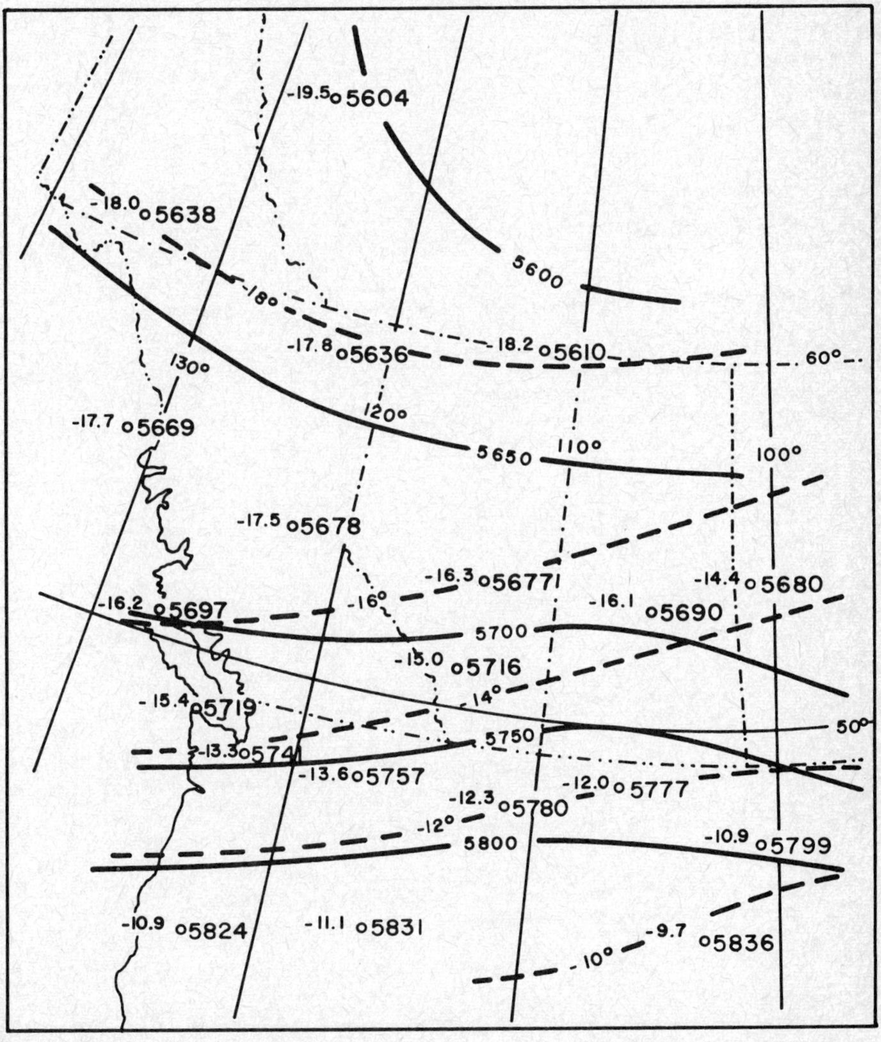

Figure 12. Mean 500-mb map, 1700 MST for days with minor hail (70 days) (Heights in metres).

that strengthened aloft. In a number of instances, this developed into a jet wind at the 300-mb level. When there was no hail in Alberta, the upper winds tended to be light. This conclusion confirms the findings of other scientists (Thompson, 1956; Ramaswamy, 1956; Beckwith, 1956; Dessens, 1960; and Müller, 1961).

When the individual days were examined, it was again found that in general, major hail and high winds went together. There have been few upper wind observations taken in southern Alberta, and so the wind had

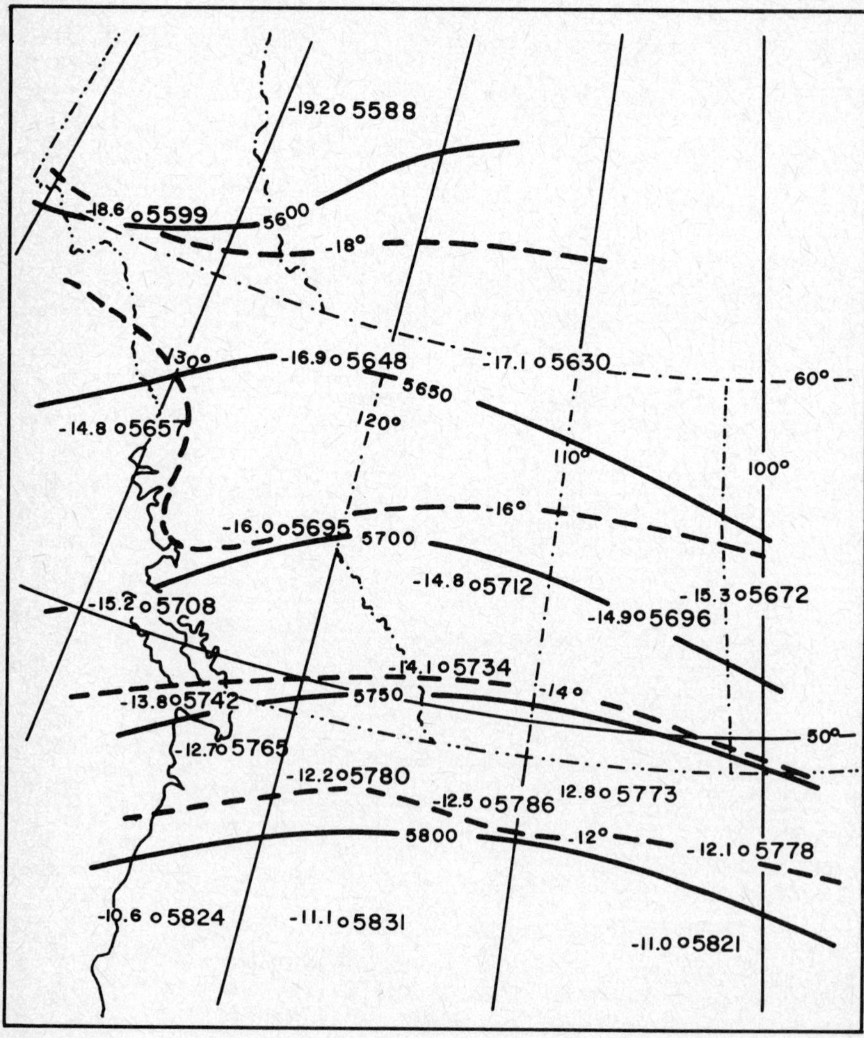

Figure 13. Mean 500-mb map, 0500 MST for days without hail (199 days) (Heights in metres).

to be estimated from the completed 500-mb chart. The estimate maximum velocity was read off for each day. The results are summarized in Table IV. Out of 83 days with major hail, 21 had maximum 500-mb winds under 30 kts (15 mps). The proportion for similar days with light winds and minor hail was slightly higher. In general, the high winds for major hail were from the southwest quadrant, but with minor hail the predominance of southwest winds was not so great. On days with no hail, the wind was under 30 kts 40 per cent of the time.

TABLE IV. RELATIONSHIP BETWEEN WINDS AT 500 MB AND WEATHER IN SOUTHERN ALBERTA

	Days with			
	Major hail	*Minor hail*	*No hail*	*Total*
Wind quadrant S-W, $\geqslant$ 30 kts	52	40	83	175
Wind $\geqslant$ 30 kts other directions	10	17	57	84
Wind $<$ 30 kts	21	26	116	163
Totals	83	83	256	422

If in the table, the columns "minor hail" and "no hail" are combined, and also the bottom two rows, one may test the hypothesis that strong southwest winds are related to major hail. The test for significance, X^2 is determined to be 19.1, thus showing that there is a significant relationship. Yule and Kendall (1950) suggested a coefficient of correlation, r, for a 2×2 contingency table where

$$r^2 = X^2/N.$$

This gives for r the value of 0.21. When the major and minor hail days are combined, computations show that $X^2 = 21.9$ and $r = 0.23$. Just as with the other parameters tested, the hail is associated with strong southwest winds aloft, but the relationship is far from perfect.

10. Vorticity

Another feature of Figures 9-13 that shows a difference is the flow over British Columbia. The mean maps for major hail days show a trough over this latter province which, as seen by comparing Figure 9 for 0500 MST and Figure 10 for 1700 MST, moved slowly eastward. The trough was also present on the mean maps for minor hail (Figures 11 and 12), but considerably less prominent. In contrast, the flow on days without hail (Figure 13) showed a broad ridge over the two western provinces with the suggestion of a trough over the eastern Pacific. Apparently cyclonic vorticity associated with a cold trough over British Columbia is conducive to the formation of hail over southern Alberta.

On individual days, as the British Columbia trough moved eastward, the jet stream associated with this trough would move southward along the Rocky Mountains. It was this phenomenon that was noted by Schleusener (1962) in his study of hail in the lee of the mountains in Colorado and Alberta.

Two parameters related to hail are associated with this trough over British Columbia. The strength of the southwest flow, already discussed, is directly related to the trough over British Columbia through the thermal wind relationship. Another, and more direct measure, was an approximation to the vorticity over British Columbia. To make the measure objective, the heights (in 10's of feet) of the 500-mb surface at Annette, Fort Nelson, Edmonton and Seattle-Olympia (see Figure 1) were added and four times the height at Prince George was substracted from the sum.

The data on vorticity for 0500 MST were related to the weather in Alberta that afternoon. The results are given in Table III. On the basis of the change over the normal range of values, the measure gives one of the best clues to the occurrence of major hail. Since it is a measure that can be computed by 0800 MST, it can therefore be used as a direct guide to forecasting. In this it differs from most of the other measures.

With the movement eastward of the trough in British Columbia, vorticity increased in Alberta on days with hail. A measure of this vorticity was obtained in a similar manner to that for British Columbia using Edmonton as center and the surrounding points of Fort Smith, Prince George, Great Falls, and Prince Albert (found by interpolation on the analyzed 500-mb chart) (see Figure 1) as the other points. Hail generally came when this measure indicated cyclonic vorticity. Results of computations are given in Table III.

11. Summary of Results

Table III provides the data from the foregoing computations. The values of *r* given in the table have very little significance in the overall picture, but give a measure of the fit of the straight line to the set of points denoting probability. If the fundamental relationship were linear, then the points would lie on a straight line after an infinite number of seasons. The degree to which the points approach the line, and hence the value of *r,* indicates the reliability of the computed trend line.

The values given in the last column give comparative measures of the different parameters. If a parameter were so related to hail that the latter occurred when and only when the parameter exceeded (or was less than) a specified value, the value in the last column would be 1.00. The larger the value found in the last column, the more is the change in the probability as the measure varies in its normal range. According to these data, major hail is most closely related to the stability index at Great

Falls, the temperature gradient between Great Falls and Seattle at 850 mb, and the vorticity over British Columbia. When all hail is considered, the best measure is again the Great Falls-Seattle temperature gradient, followed by two measures related to a cold front over Alberta, i.e., the magnitude of the temperature difference from Great Falls to Edmonton, and the difference in maximum temperature between Medicine Hat and Edmonton.

The statistics show that there is a relationship between Alberta hail and each of the phenomena that the parameters have measured: unstable warm air, a cold vortex approaching from the west, high southwest winds aloft, and a fresh outbreak of cold air from the north. On the other hand, the lack of a perfect relationship shows that no one of these phenomena alone is sufficient to produce hail.

Consideration has been given to trying to combine the results of two or more of the measures. A study of the data showed that hail was more probable when several of the parameters had significantly large (or small) values. Yet the intercorrelation among the different variables makes a simple combination difficult.

The attempt to combine data to obtain a conclusive index that would identify hail was also discouraged by an examination of some of the records. For instance, 26 July 1962[4] saw one of the worst storms on record. A marginal northwest flow, index value 10, brought a weak cold front past Edmonton. A severe storm swept southeast, starting at Whitecourt, 100 mi northwest of Edmonton, caused extensive damage near Red Deer, and gave scattered bursts of hail even southeast of Calgary. Yet, except for a probability of 0.31 based on the value of the stability index, none of the trend lines listed in Table III gave probabilities of major hail greater than 0.22. There is, then, no clue apparent to explain a storm which ruined thousands of acres of farm crops. On the other hand, the values of the variables on 27 June 1961, a day with no reported hail, gave probabilities ranging from 0.22 to 0.39, and 13 August 1962, gave values from 0.22 to 0.45.

There is one fact that may be significant in seeking an answer to the lack of hail on days when, according to the synoptic map, it should occur. Many of these days have been in the last two weeks of August. Yet there have been only two days of major hail between 17 August and 1 September. These two days were both in 1959, a year when the August precipitation was above normal. Is the hail missing at this time of year because the lack of moisture from evaporation suppresses the vertical currents?

In Section 5, mention was made of the atmospheric moisture. The data from Calgary radiosonde reports indicated that there was some dif-

[4] E. Stashko and C. E. Thompson, "Preliminary synoptic survey of the 1962 hail season in Alberta," *Tech. Cir. Series, Cir. 3774,* Tech. 443 (Toronto, Dept. of Transport, Meteor. Branch, (1962).

ference in moisture associated with hail, but the data were few. The analysis of dew points showed no significant difference. Perhaps this is not surprising. With many ponds and small lakes in contrast to large areas of grasslands, and with varying vegetation covers some of which is associated with artificial waterings near human habitations, it is probable that dew points are much more variable than is usually supposed. One can, with reason, question whether the dew points reported are representative of the total air mass.

Müller (1961) studied the incidence of hail in Switzerland. He too isolated parameters which were considered as conditions to be fulfilled for damaging hail. One parameter was the wind shear as given by the ratio of the 300-mb wind to the 850-mb wind. The second was the difference between the potential temperatures at 500 and 850 mb. This second he took as a modified Showalter index. He adjusted values of these two measures, requiring greater values of the wind shear with more stable air. He was able to show that the combination gave a statistically significant guide to the incidence of hail. Nevertheless, when one uses his data from 810 days as given in his Table I, to compute a coefficient of correlation, as described in Section 9, he obtains a value of 0.36. This value is better than similar values obtained by the current study for Alberta based upon other parameters. Yet it too is far from giving a clear-cut criterion for the incidence of hail. Müller also discusses the upper air flow, noting the prevalence of a trough through western Europe at 500-mb on days with hail, a situation which produced southwest winds over Switzerland. He does not attempt to analyze this parameter statistically.

12. Conclusion

Under what conditions does hail fall? Other scientists have shown that the phenomenon occurs with warm temperatures, high moisture content of the air, instability to high levels, active fronts moving through the district, strong upper level winds, and vorticity advection. The foregoing analysis of the incidence of hail in Alberta for the years 1959-1963 has reached no conclusion, because of lack of data, on the relationship between atmospheric moisture and hail. With the presence of one or more of the other conditions, the probability of hail has been shown to increase. Yet it has also been shown that no one of these conditions is apparently necessary for hail to occur. Nor is it a sufficient condition for the initiation of hail that many of these be present. In terms of forecasting, one can use the results of this study of five summers in Alberta to estimate in the early morning the probability of hail during that afternoon and evening. Nevertheless one is forced to conclude that the margin between hail and no hail is apparently very small, and that the study has been unable to isolate it.

References

Appleman, H., "An investigation into the formation of hail," *Nubila,* Vol. 2 (1958), pp. 28-37.

Beckwith, W. Boynton, "Hail observations in the Denver area," *Denver, United Air Lines Meteorological Circular* No. 40 (1956).

Beckwith, W. Boynton, "Analysis of hailstorms in the Denver Network 1949-1958," *Physics of Precipitation*, Washington, D.C., American Geophysical Union (1960), pp. 348-53.

Carte, A. E., "Some characteristics of Alberta hail storms," McGill University Stormy Weather Research Group, Scientific Report MW-36 (1963), pp. 1-15.

Das, P., "Influence of wind shear on the growth of hail," *J. Atmos. Sci.*, Vol. 19 (1962), pp. 407-14.

Dessens, H., "Severe hailstorms are associated with very strong winds between 6,000 and 12,000 meters," *Physics of Precipitation*, Washington, D.C., American Geophysical Union (1960), pp. 333-36.

Donaldson, R. J. Jr., A. C. Chmela and C. R. Shackford, "Some behavior patterns of New England hailstorms," *Physics of Precipitation*, Washington, D.C., American Geophysical Union (1960), pp. 354-68.

Douglas, R. H., "Alberta hail, 1958," Montreal, McGill University Stormy Weather Research Group, Scientific Report MW-30 (1959).

Douglas, R. H., "Growth by accretion in the ice phase," *Physics of Precipitation,* Washington, D.C., American Geophysical Union (1960), pp. 264-69.

Douglas, R. H., "Size distribution of Alberta hail studies," McGill University Stormy Weather Research Group, Scientific Report MW-36 (1963), pp. 55-70.

Douglas, R. H. and W. Hitschfeld, "Patterns of hail storms in Alberta," *Quart. J. R. Meteor. Soc.*, Vol. 85 (1959), pp. 105-19.

Fawbush, E. J., and R. C. Miller, "A method for forecasting hailstone size at the earth's surface," *Bull. Amer. Meteor. Soc.*, Vol. 34 (1953), pp. 235-44.

Frisby, E. M., "Relationship of ground hail damage patterns to features of the synoptic map in the upper Great Plains of the United States," *J. Appl. Meteor.*, Vol. 1 (1962), pp. 348-52.

Huff, F. A., "Correlation between summer hail patterns in Illinois and associated climatological events," Chicago, Crop-Hail Insurance Actuarial Association Research Project Report No. 10 (1961).

List, Roland, "Growth and structure of graupel and hail-stones," *Physics of Precipitation*, Washington, D.C., American Geophysical Union (1960), pp. 317-23.

Ludlam, F. H., "Hailstorm studies," *Nubila*, Vol. 1 (1958), pp. 7-27.

Ludlam, F. H., "The hailstorm," *Weather*, Vol. 16 (1961), pp. 152-60.

Mayençon, R., "Condition météorologiques, synoptiques propices aux chutes de grêle importants en Algérie, *La Meteorologie,* Vol. 61 (1961), pp. 41-54.

Müller, W., "Prévision de l'étendue des chutes de grêle," University de Clermont-Ferrand, *Bulletin de l'Observatoire du Puy de Dome,* Series 2, no. 4 (1961).

Pappas, J. J., "A simple yes-no hail-forecasting technique," *J. Appl. Meteor.,* Vol. 1 (1962), pp. 353-54.

Ramaswamy, D., "On the subtropical jet stream and its role in the development of large scale convection," *Tellus*, Vol. 8 (1956), pp. 26-60.

Ratner, B., "Do high-speed winds aloft influence the occurrence of hail?" *Bull. Amer. Meteor. Soc.*, Vol. 42 (1961), pp. 443-46.

Reed, Richard J., "A study of the physical conditions leading to an outbreak of severe convective activity," *Abstract in Bull. Amer. Meteor. Soc.*, Vol. 41 (1960), p. 139.

Schleusener, Richard A, "On the relation of the latitude and strength of the 500 mb west wind along 110° W longitude and the occurrence of hail in the lee of the Rocky Mountains," Fort Collins, Colorado, Colorado State Univ., Atmospheric Science Technical Paper no. 26 (1962).

Wilk, K. E., "An investigation of a severe local hailstorm," Abstract in *Bull. Amer. Meteor Soc.*, Vol. 41 (1960), p. 140.

Williams, G. N. and R. H. Douglas, "Continuity of hail production in Alberta storms," McGill University Stormy Weather Research Group, Scientific Report MW-36 (1963), pp. 16-54.

Yule, G. U. and M. G. Kendall, *An introduction to the theory of statistics* (London, Charles Griffin and Sons, 1950), p. 270.

12

The Structure and Movement of the Chinook in Alberta

W. Brinkmann

I. Y. Ashwell

Introduction

In those parts of Canada outside of the Prairie Provinces, the chinook is regarded as a somewhat exotic phenomenon which makes life possible in Alberta and parts of Saskatchewan during the winter months. Even in the 'chinook belt', however, the true nature of the chinook is little understood, although its effects, in some years at least, are very apparent.

Stated simply, the chinook is a warm dry wind descending the leeward side of the mountains. It is the result of the passage of a typical mid-latitude low-pressure centre at the lee of the Rocky Mountains into which air is drawn. Since the mountains lie in the path of and at right angles to the main westerly air-stream, disturbances are set up in this stream in the form of a wave-train with troughs and crests roughly parallel to the mountain ranges (Figure 1). The air descends at the dry adiabatic lapse rate bringing high temperatures and low relative humidities to the lee slopes; this is known as the 'Foehn effect'.

The amplitude of the wave system depends mainly on the strength of the upper westerly winds, besides other factors such as air mass conditions

Reprinted from *Atmosphere*, Vol. 6 (1968).

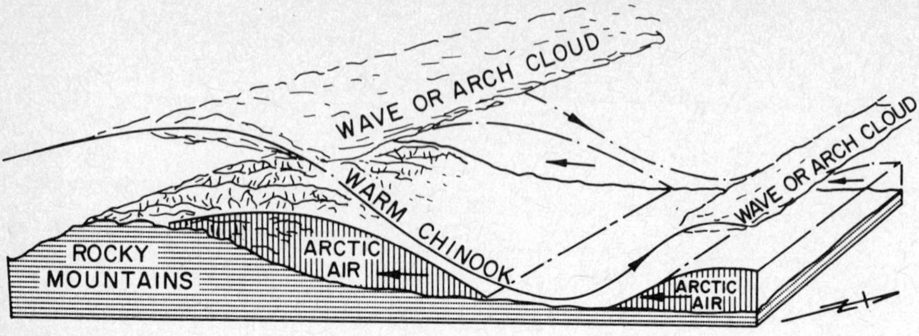

Figure 1. Diagram of the chinook waves as they displace cold Arctic air over the Rocky Mountain foothills and the Prairies, with the chinook arch at the crests of the waves.

and the topography of the area. With light winds the amplitude will be small, and the chinook current may not reach the ground. It will lie a few thousand feet above the ground as a warm layer forming an inversion, under which a light pall of pollution haze will form in the built-up areas of cities such as Calgary. In the crest of the waves the cloud formation known as the Chinook Arch may form in a linear band of lenticular altocumulus parallel to the mountains and stretching for tens or hundreds of miles in a roughly N-S direction (Thomas 1963).

Observation of the arch suggests that its western edge is close to the line of the mountains in the mornings and that the whole arch moves eastwards during the day. Winds are often strong near the mountains in the mornings under chinook conditions, while at Calgary, 45 miles from the E. edge of the mountains, the winds tend to be calm in the mornings and to strengthen towards midday. The whole wave system thus seems to be mobile, with full development starting near the mountains, where the cold Arctic air below the chinook current is thinnest, and gradually moving eastward. Upper air ascents are not normally made at present at Calgary, but a few made in November, 1966, show that at certain stages during a chinook period, the lower levels are unstable. This suggests that under normal diurnal heating conditions, convection may occur, bringing gradient wind conditions to the surface progressively further from the mountains as the day advances.

This movement of the chinook has important implications, first of which is that there is probably an area between Calgary and the mountains in which the occurrence of the chinook is at a maximum. This is of importance to any study of local hydrology and resources, especially ranching, in that the most favored area will have a lighter snow-cover and be more suitable for grazing cattle in the winter.

Method of Investigation

The Chinook Research Scheme of the University of Calgary is, therefore, investigating this movement among other aspects of the chinook. The main means is by a line of recording stations between Calgary and the Rocky Mountain front, some of which are shown on Figure 2. Continuous records of dry-bulb temperature, relative humidity, and in some cases, wind direction and speed, are made on weekly-recording charts and the temperature records are compared with a maximum-minimum thermometer when charts are changed. Instruments vary in pattern but temperature is recorded on bimetallic thermographs and relative humidity on hair hygrographs. Pressure fluctuations are obtained from a Casella micro-barograph at station 1, on the western outskirts of Calgary. All the observations for each hour are abstracted and stored on punched cards.

In the preliminary stages of the investigation a computer program depending entirely on temperature change was used for detecting chinooks, and was used in working up the results for this paper. The program selected a temperature at the sixth hour of a series and subtracted the temperature at the first hour from it. Any positive difference was stored. The process was repeated for the seventh and second hours and subsequent sets five hours apart with positive differences being stored and added. Storage ceased when negative values of difference were reached. Any total computer value of over 100 positive degrees was recorded, together with the hours over which it was attained.

This program was found unsatisfactory in some respects, in that it was sometimes grossly affected by normal diurnal temperature changes and has since been abandoned in favor of one initiated by wind direction and speed. However, it served its purpose for this investigation.

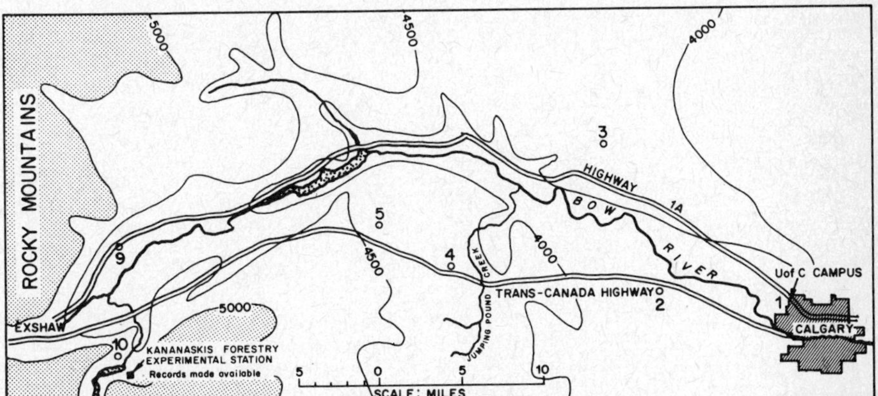

Figure 2. Stations operated by the Chinook Research Scheme in the area Calgary to the Rocky Mountains, with the Forestry Experimental Station at Kananaskis from which records are available.

The best records available were those for stations 1, 3 and 9 and these, with others, are discussed here for periods of chinooks during the winter of 1964-65. A typical synoptic situation during a period of chinooks is shown in Figure 3. The most noteworthy feature is the marked absence of precipitation associated with the warm front.

The project's technician travels through the area once a week to change charts and maintain the station equipment. At the same time, a record is made of snow cover in different parts of the area. It was considered best to concentrate on those chinooks which occurred about the time of a visit

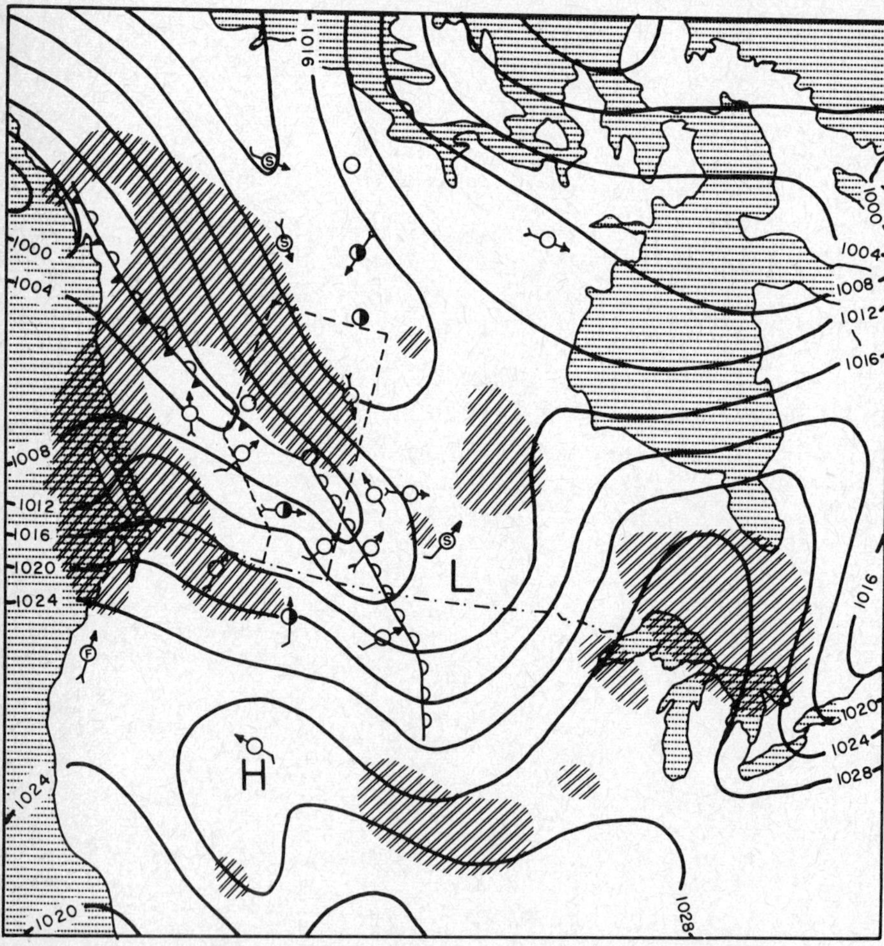

Figure 3. Synoptic situation 0600 16th February, 1964. Southern Alberta lies in the warm sector of a depression, with chinook conditions well developed. Hatched areas represent precipitation. (From Weather Map published by Department of Transport, Winnipeg.)

by the technician so that the effect of the chinook on snow cover could be investigated.

Results

Table I summarizes the rate of temperature increase at the three stations 1, 3 and 9 for those periods of chinook during the winter 1964-65 which were indicated by the computer program. Readings from intermediate stations 2, 4 and 5 were so intermittent during this period due to instrument malfunction as to be useless. The chinooks of 12-15 January and 3-6 February are discussed below and the temperature and pressure graphs are shown in Figure 4.

TABLE I.

	Date	*Station 1.*	*Station 3.*	*Station 9.*
Rate of temperature increase (°F/hr.)	Dec. 2/3, 64.	1.0.	1.4.	1.1
	Dec. 12/13, 64.	3.3.	-	3.6.
	Dec. 25, 64.	3.0.	-	4.3.
	Jan. 9/10, 65.	3.4.	-	13.7.
				2.7.
	Jan. 12/15, 65.	1.7.	1.1.	2.7.
				1.8.
				3.0.
	Jan. 25/26, 65.	2.0.	-	5.0.
				1.7.
	Jan. 27/29, 65.	2.8.	1.9.	2.1.
	Feb. 3/6, 65.	1.9.	1.6.	2.1.
	Feb. 6/7, 65.	3.9.	1.8.	3.6.
	Feb. 14/17, 65.	3.5.	1.6.	5.0.
		3.1.	2.4.	5.0.
				4.2.
	Feb. 22/23, 65.	15.0.	-	3.8.
		3.4.		
Average rate of temp. increase (°F/hr.)		3.7.	1.7.	3.8.

Chinook 12-15 January, 1965

Maximum temperatures at all stations were between 40° and 50°F. The arrival of the chinook caused a sharp increase of temperature at all stations, which was most pronounced, however, at stations 1 and 9.

The curve for station 9 shows a pattern of two temperature troughs and three temperature crests, remarkable for the fact that the two crests occurred near midnight, a day apart, and two of the troughs in the afternoons, with the second trough and crest reaching higher temperatures than the first. The fact that the period of temperature changes was exactly out of phase with the diurnal temperature pattern is an excellent indication that this was a true chinook since the diurnal temperature changes can be

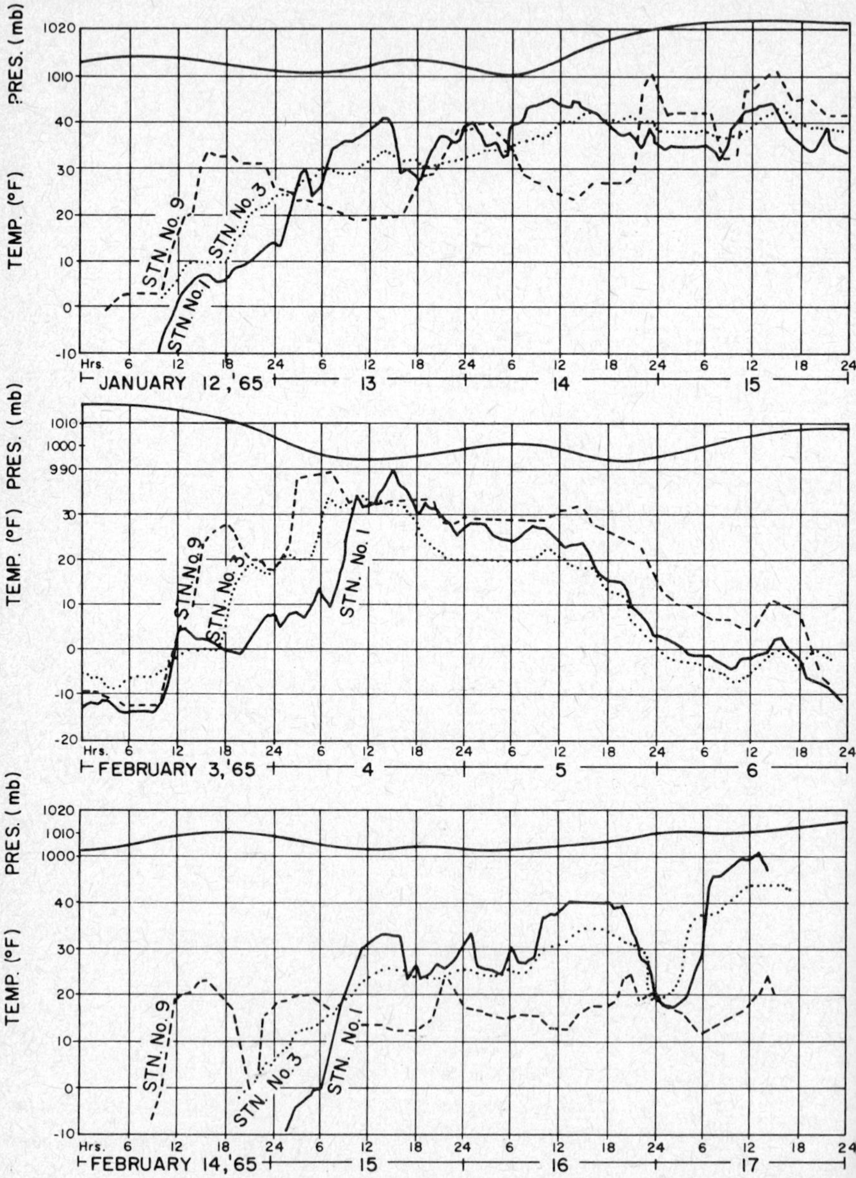

Figure 4. Pressure and temperature variations during chinook conditions. Pressures are measured at Station 1. The third case shown, that of February 14-17, is not discussed in the paper.

misleading at any other time of the day. At station 1, Calgary, the general trend was similar to that of station 9, but the troughs and crests are less noticeable. The most noteworthy fact, however, is that two of the crests at station 1 occurred around noon, possibly affected by the daily temperature cycle, when station 9 showed temperature troughs, with the other crest at station 1 coinciding with one at station 9. In the first two cases, therefore, there was a simultaneous temperature difference of about 20°F between stations 1 and 9, indicating the movement of successive chinook waves over the stations. During the 13th and 14th of January the temperature at station 1 was found to vary directly with the wind speed and inversely with the relative humidity. However, the relative humidity was generally higher around midnight of the 13th than around noon of the 13th and 14th. Therefore, the last case could be interpreted as a turbulent mixing at night which destroyed the nocturnal inversion underneath the chinook crest. This would explain the maximum which occurred at midnight of the 13th at station 1 and would indicate an almost perfect wave pattern.

As would be expected, the pressure trace from the barogram at station 1 (Calgary) shows two periods of low pressure, both coinciding with the temperature increases. It would be interesting to note pressure fluctuations nearer the mountains, to discover whether the times of pressure troughs and crests differ from those at Calgary.

On 13 January the snow cover at station 1 (Calgary) was 100 per cent, by the evening of 14 January it was 70 per cent, and on 16 January and for several days thereafter, it remained at 60 per cent. The technician on his rounds on 15 January estimated the percentage snow cover at stations 9, 5 and 4 to be 5 per cent in shaded and wooded areas, otherwise zero. At stations 5 and 4 the ground was dry, but to the East of station 4 the snow cover increased markedly, and at station 2 the cover was greater than 50 per cent. It appears, therefore, that the chinook was most intense from 13-15 January in the area including stations 4 and 9, and less intense to the East. Because of the high temperatures in both East and West, it is suggested that the main difference was in wind speed.

Since the fall in pressure at station 1 coincided with rising temperatures at that station and falling temperatures at station 9 on two occasions, it can be inferred that the chinook wave lowered at station 1 and rose at station 9 at these times. Station 3 represented a point which was almost always in the wave, with gradually rising temperatures. Indications of a similar pattern are seen for 14-17 February (Figure 4). The wavelength must have been in the order of 70 miles if the stations 1 and 9 were exactly in the trough and crest of the wave; but this is not necessarily so. Probably the wavelength was much shorter in which case the trough and

TABLE II.

	Station	Computer Value	Rate of temp. increase (°F/hr.)	Max. Temp.	Time of Max. Temp.	Observation of Snow Cover Time	Observation of Snow Cover Per cent.	Barometric Change at Calgary
Chinook 12-15 Jan., 1965	1	220	1.7.	41.	13/1 (1400)		70.	Jan. 12-13
	2	No temperature records.					50.	A rise & fall of 3 and 4 mb. occurred twice.
	3	133	1.1.	34.	13/1 (1400)			
	4	No temperature records.				15/1	5.	Jan. 14-15
	9	164	2.7.	34.	12/1 (1600)			An increase of 12 mb.
		101	1.8.	40.	14/1 (200)		5.	
		103	3.0.	51.	14/1 (2300)			
Chinook 3-6 Feb., 1965	1	170	1.9.	40.	4/2 (1500)		100.	Feb. 3-4.
	2	No temperature records.					dropping off.	Drop of 24 mb.
	3	202	1.6.	34.	4/2 (700)	5/2		
	4	No temperature records.					trace 4″ new snow	
	9	190	2.1.	40.	4/2 (700)			

crest must have been located between the two stations. Or, alternatively, another trough could have existed between the stations. It was not possible to answer this question due to lack of information from intermediate stations, although the location of station 3, which was always in the warm air indicates a fairly wide trough.

3-6 February, 1965

Maximum temperatures were in the range of 30°-40°F. Once again, station 9, near the mountains, experienced the chinook for longer than the other stations with station 1 having the shortest exposure. Station 9 also had two peaks in temperature at the beginning of the period, the first being 12 hours before the second and main peak which occurred also at the other stations. The fall in temperature indicating the end of the chinook occurred at about the same time at all stations, though station 9 was the last to experience it.

Station 1 at Calgary had a snow cover of 100 per cent throughout the period, despite the appearance of a chinook arch on 3 February. There was only a trace of snow at station 3 when the technician visited it on the 5th, but new snow had already fallen at station 9 and the cold Arctic air had spread as far as station 4 on that day (Table II). There is, unfortunately, no record of the snow conditions at stations 3, 4 and 9 before this chinook, and it is just possible that snowfalls were lighter in those areas than at Calgary.

Once again, it appears possible that the western part of the area had experienced a much greater snow clearance than the area nearer Calgary, probably due to stronger winds and a greater duration of higher temperatures. The barometer fell 24 mb during the arrival of the warmer air at station 1, and showed a further dip just before the arrival of the front showing the end of the chinook. The wave form of the chinook is apparent in the preliminary peak of temperature at station 9, with a disappearance of the wave before the re-appearance which brought all stations, to a greater or lesser extent, under its influence, and in the almost simultaneous disappearance of the warm air at the end of the chinook.

Conclusion

Although much of the area from Calgary to the Rocky Mountains experiences chinook conditions, there is an area close to the E. edge of the mountains which experiences the chinook earlier and more intensely than the Calgary area. The chinook waves move out from this area of greatest intensity towards the East, though the time-lag of occurrence may be as much as 24 hours between station 9 at the mountains and station 1 at Calgary. Snow cover is usually cleared most in the area nearer to the mountains.

References

Thomas, T. M., "Some Observations of the Chinook Arch," *Weather,* Vol. XVIII, no. 6 (June, 1963), pp. 166-70.

13

Correlation of Air Temperature Normals for the Canadian Great Plains with Latitude, Longitude, and Altitude

J. W. Hopkins

Introduction

More than 25 years ago Hopkins (1938) examined the least squares regression of monthly mean air temperatures recorded at 44 meteorological stations in central and southern Alberta and Saskatchewan on their parameters of latitude, longitude, and altitude, and concluded that linear regression equations could be used as graduation formulae to interpolate acceptable estimates of climatological averages for points between these stations. Recently, operation of the Agricultural and Rural Development Act Administration (ARDA) has generated a requirement for interpolated values in more northern parts of the Plains area. The foregoing study has accordingly been updated and extended to include the whole of the Canadian Great Plains, with the results now reported.

Reprinted from *Can. J. of Earth Sciences,* Vol. 5 (1968), pp. 199-210.

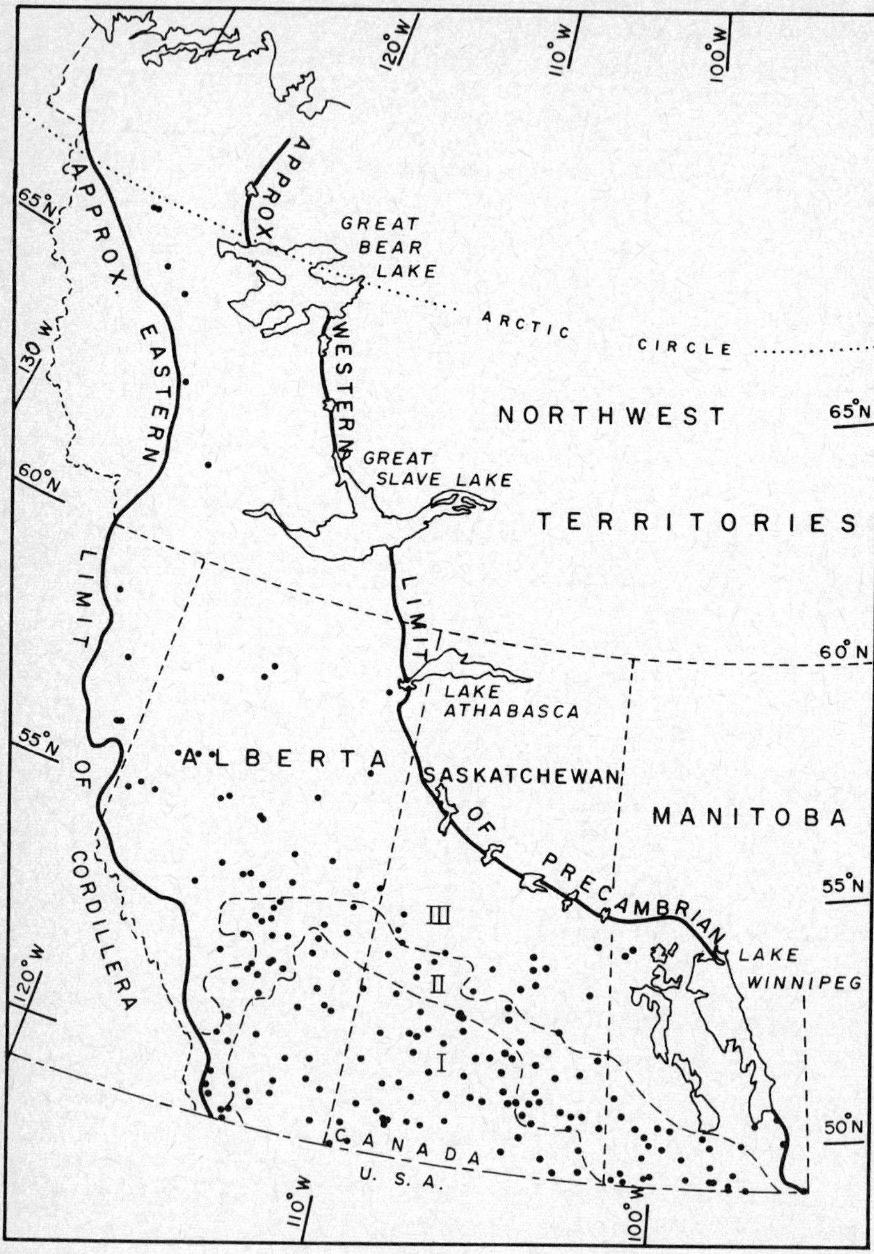

Figure 1. Location of 206 climatological stations in relation to major soil and natural vegetation zones. I, brown soil prairie grassland, II, black soil poplar-aspen parkland, III, gray etc. soil northern forest.

Data

Data comprised climatological averages ('normals') of monthly mean, monthly mean maximum, and monthly mean minimum temperatures tabulated under Project 09464 by the Meteorological Branch, Canada Department of Transport, for 206 stations located in the designated area as indicated in Figure 1, together with the latitude, longitude, and altitude of each station listed in the Branch's Climatological Station catalogue.

The Branch recognizes nine quality coded categories in its climatological tabulations, of which seven were represented in these data, namely:

Code	*No. of Stations*	*Description*
1	76	Normals were computed directly from a period of record of 25 to 30 years within the period 1931-1960. In most cases the record existed over the full 30 years.
2	28	Same as Code 1, but not as much confidence has been placed in the data. These data were considered suspect, but the overall values are mappable.
3	44	The data for these normals were from the full 10-year period 1951-1960 adjusted to the standard normal period 1931-1960.
6	37	These averages are based on the period of record of 10 to 24 years during the period 1931 to 1960. No adjustment factor has been used.
7	10	At several locations the observing station was moved from the town or city to an airport during the 1930's. At many of these locations the records were kept separate, but at those locations indicated by Code 7, the airport and town data were considered homogeneous. The resulting normals are based on the full 30-year period from 1931-1960.
8	8	These data are based on a period of record of less than 10 years.
9	3	These data are based on a period of record of less than 10 years, but adjustments have been made when an unusually warm or cold month unduly influenced the average values.

Also used for interpolation as described below were the latitude, longitude, and altitude coordinates of 1182 survey control points distributed over the area delimited in Figure 1 in an approximately 24 $\times$ 24 mile rectilinear grid. These were available through the collaboration of Mr. C. E. Hoganson, Field Surveys Section, Surveys and Mapping Branch, Canada Department of Energy, Mines and Resources.

Approximative Accuracy and Precision

It is desired to compute an approximation t_{mij} to the true normal θ_{mij} for specified location i and month j by a graduation formula

$$\hat{t}_{mij} = \textstyle\sum_k b_{mjk}\phi_{ik} \qquad [1]$$

in which the location variables ϕ_{ik} are low degree polynomial functions of latitude, longitude and altitude coordinates x_1, x_2 and x_3. Here $i = 1, \ldots\ldots, n = 206$ denote locations, $j = 1, \ldots\ldots, 12$ denote calendar months January, ..., December and $m = 1, 2$ or 3 according as normals of monthly mean, of monthly mean maximum, or of monthly mean minimum are in question.

Standard least squares calculations will yield for each j the b_{mjk} minimizing the weighted sum of squares of residuals $S^{*2}_{mj} = \Sigma_i w_i (\hat{t}_{imj} - t_{mij})^2$. These standard calculations may be conveniently summarized in matrix notation as

$$[2] \qquad \mathbf{B}_m = (\mathbf{X}'\mathbf{W}\mathbf{X})^{-1}\mathbf{X}'\mathbf{W}\mathbf{T}_m$$

In Eq. [2] $\mathbf{T}_m$ (n rows $\times$ 12 columns) is a matrix of recorded normals t_{mij}, $\mathbf{X}$ ($n \times q$) a matrix of corresponding location variables x_{ij}, $\mathbf{W}$ ($n \times n$) a diagonal matrix of weight coefficients w_i and $\mathbf{B}_m$ ($q \times 12$) the associated matrix of weighted least squares b_{mjk}.

If the residuals $(\hat{t} - t)$ are sensibly randomly distributed resultants of fortuitous siting effects and of annual fluctuations in a record of finite duration, standard statistical regression theory may be invoked as described for example by Draper and Smith (1966) to estimate the degree of statistical imprecision in the b_{mjk} from Eq. [2] and hence in individual t_{mij} from Eq. [1]. Random residuals imply an underlying functional relation $\Theta_m = \mathbf{X}\beta_m$ on which are superimposed random deviations δ_{mij} so that $\hat{\mathbf{T}}_m = \Theta_m + \Delta_m$. Then $\mathbf{B}_m$ is an unbiased statistical approximation to β_m, with statistical imprecision variance in the b_{mjk} in any monthly column vector $\mathbf{b}_{mj}$ of β_m estimated from the main diagonal elements of

$$[3] \qquad \text{cov}\,(\mathbf{b}_{mj}) = S^{*2}_{mj}\,(\mathbf{X}'\mathbf{W}\mathbf{X})^{-1}/(n - q).$$

That associated with individual $\hat{t}_{mij}$ would then be estimated as

$$[4] \qquad \text{var}\,(\hat{t}_{mij}) = S^{*2}_{mj}\,\mathbf{x}'_i\,(\mathbf{X}'\mathbf{W}\mathbf{X})^{-1}\mathbf{x}_i/(n - q)$$

with $\mathbf{x}'_i$ the ith row vector of $\mathbf{X}$.

If on the other hand $\mathbf{X}\beta$ is an appreciably inadequate formulation of the actual relation of Θ to the ϕ we shall have $\Theta_m = \mathbf{X}\beta + \Gamma_m$ with Γ_m ($n \times 12$) a matrix of systematic inaccuracies γ_{mij}, and $\mathbf{T}_m = \mathbf{X}\beta_m + \Gamma_m + \Delta_m$. In this event Eq. [2] will still give $\mathbf{B}_m$ minimizing each S^{*2}_{mj} for variations in the b_{mjk}, but these minimal S^{*2}_{mj} will be inflated by the γ_{mij} and Eq. [2] will accordingly overestimate the imprecision of $\mathbf{B}_m$. When the weights w_i are objectively determined, for example by numbers of observations, valid estimates may be possible of the variance σ^2 of the δ and of the mean square bias τ^2 of the γ in observations of unit weight. Even in this case however, because the γ_{mij} are systematic Eq. [4] cannot legitimately be used to estimate the resultant error variance of specific $\hat{t}_{mij}$.

Derivation of Approximations

Weighting

It was judged that the variance σ^2_{mij} of the deviations δ^2_{mij} associated with the t_{mij} in the foregoing quality categories 1 and 7, of those in categories 2 and 3, of those in category 6 and those in categories 8 and 9 were most likely in the ratio of 1:3/2:2:6. Weight coefficients w = 1, 2/3, 1/2, and 1/6 were accordingly used as elements of **W.**

Coordinate Datum

It was arithmetically expedient to define the coordinates x_1, x_2, and x_3 of latitude, longitude, and altitude so that

x_1 = latitude in min. north (+) or south (—) of 52° N
x_2 = longitude in min. west (+) or east (—) of 108° W
x_3 = altitude in ft above (+) or below (—) 1900 ft.

First Approximation

To provide reasonably convenient formulae for desk calculation, ϕ_k were initially restricted to the 19 distinct pairings of x_1, x_2 and x_3 specified by

[5] $$\phi_k = x_1{}^{a1}x_2{}^{a2},\ x_1{}^{a1}x_3{}^{a3},\ x_2{}^{a2}x_3{}^{a3};\ a_1,\ a_2,\ a_3, = 0,\ 1,\ 2$$

When the resulting b_{mjk} from Eq. [2] were related to the appropriate diagonal element of their imprecision matrix computed from Eq. [3], indications were that even after allowing for some bias inflation of the latter, only 10 contributed at all significantly to the approximation. Numerical values of these, recomputed from Eq. [2], omitting the other nine ϕ_k, have been recorded by Williams and Sharp (1967).

Second Approximation

For a second approximation the conditions imposed in Eq. [5] were relaxed to generate ϕ_k from the 26 distinct pairs

[6] $$\phi = x_1{}^{a1}x_2{}^{a2},\ x_1{}^{a1}x_3{}^{a3},\ x_2{}^{a2}x_3{}^{a3};\quad a_1 = 0, \ldots, 4,\quad a_2 = 0, \ldots, 3,\ a_3 = 0, \ldots, 2$$

and from the 15 triples

[6a] $$\phi = x_1{}^{a1}x_2{}^{a2}x_3{}^{a3};\quad a_1 = 1, \ldots, 4,\ a_2 = 1, \ldots, 3,\ a_3 = 1;\quad a_1 = a_2 = 1, \ldots, 3,\ a_3 = 2$$

By the putative imprecision variance criterion outlined above it was judged that only 22 of these appreciably affected the approximation. Recomputing the $\mathbf{B}_m$ to incorporate only these gave the numerical coefficients listed in Tables I, II, and III. In each of these tables the first column specifies the ϕ_k of Eq. [1] in terms of the location variables x defined above. The other 12 columns then list the corresponding coefficients b_k of Eq. [1] appropriate to each month. Also listed are the associated root mean square residuals $s_m^* = [\Sigma_i w_i(\hat{t}_{mij} - t_{mij})^2/(n - 22)]^{\frac{1}{2}}$ appropriate to observations of unit weight. These were on the average about 12% smaller than the corresponding s_{mj}^* from the ϕ_k of Eq. [5]

Computed Temperature Gradients

The coefficients in rows 2, 3, and 4 of each table specify

$$\partial \hat{t}/\partial x_1,\ \partial \hat{t}/\partial x_2 \text{ and } \partial \hat{t}/\partial x_3$$

for $x_1 = x_2 = x_3 = 0$ from Eq. [1] formulated as described. They are accordingly the computed temperature gradients appropriate to the datum location 52° N, 108° W, 1900 ft. Their seasonal variation is illustrated in Figure 2. For the monthly mean normals computed elevation gradients were greatest in summer and computed latitude and longitude gradients greatest in winter and spring. Their numerical maxima were about −0.04°F per min. N. latitude and per 10 ft elevation, and about +0.02°F per min. W. longitude. Corresponding computed gradients for monthly mean maxima in Table II were mostly numerically larger than those for minima in Table III. However this difference was also markedly seasonal, as is illustrated in Figure 2 in terms of the range (monthly mean max. *minus* monthly mean min.).

Rows 5–22 of each of Tables I to III specify computed monthly patterns of gradient modification with location. Consistent seasonal differences in these higher order coefficients are also evident. Such patterns and differences are exemplified in Figure 3. This depicts gradients $\partial \hat{t}/\partial x_1$ with latitude in monthly mean and in monthly mean range in temperature computed by substituting in the approximation formulae specified by the tables the topographical grid data for points along the parallel of 52° N. Moderation by Lakes Winnipeg and Winnipegosis of the north–south trends in both mean temperature and mean daily range east of the Manitoba Escarpment is clearly suggested in the gradients thus calculated for July, but not in those for January. For the Rocky Mountain foothills region on the other hand the calculated gradients in both mean and daily amplitude increased noticeably with westerly longitude for January but not for July.

Adequacy of Approximation

Analysis of Residuals

All of the monthly formulae gave $\hat{t}$ having coefficients of correlation of $r \geqslant 0.95$ with the recorded $\hat{t}$.

Figure 4 shows for the month of January, when temperature variations were greatest, a plot of individual $\hat{t} + u^*$ $(u^* = w_i^{\frac{1}{2}}\ (t - \hat{t}))$ in relation to

TABLE I.

Coefficients b_{1k} of regression of monthly mean temperature normals on location parameters $Ø_k$, basis deg. F/100 min., 1000 ft.

k	$Ø_k$	*Jan.*	*Feb.*	*Mar.*	*Apr.*	*May*	*June*	*July*	*Aug.*	*Sept.*	*Oct.*	*Nov.*	*Dec.*
1	1	1.80	5.91	18.42	38.52	52.20	59.00	65.77	62.49	52.42	40.68	22.06	9.36
2	x_1	—4.48	—3.35	—3.74	—3.04	—2.30	—2.69	—3.42	—3.38	—3.05	—2.57	—3.48	—5.15
3	x_2	1.75	1.59	1.63	1.24	0.78	0.62	0.65	0.55	0.73	0.71	1.20	1.72
4	x_3	—0.59	—1.10	—2.44	—3.41	—3.39	—4.01	—4.36	—4.27	—3.60	—2.35	—1.61	—1.15
5	$10^{-1}x_1^2$	5.39	5.87	3.61	—4.52	—6.96	—5.57	—4.09	—2.07	—1.21	0.97	3.91	6.71
6	$10^{-2}x_2^2$	3.67	4.47	1.55	—0.77	—2.17	—1.37	—4.43	—3.03	—1.17	0.87	2.56	0.38
7	x_3^2	1.13	1.28	0.69	—0.30	—0.39	—0.10	—0.12	—0.03	0.11	0.15	0.40	1.15
8	$10^{-1}x_1^3$	1.29	—0.35	0.99	0.67	1.50	2.03	2.31	1.91	1.43	0.58	1.45	2.11
9	$10^{-1}x_1^4$	—1.90	—1.38	—0.58	0.91	1.27	0.84	0.11	—0.21	—0.36	—0.64	—1.48	—2.25
10	$10^{-1}x_1x_2x_3$	—0.83	—0.50	0.33	—0.13	0.09	0.97	1.93	1.30	0.87	—0.80	0.07	0.65
11	$10^{-1}x_1^2x_2x_3$	0.80	0.41	0.62	0.28	0.39	0.49	0.75	0.60	0.52	0.41	1.07	1.09
12	$10^{-2}x_1^3x_2x_3$	5.14	4.85	2.15	—0.69	—4.00	—4.28	—3.73	—2.60	—1.32	1.90	3.70	5.36
13	$10^{-2}x_1^4x_2x_3$	—0.49	—0.54	—1.23	—1.38	—1.03	—0.62	0.05	0.33	0.23	—0.09	—0.38	—0.44
14	$10^{-3}x_1^3x_2^3$	—1.02	—0.76	—0.62	—0.10	0.21	0.21	0.24	0.17	0.02	—0.33	—1.09	—1.16
15	$10^{-2}x_1^4x_2$	1.19	1.00	—0.63	—1.87	—2.27	—1.75	—0.56	0.15	0.24	0.67	1.03	1.57
16	$10^{-3}x_1^4x_2^2$	1.74	1.27	1.87	1.07	0.66	0.48	—0.08	—0.36	—0.07	0.30	1.63	1.82
17	$10^{-1}x_1x_3$	—1.34	—0.48	1.57	—0.31	—0.57	—2.40	—7.16	—6.11	—3.42	—2.51	—0.29	—1.55
18	$10^{-3}x_1^4x_2^2x_3$	2.30	1.95	2.04	0.88	0.04	—0.11	—0.33	—0.38	—0.06	0.63	2.07	2.59
19	$10^{-3}x_1^3x_2^3x_3$	—1.14	—0.94	—0.74	—0.12	0.30	0.31	0.26	0.17	0.01	—0.37	—1.12	—1.32
20	$10^{-1}x_1^4x_3$	—1.12	—0.87	—0.18	0.73	1.05	0.78	0.28	0.05	—0.12	—0.35	—0.85	—1.28
21	$10^{-1}x_2x_3^2$	—0.87	—1.11	—0.65	—0.28	—1.09	—1.33	—0.44	—1.03	—0.64	—1.14	0.69	—0.32
22	$10^{-2}x_2^2x_3$	—0.02	1.78	—1.34	—0.51	2.13	0.47	—3.58	—0.28	—0.48	0.72	—4.65	—4.15
	s^*, deg. F.	±3.3	±3.1	±2.5	±1.6	±1.6	±1.6	±1.7	±1.8	±1.5	±1.8	±2.2	±3.2

TABLE II.

Coefficients b_{2k} of regression of monthly mean maximum temperature normals on location parameters $Ø_k$, basis deg. F/100 min., 1000 ft.

k	$Ø_k$	Jan.	Feb.	Mar.	Apr.	May	June	July	Aug.	Sept.	Oct.	Nov.	Dec.
1	1	11.52	16.44	28.72	50.25	65.67	71.84	79.68	76.47	65.43	53.03	31.30	18.63
2	x_1	—4.57	—2.79	—2.79	—3.44	—2.34	—2.58	—4.02	—4.03	—3.68	—3.50	—4.41	—5.69
3	x_2	1.93	1.68	1.63	1.57	0.91	0.74	0.90	0.77	1.00	1.12	1.66	2.04
4	x_3	—0.37	—0.90	—2.26	—3.96	—3.63	—4.06	—4.43	—4.26	—3.75	—2.76	—1.99	—1.67
5	$10^{-1}x_1^2$	6.68	6.00	5.98	—4.09	—7.91	—6.32	—5.49	—3.07	—1.35	—1.33	3.28	7.77
6	$10^{-2}x_2^2$	2.62	4.92	2.89	—4.15	—6.18	—5.78	—9.24	—8.60	—5.90	—3.96	—0.46	—1.89
7	x_3^2	1.25	1.45	0.53	—0.19	—0.33	—0.23	—0.32	—0.13	0.25	0.40	0.79	1.72
8	$10^{-1}x_1^3$	0.29	—2.05	—0.86	0.18	0.73	1.60	2.22	2.00	0.72	—0.21	0.93	1.30
9	$10^{-1}x_1^4$	—1.83	—0.94	—0.47	1.35	2.03	1.50	0.83	0.42	0.36	0.35	—0.92	—2.08
10	$10^{-1}x_1x_2x_3$	—0.17	—0.31	0.73	0.88	—0.69	0.60	1.68	0.67	0.41	—0.23	0.75	1.31
11	$10^{-2}x_1^2x_2x_3$	3.40	—3.32	2.66	—0.02	0.60	4.34	7.27	4.85	—0.94	—4.02	4.60	4.32
12	$10^{-2}x_1^3x_2x_3$	8.26	8.61	5.80	—0.36	—4.57	—5.61	—4.75	—3.51	—1.63	2.50	5.39	8.02
13	$10^{-2}x_1^4x_2x_3$	—0.82	—1.16	—1.73	—1.82	—1.40	—0.89	—0.46	—0.12	—0.14	—0.68	—0.91	—0.83
14	$10^{-3}x_1^3x_2^3$	—1.11	—0.83	—0.86	—0.11	0.37	0.27	0.41	0.37	0.23	—0.16	—1.06	—1.16
15	$10^{-2}x_1^4x_2$	1.23	0.66	—0.49	—2.64	—3.08	—2.57	—1.60	—0.79	—0.73	—0.48	0.33	1.32
16	$10^{-3}x_1^4x_2^2$	1.88	1.49	2.16	1.50	0.63	0.66	0.12	—0.26	0.00	0.43	1.83	1.99
17	$10^{-1}x_1x_3$	1.25	2.09	0.15	1.88	3.48	0.52	—6.55	—4.57	1.27	3.57	3.87	2.63
18	$10^{-3}x_1^4x_2^2x_3$	2.67	2.38	2.74	1.09	—0.19	—0.30	—0.41	—0.54	—0.39	0.49	2.21	2.84
19	$10^{-3}x_1^3x_2^3x_3$	—1.31	—1.10	—1.09	—0.11	0.51	0.49	0.46	0.40	0.32	—0.16	—1.10	—1.39
20	$10^{-1}x_1^4x_3$	—1.24	—0.81	—0.27	0.96	1.51	1.21	0.77	0.51	0.32	0.19	—0.59	—1.30
21	$10^{-1}x_2x_3^2$	1.35	1.15	2.03	1.04	—1.93	—1.49	—0.39	—1.71	—1.28	—0.35	1.72	0.92
22	$10^{-2}x_2^2x_3$	—8.33	—5.91	—9.78	—6.00	4.56	0.86	—4.09	1.40	1.58	—2.31	—9.19	—8.97
	s^*, deg. F.	±3.6	±3.4	±3.3	±2.1	±2.2	±2.2	±2.3	±2.6	±2.2	±2.0	±2.1	±3.4

TABLE III.

Cofficient b_{3k} of regression of monthly mean minimum temperature normals on location parameters $Ø_k$, basis deg. F/100 min., 1000 ft.

k	$Ø_k$	*Jan.*	*Feb.*	*Mar.*	*Apr.*	*May*	*June*	*July*	*Aug.*	*Sept.*	*Oct.*	*Nov.*	*Dec.*
1	1	−7.92	−4.65	8.08	26.75	38.69	46.12	51.81	48.48	39.37	28.28	12.78	0.06
2	x_1	−4.36	−3.91	−4.66	−2.66	−2.26	−2.79	−2.83	−2.75	−2.42	−1.64	−2.56	−4.66
3	x_2	1.55	1.50	1.63	0.92	0.65	0.50	0.39	0.34	0.46	0.30	0.75	1.41
4	x_3	−0.81	−1.30	−2.58	−2.88	−3.16	−3.95	−4.31	−4.28	−3.47	−1.96	−1.26	−0.71
5	$10^{-1}x_1^2$	3.94	5.55	1.22	−4.93	−5.99	−4.95	−2.58	−0.93	−1.14	3.19	4.52	5.58
6	$10^{-2}x_2^2$	4.69	4.03	0.30	2.57	1.85	3.05	0.33	2.28	3.57	5.83	5.65	2.61
7	x_3^2	1.01	1.13	0.81	−0.40	−0.47	0.02	0.09	0.07	−0.04	−0.11	0.04	0.62
8	$10^{-1}x_1^3$	2.21	1.36	2.77	1.18	2.23	2.43	2.40	1.81	2.14	1.32	1.93	3.01
9	$10^{-1}x_1^4$	−1.92	−1.77	−0.67	0.49	0.53	0.20	−0.63	−0.83	−1.09	−1.60	−2.02	−2.42
10	$10^{-1}x_1x_2x_3$	−1.47	−0.59	−0.05	−1.11	0.90	1.36	2.26	2.06	1.33	−1.34	−0.59	−0.01
11	$10^{-1}x_1^2x_2x_3$	1.25	1.14	0.96	0.56	0.73	0.54	0.76	0.72	1.14	1.21	1.66	1.74
12	$10^{-2}x_1^3x_2x_3$	1.98	0.96	−1.46	−1.05	−3.49	−3.02	−2.65	−1.68	−1.04	1.22	2.05	2.62
13	$10^{-3}x_1^4x_2x_3$	−1.77	0.52	−7.33	−9.41	−6.41	−3.38	5.58	7.83	6.27	4.98	1.31	−0.70
14	$10^{-3}x_1^3x_2^3$	−0.90	−0.67	−0.37	−0.09	0.04	0.17	0.08	−0.05	−0.19	−0.51	−1.12	−1.16
15	$10^{-2}x_1^4x_2$	1.11	1.26	−0.78	−1.11	−1.45	−0.96	0.51	1.08	1.23	1.79	1.70	1.78
16	$10^{-3}x_1^4x_2^2$	1.58	1.08	1.57	0.66	0.66	0.28	−0.26	−0.42	−0.17	0.18	1.44	1.67
17	$10^{-1}x_1x_3$	−3.95	−3.07	2.93	−2.41	−4.67	−5.27	−7.66	−7.36	−8.17	−8.65	−4.52	−5.59
18	$10^{-3}x_1^4x_2^2x_3$	1.91	1.51	1.34	0.67	0.25	0.05	−0.24	−0.20	0.24	0.76	1.95	2.33
19	$10^{-3}x_1^3x_2^2x_3$	−0.94	−0.76	−0.39	−0.13	0.10	0.15	0.06	−0.08	−0.28	−0.58	−1.13	−1.26
20	$10^{-1}x_1^4x_3$	−0.98	−0.91	−0.09	0.52	0.59	0.36	−0.22	−0.42	−0.57	−0.87	−1.11	−1.24
21	$10^{-1}x_2x_3^2$	−3.10	−3.40	−3.17	−1.59	−0.21	−1.14	−0.38	−0.10	−0.03	−1.91	−0.38	−1.57
22	$10^{-2}x_2^2x_3$	8.23	9.26	6.69	4.90	−0.32	−0.03	−3.36	−2.77	−2.44	3.81	−0.06	0.83
	s^*, deg. F.	±4.2	±4.2	±3.7	±2.5	±2.8	±2.9	±3.3	±3.5	±3.2	±3.5	±3.3	±4.2

the $\hat{t} = \sum_k b_k \phi_k$ obtained from the first column of Table I and the 206 sets of station coordinates x_1, x_2, and x_3. So ranked, this ensemble of u^* exhibits no evident systematic pattern. Also the size distribution of individual u^* within such monthly ensembles did not differ radically from that

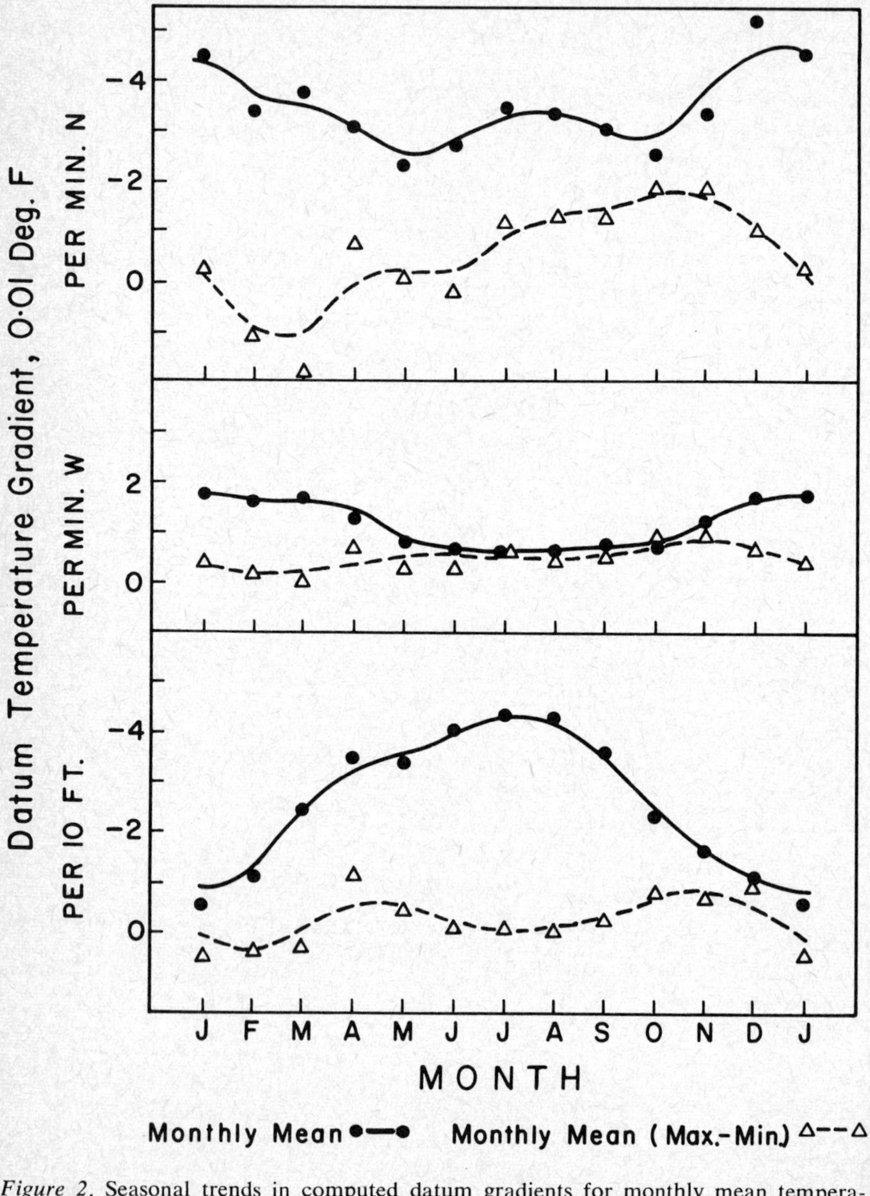

Figure 2. Seasonal trends in computed datum gradients for monthly mean temperature and for monthly mean range (mean maximum *minus* mean minimum).

characteristic of 'normal', i.e. Gaussian frequencies. For January and July monthly mean t, percentages of the 206 u^* comprised within the limits $\pm s^*$ and $\pm 2s^*$ were

Limits	January	July	Gaussian	
$\pm s^*$	71%	70%	$\pm\ \sigma$	68¼%
$\pm 2s^*$	96%	96%	$\pm 2\sigma$	95½%

Some heterogeneities were however revealed when the stations were classified according to location. For example Figure 5 (upper panel) portrays the average u^* associated with the monthly mean t computed for the 75, 66, and 65 stations respectively comprised within the major black, brown, and gray etc. soil and natural vegetation zones outlined in Figure 1. Indications from this categorization were that for the months of November through March the approximation formulae tended on the whole slightly to under-estimate the normals actually recorded at gray soil, northern forest zone stations, and correspondingly to overestimate those for brown soil, prairie grassland zone stations. For May through September this indicated differential was reversed.

Also indicated was a tendency for most of the monthly formulae slightly to underestimate monthly mean normals recorded at stations in the black

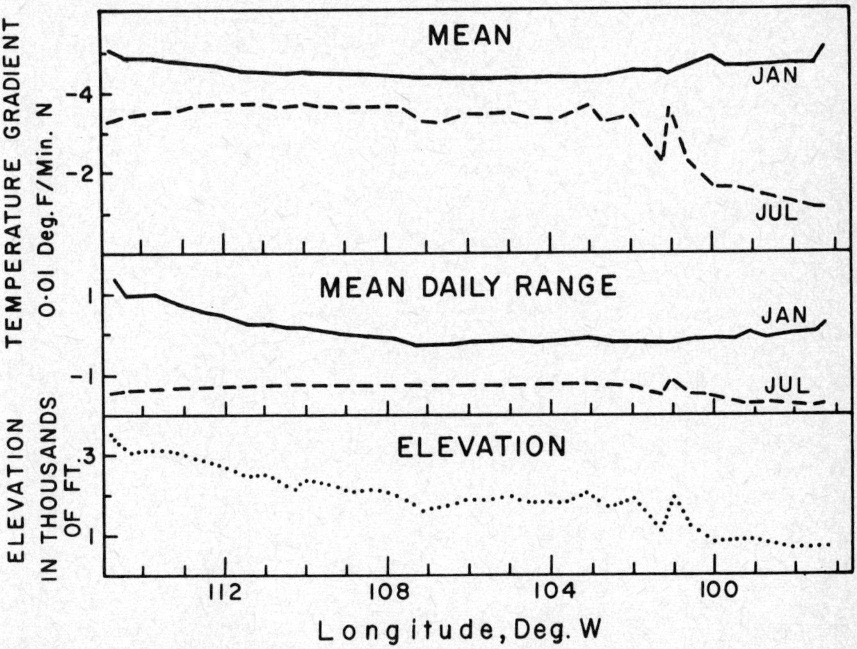

Figure 3. Top, longitudinal trends in estimated temperature gradients along parallel of 52°N. Bottom, topographical profile.

soil, poplar–aspen parkland zone, particularly for the winter months of November, December, and January. This discrepancy was most pronounced for stations in the North Saskatchewan River Basin and Red Deer River Basin districts, for which the average u^* are shown separately in the lower panel of Figure 5. Those for the Red Deer River Basin manifested an especially marked seasonal trend.

Estimated Root Mean Square Errors of Approximation

Both random annual fluctuations and statistically independent local siting effects will tend to be smoothed out by the approximation formulae. These may therefore in some instances actually provide more precise neighborhood climatological values than do the recorded station normals. This random component of error variance in $\hat{t}$, minimal at the centroid of station location coordinates, will vary with location in accordance with Eq. [4].

Systematic inaccuracies such as those noted above cannot however be assumed to be distributed in this way, or indeed to be smaller near the centroid than elsewhere. The most facile assumption to be made in respect of them would be to suppose that $\tau_{mj}{}^2 = \sigma_{mj}{}^2 = s^*{}_{mj}{}^2$, namely that the mean square bias and mean square random deviation were of equal magnitude and could additively be equated to s^{*2}, the computed mean square residual of unit weight. On this and on the further supposition that the point biasses γ_{mij} were representative of those affecting interpolation throughout the region the resultant root mean square error to be apprehended in $\hat{t}$ calculated for an arbitrary location would be

$$[7] \qquad \text{r.m.s.e.}(\hat{t}) = \pm[\tfrac{1}{2}s^*{}_{mj}{}^2(1 + (\mathbf{x}'\mathbf{Z}^{-1}\mathbf{x}))]^{\frac{1}{2}}$$

with $\mathbf{Z} = \mathbf{X}'\mathbf{W}\mathbf{X}$ and $\mathbf{x}$ the appropriate 22-element vector of location variables.

Applied to hypothetical locations 56° N, 118° W, 2150 ft (Grimshaw vicinity, Alberta) and 52° N, 106° W, 1850 ft (Saskatoon vicinity, Saskatchewan), Eq. [7] gave estimated ± °F r.m.s. errors in approximations of climatological monthly mean temperatures of

	Grimshaw	Saskatoon
January	1.4	1.2
February	1.3	1.1
March	1.1	0.9
April	0.7	0.6
May	0.7	0.6
June	0.7	0.6
July	0.8	0.6

August	0.8	0.6
September	0.7	0.5
October	0.8	0.6
November	1.0	0.8
December	1.4	1.1

A further application of Eq. [7] to each point of the topographical survey grid permitted location of equal estimated approximation error by inverse interpolation. Figure 6 shows some contours so deduced for calculated September mean temperatures. They suggest r.m.s.e. ($\hat{t}$) of less than $\pm 1°$ F for most of the central plains south of Great Slave Lake and of the valleys of the Hay and Fort Nelson Rivers, excepting the Caribou and Birch Mountain highlands bordering the Peace River Valley east of Fort Vermilian.

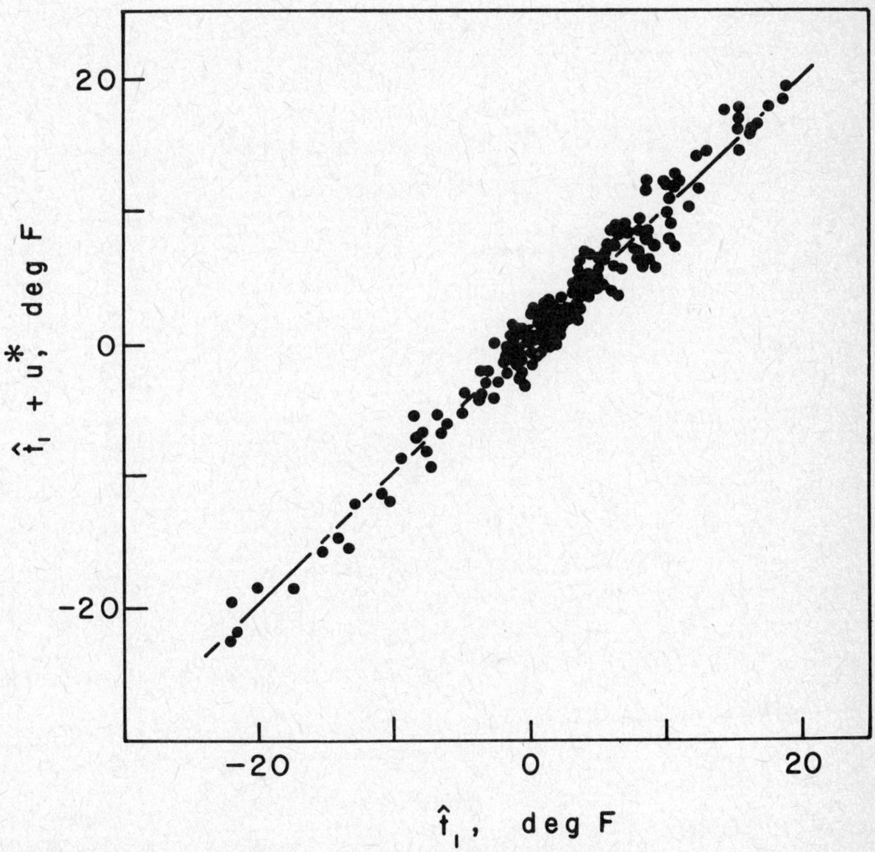

Figure 4. Weighted residuals u^*, observed *minus* calculated, in relation to calculated January mean temperatures $\hat{t}_1$.

Isotherm Approximations

Inverse interpolation between $\hat{t}$ calculated for survey grid points may also be used to compute estimated loci of isotherms. Figure 7 outlines such estimates of 30, 40, and 50 °F September mean temperature isotherms.

Of course these putative T deg. isotherms are actually loci of T + ΔT, with ΔT a variable quantity of order r.m.s.e. ($\hat{t}$). Along the depicted putative 50 °F. isotherm r.m.s.e. ($\hat{t}$) estimated from Eq. [7] averaged ±0.6 °F., while along that for 40 °F shown extending northwesterly from the vicinity of Great Slave Lake it averaged ±1.9 °F. Such putative contours will deviate from the true isotherm loci to an extent varying directly with ΔT and inversely with $\partial\hat{t}/\partial x_1$, $\partial\hat{t}/\partial x_2$, and $\partial\hat{t}/\partial x_3$ at each point.

Discussion

The approximation formulae specified in Tables I–III are readily amendable to automatic calculation. In fact, the 1182-point grid of $\hat{t}$ used in Figure 7 was produced by the NRCC IBM 360/50 in about 5 minutes.

Although estimates of r.m.s.e. ($\hat{t}$) as a specific function of location are subject to the reservations noted, they nevertheless indicate the general order of accuracy attained. This suggests that calculated spring and summer monthly $\hat{t}$ may be of practical utility for appraisals of potential land utili-

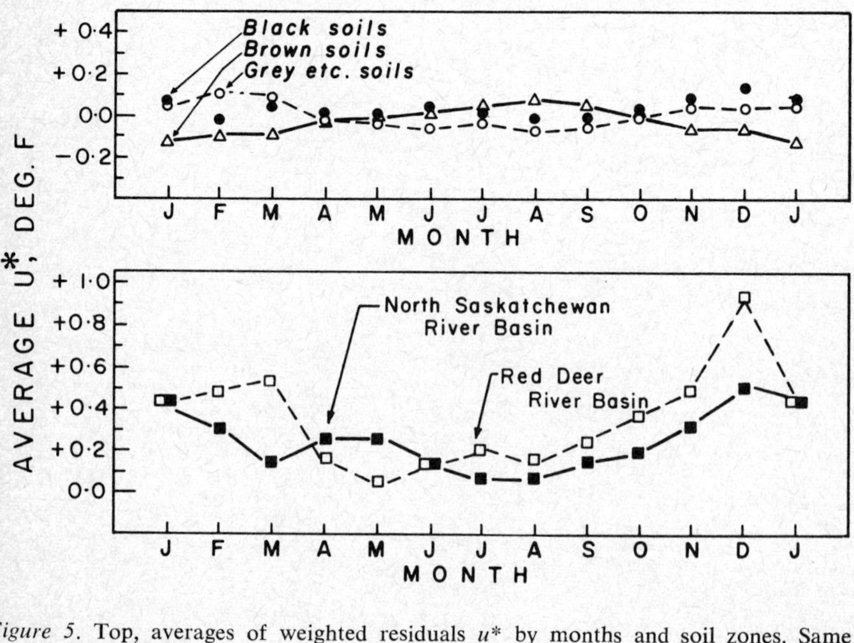

Figure 5. Top, averages of weighted residuals u^* by months and soil zones. Same for North Saskatchewan and Red Deer River Basin districts of black soil, poplar-aspen parkland zone.

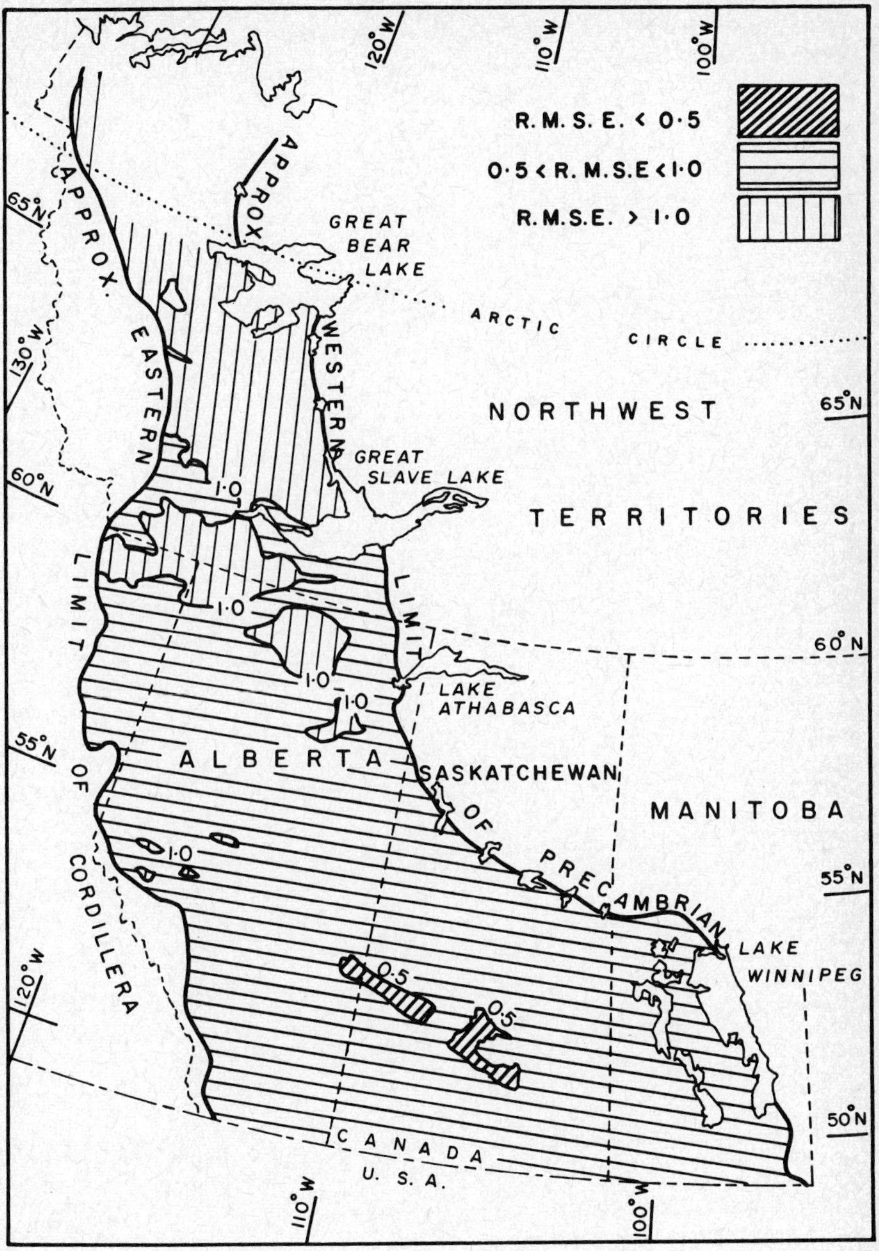

Figure 6. Estimated contours of root mean square error of computed September normal mean temperatures.

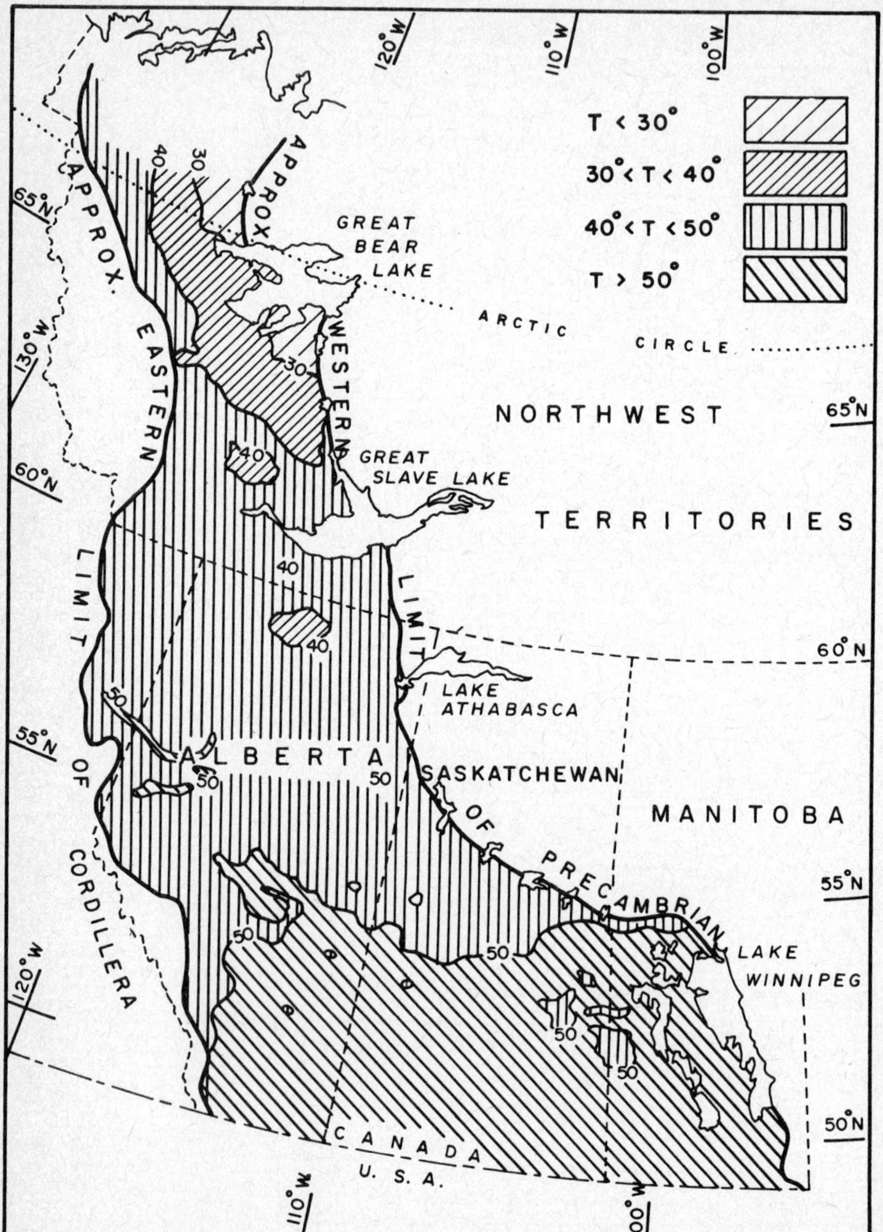

Figure 7. Approximations to isotherms of September normal mean temperature located by inverse interpolation.

zation in some areas only sparsely documented by established climatological normals. Extrapolation beyond the area limits defined in Figure 1 is inadvisable, and even within these limits the margin of uncertainty in $\hat{t}$ for high latitudes, say beyond the Great Slave Lake vicinity, becomes considerable.

The degree of approximation achieved reflects the infrequency of major geographical discontinuities in the Plains area. It is nonetheless hardly surprising that these purely empirical formulae did not sensibly account for all systematic features of the data. From the foregoing analyses of residuals it may be conjectured for example that aspect as well as relief may have influenced temperatures recorded in the predominantly west–east oriented North Saskatchewan River basin and in the predominantly north–south oriented Red Deer River basin. Figure 3 suggests that the formulae provided at least an approximation to temperature gradient inflexions occasioned by large bodies of water. If, however, this was not completely adequate, the average differentials between residuals classified by soil zones illustrated in Figure 4 may have resulted in part from the several large water masses lying in the gray soil, northern forest region. In this event approximations might be improved by incorporation in the formulae of an additional variable x_4 quantifying the distance of each location from major lakes jointly with the size of the latter.

References

Draper, N. R. and Smith, H., *Applied Regression Analysis* (New York, John Wiley and Sons Inc.).

Hopkins, J. W., "Agricultural meteorology: correlation of air temperatures in central and southern Alberta and Saskatchewan with latitude, longitude and altitude," *Can. J. Res.*, C. 16, pp. 16-26.

Williams, G. D. V. and Sharp, W. R., "A program to estimate normals of temperature and related agroclimatic elements for locations on the Canadian Great Plains," Can. Dept. Agr., Res. Br., Agr. Met. Tech. Bull. 11.

14

Great Lakes Water Temperatures by Aerial Survey

T. L. Richards

Introduction

The Great Lakes of the Laurentian system, situated in the heart of the North American continent, represent 250,000 km^2 of fresh water that is of utmost importance to the economic and social life of some thirty to forty million Canadian and United States residents who live near their shores. These large bodies of water present many problems to the hydrologists and meteorologists of the area, and not the least of these are associated with the surface water temperatures. The lakes significantly influence the climate of the region making a knowledge of water temperatures a necessity for understanding and forecasting meteorological conditions over the lakes and the adjacent land areas. In addition, water temperatures are becoming increasingly important for evaluating evaporation losses and the resultant changes in lake levels, and in studies of the formation and dissipation of ice.

Until recently water temperature data for the Great Lakes were obtained through a modest, but expanding network of thermographs located mainly at lighthouse weather observing stations and by means of ship's surveys. The most regular of these surveys have been conducted by the

Reprinted from Pubn. No. 70 of the International Association for Scientific Hydrology, Symposium of Garda (1965), pp. 406-19.

Canadian Coast Guard Ship "Porte Dauphine" a research vessel and weather ship operated in support of the Great Lakes Institute, University of Toronto by the Canada Department of Transport. Both methods have recognized shortcomings. The exposure of each thermograph is critical and the data at each site are really representative of only a single fixed point near land. The ship's surveys on the other hand, take 4 to 5 days per lake and so are not truly synoptic in nature.

It has long been recognized that for best results surface water temperatures should be obtained by surveying as large an area as possible in as short a time as possible. Because of such requirements the Meteorological Service of Canada (Meteorological Branch, Department of Transport) decided to investigate the usefulness of the infra-red thermometer (IRT) and the airborne radiation thermometer (ART) technique over the Great Lakes.

Instrumentation and Technique

History

The first use of the ART technique was by Stommel[1] in 1953 to trace the edge of the Gulf Stream. In a more sophisticated model developed by Richardson and Wilkins[2] in 1957 the original Golay detector was replaced by a thermistor bolometer. During the past four or five years several commercial models of the IRT have become available and have been subjected to a variety of evaluation programs; Pirart,[3] Snowden,[4] McFadden and Ragotzkie,[5] U.S. Department of Interior Workshop Report 202,[6] Trites.[7] As a result of these and other programs substantial improvements have been made in the basic instrument and in airborne techniques.

It was in order to exploit these recent advances that the Lakes Investigation Unit (Hydrometeorology Section, Climatology Division) of the Canadian Meteorological Service acquired a portable IRT for research and eventual operational purposes and undertook a flight test and evaluation program in the Great Lakes area early in 1965. The initial program was later extended to include a study of the environmental errors inherent in the technique with a view to making corrections based on meteorological conditions.

Basic Principles

In brief, the IRT senses the radiant energy from the surface over which it is transported. The detector, a thermistor bolometer, looks through an optical system alternately at the water surface whose temperature is desired and at an accurately calibrated reference black body. The difference between the two radiations is converted to an electrical signal that varies with the temperature of the water surface. The radiant energy is passed through optical filters which limit transmission to between 8.0 and 14.0

microns. Because of the filter system the detector is exposed to infrared emissions in the region where the attenuation due to carbon dioxide and water vapour is small. For this reason the measurements have been judged to be relatively independent of altitude although it will be shown later that altitude does produce a significant and measureable effect.

The accuracy of such instruments depends very much on their particular design. The larger and more sensitive units are reported to be effective over a range of —2 to +35°C with an absolute accuracy of ±0.2°C. Smaller, portable units are available with essentially the same range but with an accuracy more of the order of ±1.0°C.

Instrumentation

An instrument manufactured by the Barnes Engineering Company (Model IT-2S) was selected on the basis that it was readily available, conveniently portable and reasonably priced. This particular model has an advertised mid-scale sensitivity of 3/4°F (.4°C) and an absolute accuracy of 1½°F (.8°C). The field of view is a 3° cone and the temperature is read directly from a meter on an electronics console to which a recorder may be attached. The power requirement is 105-125 volts, 60 cycles per second and 25 watts.

Installation

The instrument was installed on a chartered Lockheed 14 aerial survey aircraft based at Toronto International Airport. The complete unit consisted of a sensor or sensing head, an electronics console with temperature meter, a recorder and power inverters. With the exception of the sensor all the equipment was shock-mounted on an instrument rack in the aircraft cabin. The sensor itself was mounted on a bracket on the floor of the aircraft with a field of view through an opening in the camera hatch.

The Technique

The accepted ART survey technique is to fly over the water body to be surveyed at a pre-determined height usually 300 to 1000 feet, (90 to 300 m) above water level and to follow a fixed flight pattern. The data are collected by reading the temperature meter on the electronics console at selected time intervals, or better, by recording an analogue output on an attached recorder. These raw data are corrected by applying a correction factor established empirically by flying over a known water temperature (usually from a ship) at the pre-determined survey height preferably both before and after the survey.

This technique assumes that the meteorological conditions remain relatively constant during the flight i.e., no precipitation, no clouds below the aircraft, good visibility, cloud cover over the aircraft not changing, and little or no change in temperature, humidity and visibility in the layer of air below. Under these conditions an empirical correction for the environmental error due to the optical path through the air below the aircraft is considered valid for a flight of three or four hours.

In comparing the ART readings and ship's water temperature it is important to keep in mind that the infra-red thermometer senses the temperature of the microsurface or "skin" of the water whereas the ship's bucket thermometer is more likely to read the average temperature of the top 6-12 inches (15-30 cm) of water. Recent studies[6] have reported an average difference between skin and bucket temperatures of 0.6°C. As part of the test program an IRT was installed on the research vessel "Porte Dauphine" for two of the flights. For this period the ship's IRT readings averaged 0.4°C lower than the bucket temperatures with individual differences ranging from 0.8°C to zero. More extensive studies of this phenomenon are planned for a later date.

The Test and Evaluation Program

The initial test and evaluation program was conducted between February and September of 1965 and has been described in detail by Richards and Massey.[8] The program was designed:

(i) to determine the most effective mounting for the sensor,
(ii) to establish the optimum height for aerial surveys,
(iii) to evaluate the technique as a method of making full lake temperature surveys and for detecting small scale horizontal temperature changes,
(iv) to assess the size and nature of the various factors contributing to the environmental error with the end in view of making corrections to the IRT readings based on observed or forecast meteorological conditions.

The Program—Phase I

The first phase of the initial program was in conjunction with four regular Meteorological Branch ice reconnaissance flights over the Great Lakes early in 1965 and was designed to determine the most effective mounting of the sensing head. It was soon established that the mounting of the sensor is critical and a review of the literature indicates that there is likely a different and specific requirement for each type of aircraft. It was found that to avoid excessive fluctuations in the temperature record there must be a gentle outflow of air from the cabin with a minimum of small scale turbulence at the opening through which the sensor views the water. For the Lockheed 14 aircraft the high frequency fluctuations (or "noise") in the ART temperature trace were reduced to an acceptable ±1°C by mounting the sensor 7 to 10 cm inside the cabin and viewing the water through a stream-lined pipe designed as an air diverter.

The Program—Phase II

The second phase of the test and evaluation program consisted of a series of five flights in co-operation with the research vessel C.C.G.S. "Porte Daphine". The aircraft's program was to fly upwind and downwind over the ship at altitudes of 300, 500, 1000, 1500 and 2000 feet (90, 150, 300, 450, 600 m) above lake level and then to proceed on a survey of a lake. The surveys were designed to criss-cross the lake to provide a check on ART temperatures at the cross points.

The ship's program included the taking of water temperature observations by means of the conventional bucket thermometer and for two flights this program was augmented by an IRT based on the bridge. In addition to the water temperature measurements and the regular meteorological observations the technician staff also took micro-profile measurements of the atmosphere up to 1000 feet (300 m) by means of the ship's meteorological boom and wiresonde facility. These latter observations were in support of the study of environmental errors.

Tables I and II are samples of the data collected during Phase II of the program. Table I tabulates results of the first flight of the series on March 25, 1965 when the air temperature was slightly lower than the

TABLE I.

Aircraft and ship's observations

Flight 2.1 : 25 March 1965

Aircraft data			*Water temperature (°C)*				
Height (ft.) above lake	*Heading*	*Ambient air temp. °C*	*A R T aircraft*	*Bucket temp. ship*	*Difference A R T- bucket*	*I R T ship*	*Difference ART-IRT*
300	West	—	1.2	1.7	−0.5	—	—
300	East	—	1.2	1.7	−0.5	—	—
500	West	—	1.7	1.7	0.0	—	—
500	East	—	1.7	1.7	0.0	—	—
1000	West	—	1.0	1.6	−0.6	—	—
1050	East	—	0.9	1.6	−0.7	—	—
1500	West	—	0.9	1.6	−0.7	—	—
1550	East	—	0.3	1.6	−1.3	—	—
2000	West	—	0.3	1.6	−1.3	—	—
2000	East	—	0.3	1.6	−1.3	—	—

Ship weather
Cloud: 4/8 of Altostratus at 8,000 ft. and 4/8 of Cirrostratus.
Visibility: 15 miles.
Air temperature: 0°C.
Dew point: — 1°C.
Wind: 110° at 19 knots.
Wave height: 1 to 2 feet.

surface water temperature. Table II shows the results of the flight of August 11, 1965 when the air temperature was higher than the water temperature. The ship-board IRT was also in operation for this flight.

TABLE II.

Aircraft and ship's observations
Flight 2.3: 11 August 1965

Aircraft data				*Water temperature (°C)*			
Height (ft.) above lake	*Heading*	*Ambient air temp. °C*	*A R T aircraft*	*Bucket temp. ship*	*Difference A R T-bucket*	*I R T ship*	*Difference ART-IRT*
350	West	15	15.3	15.0	0.3	14.6	0.7
350	East	15	15.4	15.0	0.4	14.5	0.9
500	West	14	16.0	15.1	0.9	14.8	1.2
500	East	15	15.7	15.1	0.6	14.7	1.0
1000	West	13	16.0	15.1	0.9	14.7	1.3
1000	East	14	16.0	15.1	0.9	14.7	1.3
1500	West	13	15.7	15.2	0.5	14.9	0.8
1500	East	13	15.5	15.2	0.3	15.0	0.5
2000	West	12	15.9	15.2	0.7	14.8	1.1
2000	East	12	15.7	15.3	0.4	14.7	1.0

Ship's weather
Clouds: 1/8 of Cumulus at 1500 ft. and 1/8 of Cirrus.
Visibility: 15 miles.
Air temperature: 18°C.
Dew point: 11°C.
Wind: Calm.
Wave height: Less than 1 foot.

Optimum Flying Height

Table III is a summary of the data from the five flights in Phase II of the program. The table indicates that smallest temperature corrections were required for flights of 300 to 500 feet (90 to 150 m). At these altitudes there was an average error of about 1°C within a range of −1.2°C to

TABLE III. TEMPERATURE DIFFERENCES WITH HEIGHT (ART - SHIP'S BUCKET)

Aircraft Height (feet)	*Average Difference °C (regardless of sign)*	*Range of Difference °C* Negative	Positive
300	0.9	−1.0	2.2
500	1.0	−1.2	2.9
1000	1.7	−1.5	3.6
1500	1.6	−1.7	3.4
2000	1.7	−2.1	3.3

+2.9°C. Flights at one to two thousand feet (300 to 600 m) required larger corrections but these were constant enough that flights at these heights appear feasible if safety or crew comfort indicates a higher altitude.

Temperature Corrections

From the data collected on the five air and ship trials correction required for the ART varied from −2.1°C to +3.6°C. It was evident that these differences were related to the height of the aircraft and to day-to-day changes in the atmosphere. For any given aircraft height there was a close relationship between the temperature difference (ART — ship) and the temperature of the air column through which the sensor was viewing. Table IV establishes this relationship in general terms, i.e., the ART reads lower when the air column is colder than the water and higher when the air column is warmer.

TABLE IV. EFFECT OF AIR TEMPERATURE ON ART READINGS

Flight	*Temperature differences (A R T-ship)*	*Air temperature (relative to water)*	*Lapse rate*	*Visibility (miles)*
2.1	small negative	air colder than water	estimated unstable	15
2.2	large positive	air much warmer	marked inversion	15
2.3	small positive	air a little warmer	slightly unstable	15
2.5	negative	air a little colder	unstable	6
2.6	negative	air and water about same	very unstable	15

Calibration

For best results the infra-red thermometer requires frequent laboratory calibration checks. Until more is learned of the environmental errors inherent in the ART technique an in-flight calibration check over a known water temperature is also required, preferably both prior to, and after, each survey. Frequent in-flight in-cabin calibration checks with a well-stirred water bath are recommended although a drift in excess of 0.5°C was detected on only one flight. A monitor of the power input to the ART is also advisable to guard against errors introduced by frequency changes to which the instrument (AT2S) is extremely sensitive. A newly acquired model (AT3S), said to be much less sensitive to frequency variations is currently undergoing tests.

Temperature Surveys

The results of the two full-lake temperature surveys of Lake Ontario are shown in Figure 1 and Figure 2. Figure 1 shows the survey of March 25, 1965 with comparative temperatures from a limited ship's survey on

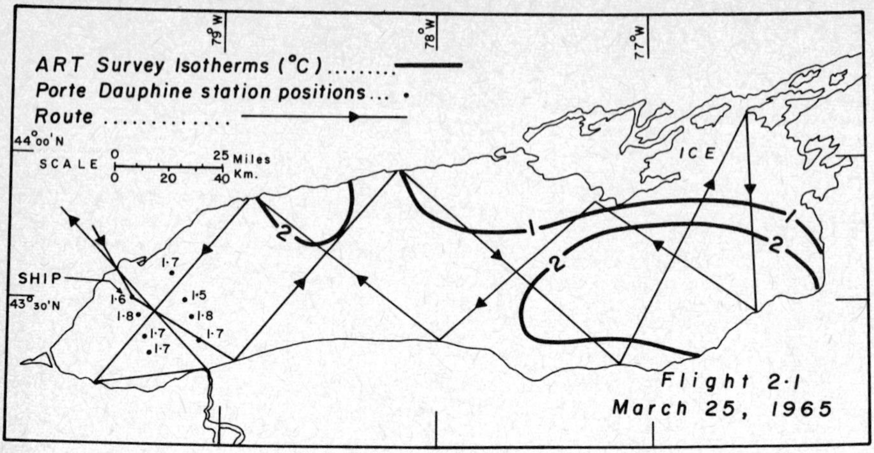

Figure 1. ART survey, Lake Ontario, March 25, 1965 showing flat temperature gradient typical of early spring. Limited Porte Dauphine survey in figures °C.

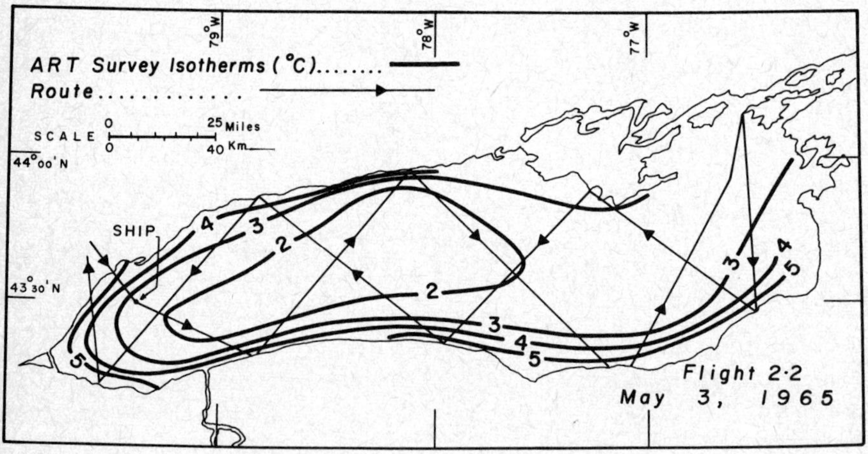

Figure 2. ART survey, Lake Ontario, May 3, 1965 showing shoreline warming.

the same day in the west end of the lake. Figure 3 shows the results of the survey of May 3, 1965. From these and later surveys compared with ship's observations the tentative conclusion has been reached that with fly-over and in-cabin calibration the isotherms have an absolute accuracy of about 1.0°C and a relative accuracy (between isotherms) of at least 0.5°C.

Small Scale Variations

The ART temperature traces from the survey flights have revealed

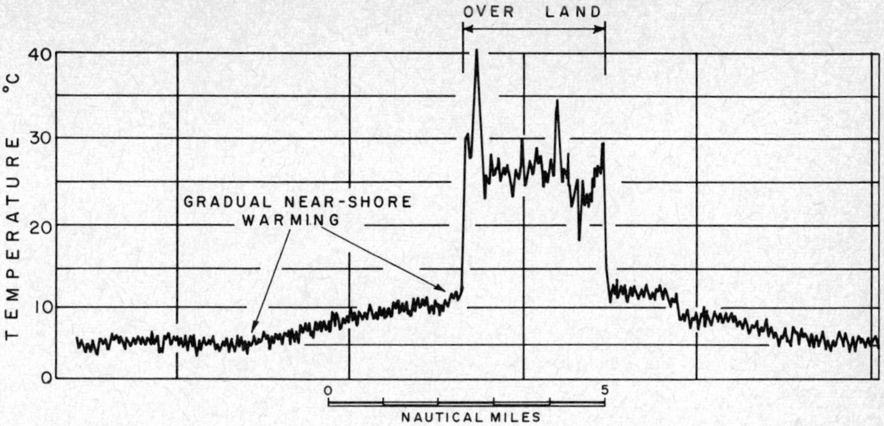

Figure 3. ART trace, Flight 2.2, May 3, 1965 from 500 feet (150m.). Example of gradual warming towards shore.

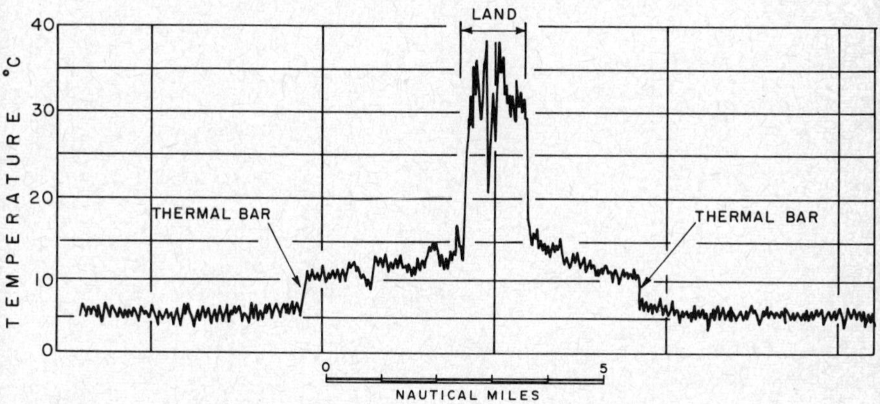

Figure 4. ART trace, Flight, 2.2, May 3, 1965 from 500 feet (150m.). Example of abrupt temperature change near shore – the "Thermal Bar".

some good examples of the ability of the instrument and the technique to detect surface water temperature changes over short horizontal distances.

Figure 3 (May 3, 1965) shows the gradual near-shore warming typical of any of the lakes in the spring.

Figure 4 (May 3, 1965) illustrates an abrupt change of upwards to 5°C with a "thermal bar" in Lake Ontario as described by Rodgers.[9] This bar is a phenomenon recently discovered in several of the Great Lakes. It is particularly evident in the spring and to a lesser degree in the autumn.

Figure 5 (Sept. 2, 1965) shows the temperature of the Niagara River and follows the outflow of the river three miles into Lake Ontario before

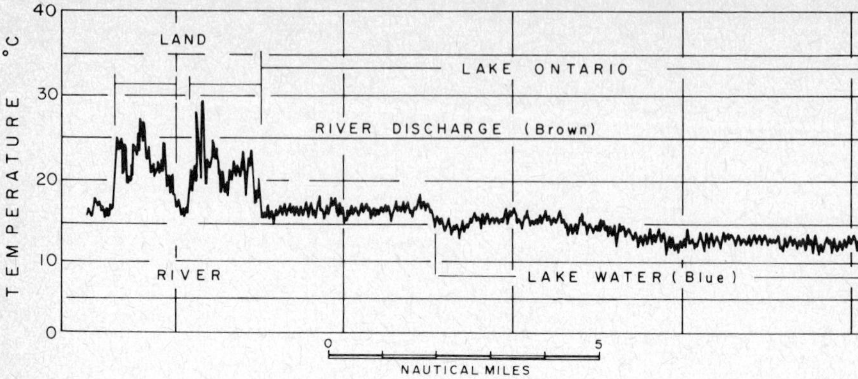

Figure 5. ART trace, Flight 2.6, September 2, 1965 from 1000 feet (300m.), showing warming in Lake Ontario at outflow of Niagara River.

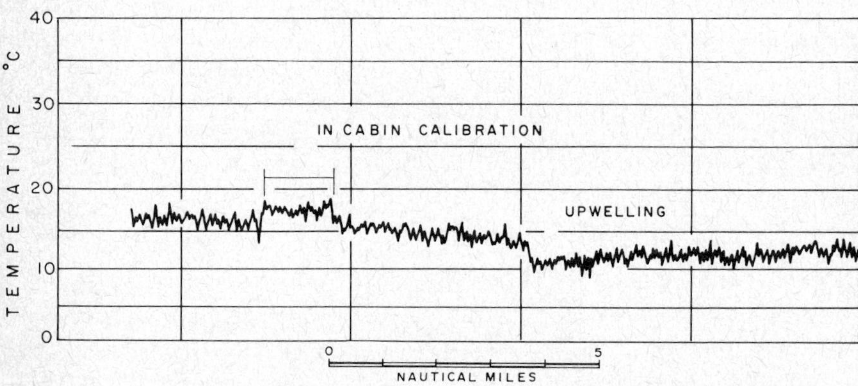

Figure 6. ART trace, Flight 2.6, September 2, 1965 from 1000 feet (300m.). Example of upwelling cold water, Lake Ontario.

it begins to mix with the colder water of the lake. A colour photograph taken on the flight confirms the pattern showing a sharp change in colour and texture of the water at the point of temperature change.

Figure 6 (Sept. 2, 1965) reveals a 7°C temperature change in north central Lake Ontario associated with a cold upwelling characteristic of the lake in the summer and autumn.

As a result of these and many other examples the ART, with a relative accuracy of ±0.5°C, has been proven to be an excellent technique for determining small scale horizontal gradients in surface water temperatures.

Environmental Errors

The second phase of the evaluation program was developed to assess the nature and size of the various factors contributing to the environmental error with the purpose of making corrections to the IRT readings based on observed or forecast meteorological conditions. As indicated earlier the ART technique is based on the measurement of the infra-red emissions from the surface of the water body by the sensor located on the aircraft. Since the emissions received by the sensor are not solely from the water surface and since they may also be affected by the atmosphere any investigation of the environmental errors inherent in the technique should be preceded by a consideration of the basic theories involved.

Blackbody Radiation

All objects in the universe whose temperature are above absolute zero emit radiation. The perfect or "blackbody" radiator has an emissivity of 1 with zero reflectivity and zero transmissivity. A pure water surface with a depth of 1 mm or more has a known transmissivity of zero and when viewed normally in the infra-red wavelength band of 8 μ to 14 μ has an average emissivity of .98 and a reflectivity of .02. When viewed at an angle of 60° from normal the emissivity becomes .96 and the reflectivity .04.

A spectral window of 8 μ to 14 μ was chosen for the IRT because it covers the band of maximum radiation and is that part of the spectrum where attenuation or absorption by water vapour and carbon dioxide is at a minimum (Figure 7).

Radiation Equation

In theory, the sensor receives infra-red emissions from the water surface and the IRT translates these to readings of surface water temperatures.

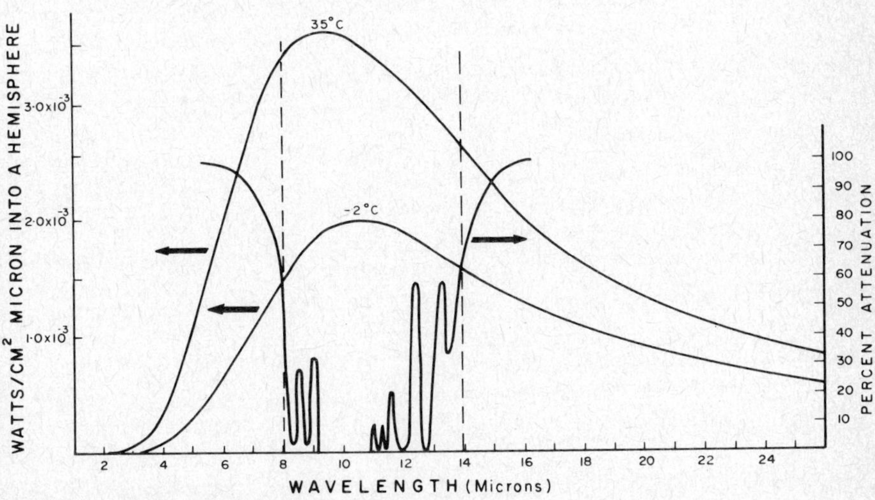

Figure 7. Blackbody distribution (Planck Law) for −2°C and +35°C and atmospheric attenuation of 1000 feet of atmosphere near sea level.

In fact, the emissions are partially attenuated by the layer of atmosphere between the aircraft and the water and at the same time are increased by re-emissions from the atmosphere and by reflected emission from the sky or cloud cover above the aircraft. To correct for the environmental error by theoretical considerations each contributing factor must be isolated and the total effect of all the factors must be assessed.

The following equation summarizes the major gains and losses affecting the radiation emission from the water surface before it reaches the IRT sensor.

tRadiation measured by IRT

  Radiation from water surface
+ Radiation reflected from the sky or cloud cover above aircraft.
+ Radiation re-emitted by atmosphere.
— Radiation attenuated by atmospheric water vapour.
— Radiation attenuated by carbon dioxide (and other gases).
— Radiation attenuated by aerosols.

(Equation 1)

These are represented schematically in Figure 8.

Computer Program

In his M.A. Thesis on the Environmental Errors in the Use of the Airborne Radiation Thermometer, Shaw[10] working with the Hydrometeorology Section developed a computer program for the IBM 7094 which calculates the total effect of the contributing factors in Equation 1, based on a technique and emission equations described by Godson.[11] The program divides the atmosphere into horizontal layers of 10 mb thickness from the aircraft to the water and back up to the top of the atmosphere or to an intervening cloud layer above the aircraft. In addition, the spectral window of 8 μ to 14 μ is divided into 18 bands of wavelength. The program then computes the contributions of each layer of atmosphere for each of the 18 bands of wavelength taking into account each of the contributing factors. The computer integrates these effects to evaluate the "true" infrared radiation from the water surface and hence the "true" temperature of the water surface.

Preliminary tests of the program found that the computer correction substantially improved the original or "raw" data and, in general the "corrected" temperatures were within 1°C of the surface water temperatures measured by conventional methods. However, since the attenuation and re-emission by the atmosphere below the aircraft is highly dependent on the water vapour content and the temperature of the atmosphere, the accuracy of the "corrected" temperature can be no better than the accuracy to which these meteorological parameters can be measured or forecast.

Based on the observations obtained on five special flights over the research vessel "Porte Dauphine" the program provided some preliminary estimates of the relative importance of each of the contributions noted in Equation 1. Radiation re-emitted from the gaseous atmosphere (including water vapour and carbon dioxide) varied from 10 to 25% of the total radiation received by the sensor and as indicated earlier is dependent on the water vapour content and temperature of the air. This gaseous re-emission is particularly affected by temperature changes in the air column below the aircraft—the colder the air column relative to the surface water the greater the negative error in the temperature readout and vice-versa. The effect of the reflected radiation from a clear sky varied from 0.1 to 0.6% of the total radiation. Overcast cloud above the aircraft increased the reflected radiation by 1.0 to 1.5% of the total received, i.e., an over-

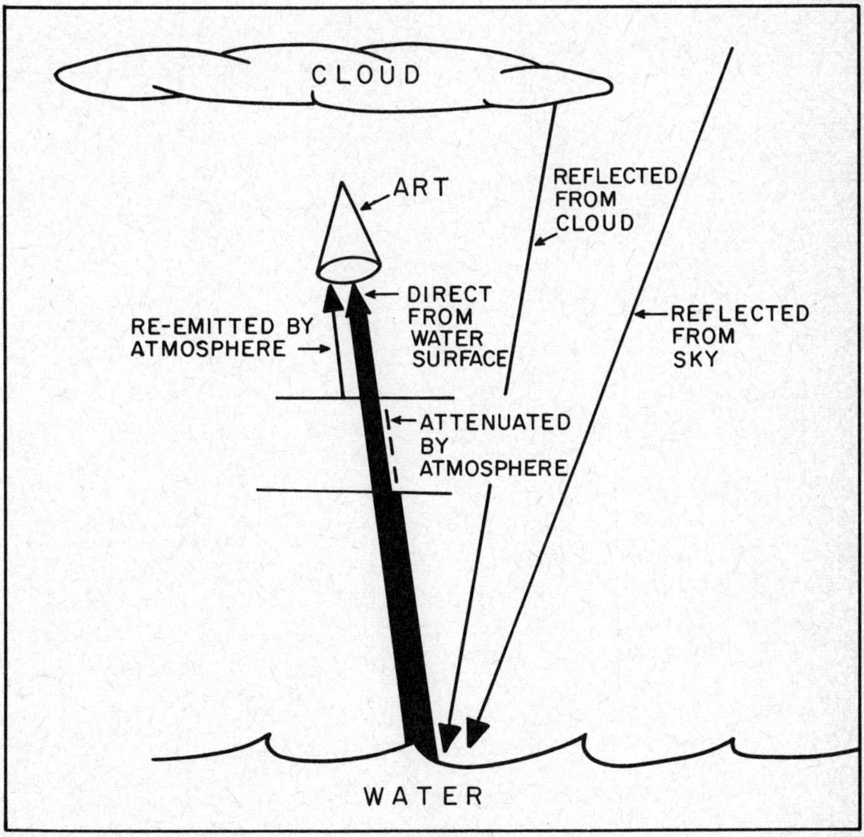

Figure 8. Schematic diagram of factors contributing to environmental errors in the ART technique.

cast sky can increase the uncorrected temperature by about 0.5°C. The program did not directly calculate the specific attenuation by atmospheric water vapour and carbon dioxide but by testing it with data representing extreme conditions it is apparent that the effects of gaseous attenuation must range from about 5-10% of the total radiation received from the water for dry air to 10-20% for relatively humid air. The attenuation by aerosols was not considered since it has been established that in visibilities of 6 miles (10 km) or better the contributions (negative or positive) would be a full order of magnitude smaller than the other contributions, (Deirmendjian).[12] Both of these latter effects are to receive further consideration by Shaw as he will remain with the Hydrometeorology Section for a limited period to develop simplified graphical techniques to estimate the environmental error from theoretical considerations.

Conclusions

Although the exposure of the sensor is critical, the airborne radiation thermometer (ART), when used under suitable meteorological conditions and with proper calibration checks, provides a good technique for making fast and reliable water temperature surveys of the Great Lakes.

The optimum (i.e. smallest) correction required appears to be at 500 feet (150 m) or lower, although, if necessary, flights at one to two thousand feet (300-600 m) appear feasible with proper calibration.

With the present instrumentation and using a correction factor obtained by flying over a known water temperature the technique provides an absolute accuracy of the order of 1.0°C with a relative accuracy within 0.5°C.

A computer program, based on theoretical considerations, has been developed which provides a "corrected" temperature that in preliminary tests substantially improved the original data, generally to within 1.0°C of the measured surface water temperature. However, since attenuation varies as the water vapour content of the atmosphere and re-emission varies as the temperature and water vapour content, the "corrected" temperature can only have the same order of accuracy as the observed or forecast meteorological conditions.

Although the computer program integrates the effects of the various factors influencing the total infra-red radiation received by the sensor it was also possible to make estimates of the relative values of each contribution (Equation 1) which has led to an improved understanding of the technique.

The experience gained with the instrumentation and the aircraft has been invaluable and results confirm the usefulness of the technique for both research and operational purposes in the Great Lakes area. As a result, monthly ART flights are being conducted in 1966 over Lakes

Ontario and Erie with less frequent flights planned for Lakes Huron and Superior. In addition, preliminary plans call for a program of monthly surveys of all the Great Lakes bordering on Canada in 1967.

References

[1] H. Stommel, "Rapid Aerial Survey of the Gulf Stream with Camera and Radiation Thermometer," *Science,* Vol. 117 (1953), pp. 639-40.

[2] W. S. Richardson and C. H. Wilkins, "An Airborne Radiation Thermometer," *Deep Sea Res.*, Vol. 5 (1957), pp. 62-71.

[3] M. Pirart, "Airborne Radiation Thermometer (FRB 1)," *Fish. Res. Bd.*, Canada, MRS no. 102 (1961).

[4] W. J. Snowden, "Report on the Flying of the Airborne Radiation Thermometer," Canada, Department of Transport, Meteorological Branch, Technical Circular Series CIR-3778, TEC-445 (1962).

[5] J. D. McFadden and R. A. Ragotzkie, "Aerial Mapping of Surface Temperature Pattern of Lake Michigan," Publication no. 10, Great Lakes Res. Div., University of Michigan (1963), pp. 55-58.

[6] "Techniques for infra-red Survey of Sea Temperatures," Report of a Workshop, April 27-28, 1964, U.S. Dept. of Interior, Bureau of Sport Fisheries and Wildlife, CIR. no. 202 (1964).

[7] R. W. Trites, "Some Aspects of the Airborne Radiation Thermometer Trial Conducted by the RCAF and AOG," Fish. Res. Bd., Canada, MRS no. 167 (1964).

[8] T. L. Richards and D. G. Massey, "An Evaluation of the Infra-red Thermometer as an Indicator of Surface Water Temperature," Canada, Dept. of Transport, Meteorological Branch, Technical Circular Series, CIR-4354, TEC-592 (1966).

[9] G. K. Rodgers, "The Thermal Bar in the Laurential Great Lakes," Publ. no. 13, Great Lakes Res. Div., University of Michigan (1965).

[10] R. W. Shaw, "Environmental Errors in the Use of the Airborne Radiation Thermometer," *M.A. Thesis* (Meteorology), University of Toronto (1966).

[11] W. L. Godson, "Radiation Notes," Research and Training Division, Meteorological Branch, Canada Dept. of Transport (unpublished) (1959).

[12] D. Deirmendjian, "Atmospheric Extinction of Infra-red Radiation," *Quarterly Journal, Royal Met. Soc.*, Vol. 86 (1960), pp. 371.

15

Correlating Freeze-up and Break-up with Weather Conditions

G. P. Williams

The first ice cover of freeze-up in early winter marks the end of water navigation and the end of the season when run-off from rainfall will increase the water supply in reservoirs. Spring break-up (defined as the complete clearing of ice covers from rivers, lakes, and harbours) marks not only the beginning of the navigation season, but also the time when depleted water reservoirs will be filled by spring run-off. Although the prediction of freeze-up and break-up would be of great value in planning construction, logging, navigation, and hydro-electric operations, it is only in very recent years that organized efforts have been made to assemble the basic information available in Canada on this problem (Meteorological Branch, DOT, 1959).

One approach to the prediction of free-up and break-up is through a statistical analysis of past records (MacKay, 1962a; Cote, 1950). With such an analysis it is possible to use past freeze-up or break-up records as a guide to what can be expected in the future. This approach has limitations, however, which can best be illustrated by an analysis of some available freeze-up and break-up records.

The results of statistical analysis of relatively long-term records from

Reprinted from *Can. Geotechnical J.*, Vol. 2, no. 4 (1965), pp. 313-26.

TABLE I. DIFFERENCE IN DAYS BETWEEN THE 10 PER CENT LEVEL EARLY AND LATE FREEZE-UP AND EARLY AND LATE BREAK-UP FOR SOME LAKES AND RIVERS

Body of water	*Years of record*	*Difference between dates of earliest and latest freeze-up*	*Difference between dates of earliest and latest break-up*	*Reference*
		(10 per cent levels—days)		
Charlottetown Harbour	37	22	37	Met Branch 1959
Christina L. BC	37	37	45	Met Branch 1959
L. Kenogami	31	25	20	Met Branch 1959
L. Mendota	107	28	31	Ragotzkie 1960
Port Arthur and Fort William	59	14	25	Met Branch 1959
South Bay (Coral Harbour)	29	16	22	Met Branch 1959
Miramichi R.	128	27	20	Met Branch 1959
North Saskatchewan R. (Edmonton)	56	20	21	Met Branch 1959
Coppermine R.	28	18	20	Met Branch 1959

several lakes and rivers are tabulated in Table I. The difference, in days, between the dates of earliest and latest freeze-up and between the dates of earliest and latest break-up are shown. The 10 per cent level (one year out of ten) was used to define these dates. The difference for freeze-up varies from 14 to 37 days for the rivers and lakes analysed; that for break-up varies from 20 to 45 days. This type of statistical analysis of past records enables one to estimate only that freeze-up or break-up will probably occur within a period of several weeks.

Freeze-up and break-up records seldom show a cyclic pattern from which future trends can be established. For example, an analysis of 106 years of break-up records for Lake Mendota, Wisconsin (Ragotzkie, 1960), shows that in 1880-81 freeze-up occurred on 23 November and break-up on 3 May, a duration of ice cover of 161 days and one of the longest on record for that lake. The next winter season, 1881-82, freeze-up occurred on 2 January and break-up on 21 March, a duration of ice cover of 78 days and one of the shortest on record.

Attempts have been made to improve on this statistical type of prediction by correlating meteorological elements with ice conditions so that weather forecasts can be translated into ice forecasts. The usual procedure is to correlate accumulated degree days (the differences between air temperature and a given base temperature for a given period) with freeze-up and break-up dates. Stankiewicz (1947) used this degree-day technique for estimating from air temperature records the date when a river would be open for driving logs. Mackay (1961), in a study of freeze-up and break-up of the lower Mackenzie River, related a weighted air temperature, based on the estimated flow time of the river, to freeze-up. Burbidge and Lauder (1957), in a preliminary investigation of freeze-up and break-up, presented some relationships between degree days and freeze-up and break-up. More recently, Bilello (1964) presented an improved technique for forecasting river and lake ice formation using air temperature records, and MacKay (1962b) studied in some detail the factors that affect freeze-up and break-up on the Nelson River drainage system. The use of weather information for predicting freeze-up and break-up has limitations, however, which are considered in this paper.

In 1956 observations on initial ice formation, on the rate of ice growth in winter, and on the rate of ice in spring were initiated at McKay Lake, a small, sheltered lake located within the city of Ottawa. Ice thickness observations were used, along with ice thickness records from other lakes, in a statistical analysis of ice growth rates and maximum ice thicknesses (Williams, 1963b). In 1959 observations on water temperature in the fall cooling period were included in this program. Heat losses determined from these temperature measurements were shown to be approximately proportional to the difference in temperature between the water surface and the air, indicating that the energy gained by solar radiation equals that lost by evaporation (Williams, 1963a). These water temperature observations, together with observations on initial ice formation and ice melt, are analysed and used in this paper to establish relationships between freeze-up and break-up and weather.

Environmental Factors Affecting Freeze-up and Break-up

Some background information on the environmental factors that affect freeze-up and break-up may be useful.

The formation of a solid ice cover at freeze-up is determined by the following factors: heat exchange with the atmosphere, heat stored in a water body, and inflow of warm or more saline water to a site.

1. Atmospheric heat exchange (the rate of heat loss from the water surface to the surrounding air) is determined by the weather conditions that prevail during the period prior to freeze-up. Air temperature, wind veloc-

ity, and solar radiation are the most important weather elements affecting this rate of heat loss.

2. The heat stored in a water body is a function of water depth. In general, the deeper the lake, the deeper the convective mixing and the slower the rate of cooling for a given surface heat loss. In rivers the water is kept at nearly the same temperature by turbulent mixing, but the deeper the river the slower the rate of cooling for a given rate of heat loss.

3. Inflow of heat from upstream sources such as water flowing from a deep reservoir can delay or prevent sheet ice formation on a river. In salt-water harbours inflow of more saline water because of density currents can also retard the formation of an ice cover.

The clearing of ice from bodies of water at break-up is affected by the following factors: heat gain from the atmosphere, snow and ice conditions, wind and currents, and inflow of warm or more saline water to a site.

1. The heat gained from the atmosphere, principally from solar radiation, weakens and melts the ice; the reflectivity of the snow or ice surface determines the amount of solar radiation absorbed.

2. The thickness of the snow and ice determines the amount of ice to be melted, and this affects the duration of the break-up period.

3. Of great importance is the mechanical action of wind and currents in breaking up the ice and mixing it with warmer sub-surface water.

4. In salt-water harbours wind and current can break up the ice so that it is mixed with more saline water. Run-off water also from surrounding land areas can accelerate break-up, as can the flow from southern portions of a northward-flowing river affect the break-up of the northern section.

The mechanical breaking of solid cover by wind and currents can sometimes be more important in the break-up process than the heat gained from the atmosphere. To illustrate the difference between break-up due almost entirely to heat gained from the atmosphere and break-up resulting from wind or currents, some available break-up records for several lakes, rivers, and salt-water harbours in Canada were analysed (Meteorological Branch, DOT, 1959). The number of days from the time the ice thickness was last measured in the spring to the time when it was completely cleared was related to the total ice thickness at the time of the last measurement (Figure 1). The comparison shows the great variation in the rate at which ice disappears from different bodies of water. The lower limit, where the ice melt is due entirely to the heat from the atmosphere, is about 1 in./day. This is less than one tenth the upper limit where the ice is cleared from a river or harbour largely by wind and current effects.

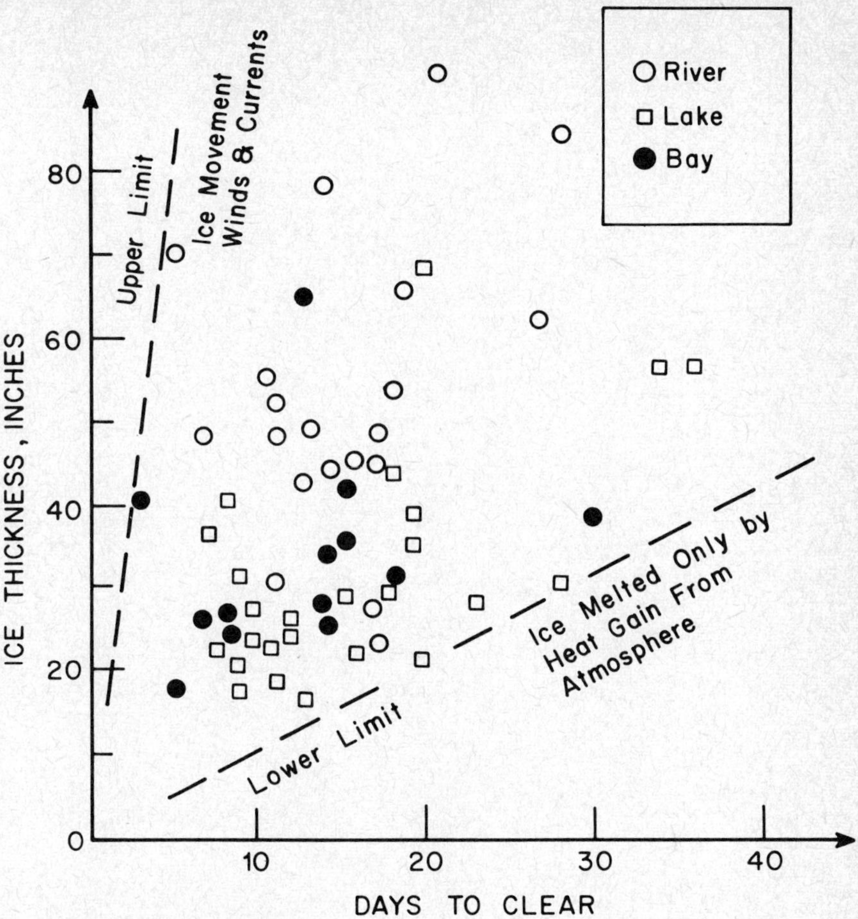

Figure 1. Comparison of number of days for ice to disappear with ice thickness at time of last measurement.

Ice break-up caused only by heat from the atmosphere will probably occur for a relatively small percentage of lakes, and harbours. Even for quite small lakes, wind effects can be appreciable. After observing ice break-up for several years on lakes in New Brunswick, Smith (1951) stated: "With strong winds the ice disappeared simultaneously from several small lakes, but with calm weather, such as in 1943, a number of days may intervene between their break-up dates."

The factors that determine when an ice cover will form on a water body or when it will melt, weaken, and disappear are complicated and interrelated. Of all these factors, however, the rate at which heat is gained or lost at the surface because of prevailing weather conditions is usually dominant. Even when wind and currents accelerate break-up by mechanically breaking the solid cover the ice has often been weakened previously by mild weather. A large part of forecasting freeze-up or break-up dates at any site is thus a problem of relating weather conditions to ice conditions. It is for this reason that the relation between freeze-up and break-up and weather conditions was investigated on a sheltered lake where atmospheric heat exchange was dominant and effects of wind and current were minor.

McKay Lake Freeze-up

Figure 2 shows typical water temperature profiles at McKay Lake from the beginning of the fall cooling period in 1961 until freeze-up. The profiles obtained for other years were similar, although there were some differences in the thickness of the thermocline at any particular time. After the water became isothermal at 39 to 40°F, the surface water remained at the surface and quickly cooled to 32°F. With further cooling sheet ice started to form.

The observations show that the cooling of a fresh-water lake is divided into two stages: gradual cooling until all the water is at a temperature of about 39 °F, and cooling of a shallow layer of surface water from the time the water is isothermal at 39 °F until sheet ice forms. Predicting

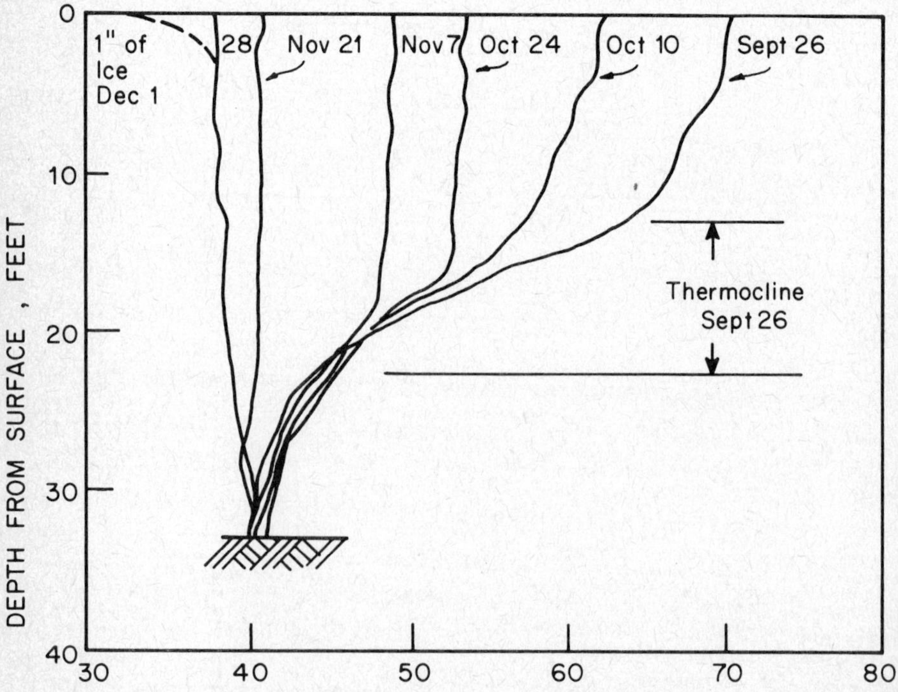

Figure 2. Typical water temperature profiles, McKay Lake, 1961.

freeze-up dates is a two-fold problem: to predict when the water will become isothermal at 39 °F, and when ice will form after the water temperature has reached 39 °F.

These two aspects can be illustrated by plotting changes in surface water temperature during several fall cooling periods (Figure 3). While the lake is cooling to 39 °F the rate of change of the water surface temperature parellels the rate of change of mean air temperature. During this period the lake has a large thermal inertia and surface heat loss cools it to a considerable depth. In McKay Lake the effects of cooling extend 15 to 20 feet below the surface and it thus takes considerable surface cooling to change the surface temperature appreciably. When the lake water becomes isothermal at 39 °F, however, surface heat loss affects only a thin surface layer of water so that surface temperature is much more sensitive to the weather and fluctuations in it. Surface water may be quickly cooled to 32 °F and ice will form if cooling continues. This ice may melt again or freeze into a permanent cover, depending on the weather conditions.

The sensitivity of surface ice formation to variable weather conditions during this final stage of cooling is one reason why freeze-up dates fluctuate within rather wide limits (Table I). In 1962 the duration of the second stage of cooling was 28 days (14 November to 13 December); in 1963 it lasted only 5 days (30 November to 4 December).

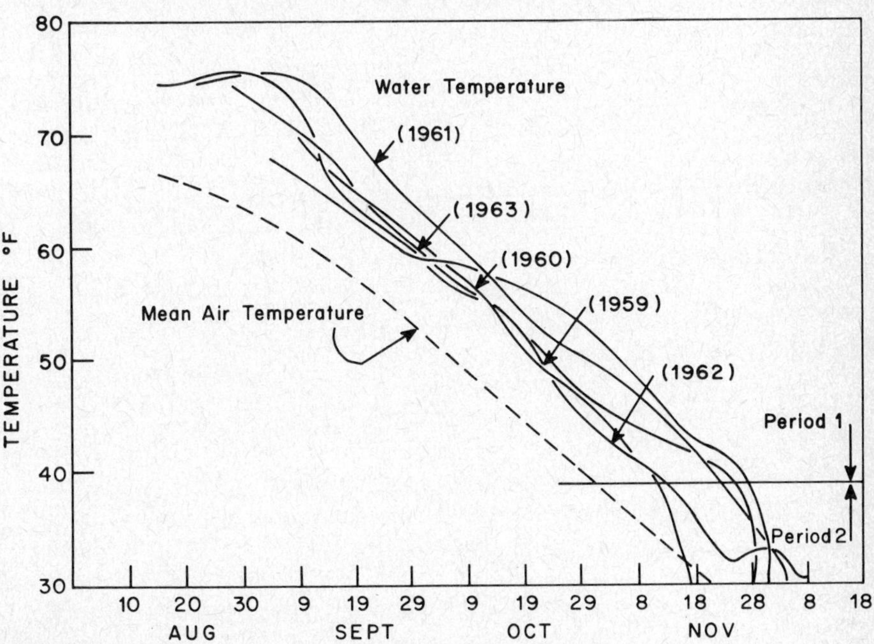

Figure 3. Surface water temperature, McKay Lake.

Because surface water temperature change parallels air temperature change during the first stage of cooling, it should be possible to establish a relationship between air and water temperature. Such a relationship has been established for sea water where, because the maximum density of salt water depends on salt concentration, convective mixing is maintained until it cools to the freezing point, and there is generally no second stage of cooling. For example, Rodhe (1952) used a relationship between mean daily air temperature and water temperature for predicting ice formation in the Baltic.

The mathematical expression, based on the assumption that the rate of heat loss from a water surface depends on air-water temperature difference, is:

$$h_0 A(T_W - T_A)dt = ADsdT_W \qquad (1)$$

where T_W = surface water temperature, which remains essentially constant to depth D (depth of convective mixing),
T_A = *air temperature,*
A = area of water,
s = specific heat of water,
h_0 = coefficient of heat transfer from a water surface.

This equation can be reduced to that used by Rodhe by letting $K = h^0/Ds$, which can for practical purposes be treated as a constant with the inverse dimensions of time. Equation (1) then becomes:

$$dT_W/dt = K(T_W - T_A) \qquad (2)$$

Bilello (1961) used the technique developed by Rodhe to compute from daily air temperatures the date of ice formation at several locations in the Arctic. A value of K^{-1} was found by trial and error for each location. Once K was determined, it was possible to make a reasonably accurate estimate of the date of ice formation from daily air temperature records.

Using the computation procedure outlined by Bilello, K^{-1} was calculated to be 17.5 days for the period 1 October to 28 November 1961 for McKay Lake. With this value of K^{-1} mean air temperatures were used to compute synthetic water temperatures for the first stage of fall cooling for McKay Lake for the years 1959, 1960, 1962, 1963. Figure 4 shows the result of plotting the calculated water temperature against the measured water temperature. By this procedure the date when McKay Lake would become isothermal at 39 °F (obtained from synthetic water temperatures) agreed with the actual date within ±3 days, an error about the same as that obtained by Bilello for estimating freeze-up dates in the Arctic.

The relationship between air temperature and surface water temperature cannot be applied after a sheltered fresh-water lake becomes isothermal at 39 °F because the water surface temperature is much more

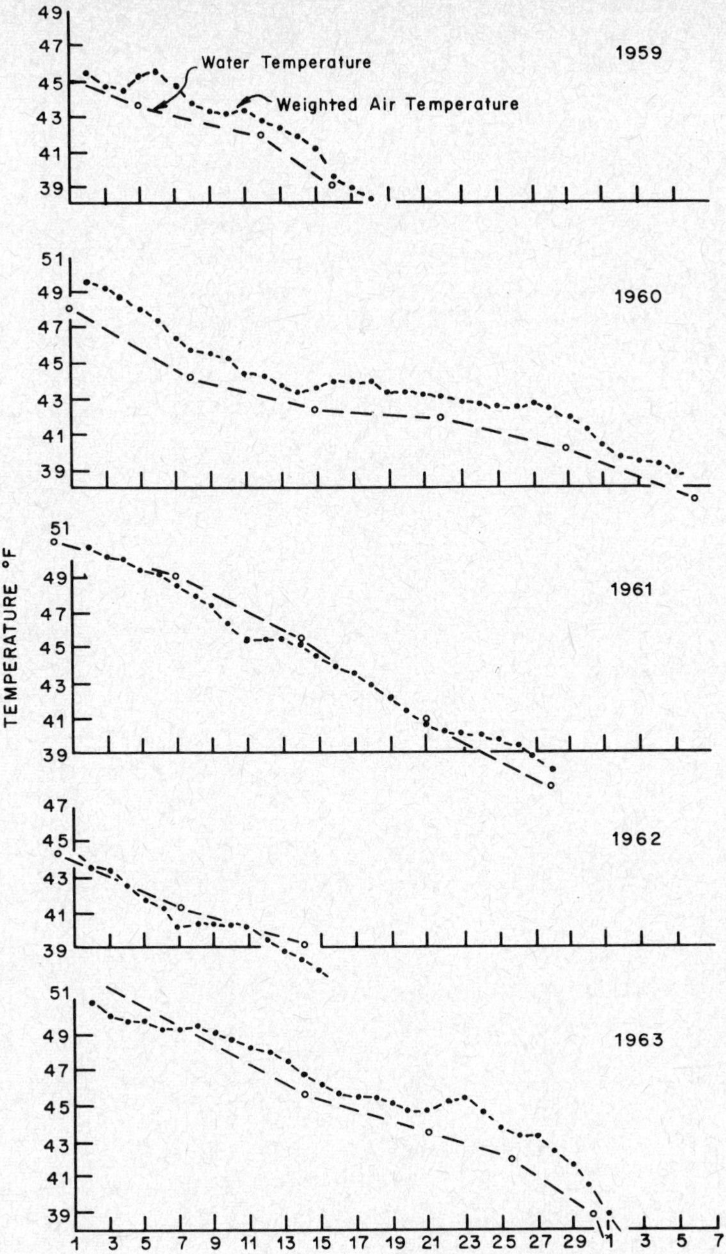

Figure 4. Relation between water and weighted air temperature, McKay Lake, 1959-63.

sensitive to the weather and may be quickly cooled to 32 °F. During this second cooling stage it was found that accumulated degree days, using 32 °F as a reference temperature, could be correlated with the time for a solid ice cover to form on the lake. The basis for this relationship is that during this final cooling stage the lake-water surface temperature is close to 32 °F, and the rate of heat loss is approximately proportional to $(T_A - 32)$ °F.

Accumulated degree days for McKay Lake were calculated for 1959 to 1963 from the time the lake water was isothermal at 39 °F until 4 to 6 in. of ice had formed. A solid cover of 4 to 6 in. was used as a reference because this thickness of ice will generally not melt during subsequent mild weather. Figure 5 shows that the accumulated degree days during this second stage of lake cooling varied from 70 to 100 for the five years studied. After the lake water became isothermal at 32 °F the first sustained period of cold weather, usually from 3 to 5 days, would result in an ice growth of 4 to 6 in. By using this relationship the time taken for a solid ice cover of 4 to 6 in. to form could be estimated from air temperature records to within a day or two.

McKay Lake Break-up Relationships

Predicting the date when a sheltered lake will be completely free of ice is somewhat different from predicting when a solid ice cover will first form. Prediction of freeze-up depends almost entirely on prediction of weather conditions, but prediction of break-up depends not only on weather

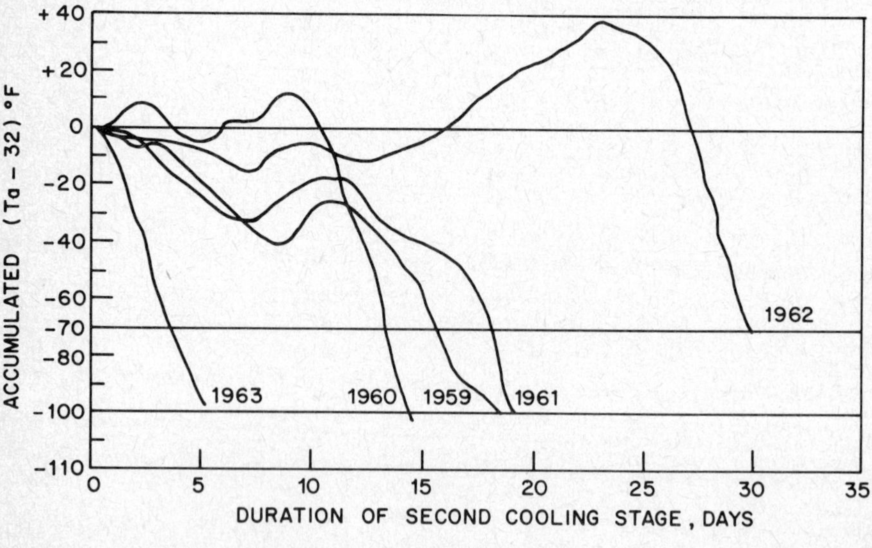

Figure 5. Accumulated degree days during period from when water temperature is 39°F until 4 to 6 in. of ice forms, McKay Lake.

conditions but also on the thickness of ice to be melted and the condition of the ice surface, including the depth of snow cover. In the fall the rate of heat loss from open water can be related to the air-water temperature difference because solar radiation approximately equals evaporation. During the spring melt period the heat from solar radiation is usually much greater than evaporative heat loss and, consequently, calculations of the heat gained at melting ice surfaces must include heat gained from solar radiation.

In a previous study on the use of dust to advance the break-up of ice on lakes and rivers (Williams and Gold, 1963) a simplified heat balance equation was used for estimating the heat gained at a melting ice surface. A modified form of this equation is:

$$(3)\quad Qm = [\{(100 - \alpha)\, 3.69\ \mathrm{Rsw}\}/100] + 85\, \Delta e + 37\, \Delta T \ \text{Btu/sq. ft./24 hr}$$

where

Qm = net heat gain at melting ice surface,

α = albedo or reflectivity of surface (per cent),

$\Delta e = (e_s - e_a)$, e_s equals saturated vapour pressure of air at temperature of surface and e_a equals vapour pressure of air (mbs),

$\Delta T = (Ts - Ta)$. Ts temperature of surface, and Ta equals temperature of air (°F),

Rsw = Incoming short-wave radiation (cal./sq. cm./24 hr).

Equation (3) was used to calculate the heat available for melting ice in each spring melt period at McKay Lake for the ten years from 1952 to 1961. In making the calculations the following assumptions were made, based on field observations in the Ottawa area from 1956 to 1963:

1. If the daily air temperature was above freezing the surface temperature was assumed equal to 32 °F.
2. If the daily air temperature was below freezing, the surface temperature was assumed equal to the daily air temperature.
3. Values of albedo or reflectivity of the melting surface were assumed as follows:

(*a*) if air temperature was above 32°F, albedo equalled 50 per cent;
(*b*) if rain occurred, the albedo was assumed to be 40 per cent for the two days following the rain, then 50 per cent if air temperature was above 32 °F;
(*c*) if new snow occurred, the albedo was assumed to be 80 per cent on the first day after the snowfall, 70 per cent after the second day, and 50 per cent if air temperature was 32 °F by the third day;

(*d*) in the late stages of melting, if the air temperature was above 40 °F for several days, the albedo was assumed to equal 40 per cent.
4. The beginning of the effective melt period was assumed to start when the five-day average air temperature was above 35 °F.
5. The end of the melting period was taken as the time when ice completely melted from the lake.

A check on the accuracy of the calculated total heat loss was obtained by comparing it with the heat required to melt the thickness of ice recorded at McKay Lake at the beginning of the break-up periods (prior to 1956 maximum ice thickness was estimated from ice thickness records on other lakes in the Ottawa area). The calculated heat is compared with the heat required to melt the observed thickness of ice in Table II. Considering the many assumptions made, the agreement is surprisingly good.

TABLE II. COMPARISON OF CALCULATED HEAT GAIN TO HEAT REQUIRED TO MELT ICE BREAK-UP PERIOD 1952-61, MCKAY LAKE

Year	*Calculated heat (Btu)*	*Heat required melt ice (Btu)*	*Per cent difference*
1952	14,700	14,200*	−3.5
1953	12,900	12,700*	−1.5
1954	14,900	14,900*	0
1955	22,600	25,400*	+11.0
1956	16,200	17,200	+6.0
1957	14,600	14,200	−2.5
1958	13,500	14,200	+5.0
1959	15,700	17,200	+9.0
1960	18,400	20,200	+9.0
1961	19,000	21,000	+9.5

*Estimated.

To show the relative importance of the different components of the heat exchange, these are tabulated in Table III as percentages of the total calculated melt. The importance of each component varied considerably from year to year. The short-wave radiative heat gain varied from 65 to 109 per cent, the evaporative heat loss from 2 to 29 per cent, and the combined long-wave and convective heat gain from 20 to 40 per cent of the total melt. It is considered that these percentages are representative of the relative importance of the different components of the heat balance in melting the ice cover of a sheltered lake in the Ottawa area.

These calculations illustrate the importance of solar radiation in the ice melting process. Because the albedo of the ice surface is highly variable, it will always be difficult to estimate accurately how much incoming solar

TABLE III. CALCULATED PERCENTAGE OF TOTAL MELT CAUSED BY DIFFERENT COMPONENTS OF HEAT EXCHANGE BREAK-UP PERIOD 1952-61, MCKAY LAKE

	Percentage of Total Melt		
Year	*Short-wave radiation*	*Long-wave radiation and convection*	*Evaporation*
1952	+86	+25	−11
1953	+65	+40	−5
1954	+67	+35	−2
1955	+77	+28	−5
1956	+81	+30	−11
1957	+109	+20	−29
1958	+92	+25	−17
1959	+82	+28	−10
1960	+84	+23	−7
1961	+77	+32	−9

energy goes to melt ice, unless actual measurements of albedo are made at several representative locations on a lake.

Development of a Simplified Ice-Melting Formula

As an outcome of the preceding analysis a simplified exchange formula was developed combining a solar radiation term with a degree-day term:

$$(4)\quad Qm = 0.5\ \Sigma \mathrm{Rsw} + 10\ \Sigma \Delta T \text{ cal./sq. cm.}$$

where

Qm = heat available for melting lake ice,

ΣRsw = accumulated short-wave radiation over break-up period (cal./sq. cm.),

$\Sigma \Delta T = (Ta - 32)$ accumulated degree days over break-up period (°F).

This formula assumes that the albedo of the ice cover equals 50 per cent during the entire break-up period. The term 10 ΔT in equation (4) was derived from the combined long-wave radiation and convection equations previously developed in equation (3). The evaporation term is not included in this simplified formula. The units (°F and cal./sq. cm.) are used because air temperature and solar radiation are normally recorded at Canadian meteorological stations in these units.

In using this formula the active melt period is assumed to begin when the five-day average air temperature is 35 °F or above. (Thus degree-day accumulations and solar radiation accumulations are started at this time.) This arbitrary rule has some physical basis as, judging from the

McKay Lake observations, unless average air temperature is well above 32°F, melting of lake ice during the day is offset by freezing during the night.

The formula was checked by comparing the melt estimated from the reported decrease in ice thickness at several Canadian stations with the melt calculated by formula. Only those periods during which decrease of ice thickness was estimated to be entirely due to heat exchange from the atmosphere were considered; in most cases the period of calculation was for only part of the melt period. Solar radiation and air temperature records used in the formula were obtained from meteorological stations located near where ice thickness observations are made. Unfortunately, there are few stations in Canada where both radiation and ice thickness records

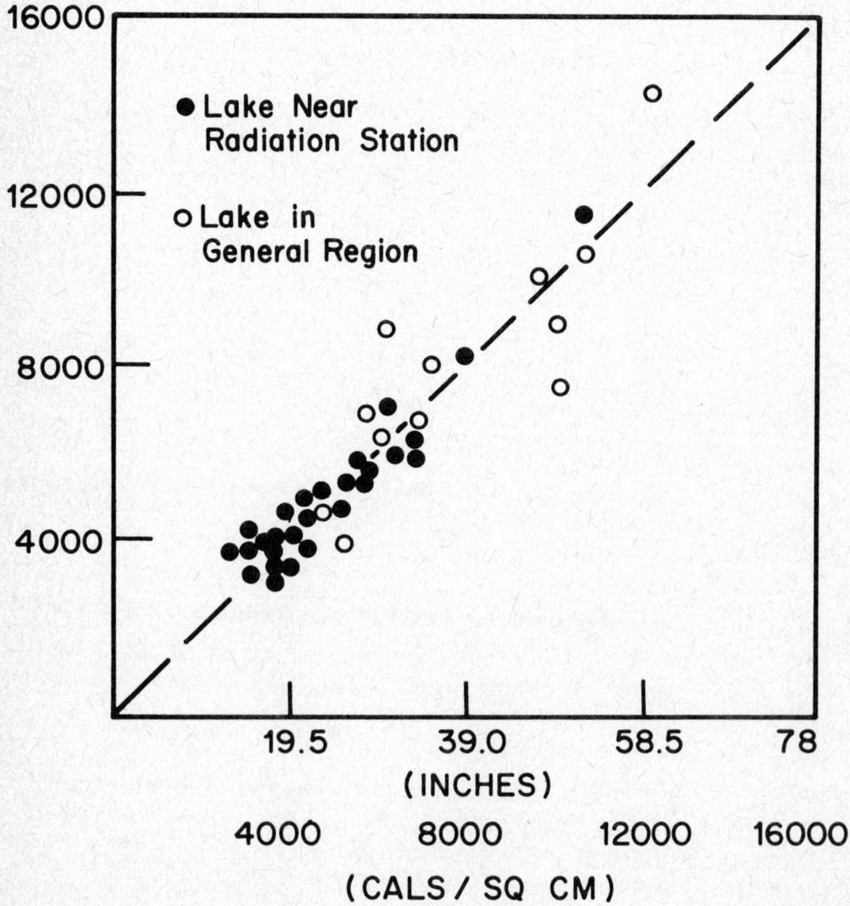

Figure 6. Heat required to melt ice *vs.* calculated heat.

are maintained. Those available, usually with quite short records, were Knob Lake, Churchill, Resolute, Moncton (McLaughlin Reservoir), and Ottawa (McKay Lake). In addition, radiation observations at Churchill were used to calculate the rate of melting at Brochet and Ennadai Lakes, two fresh-water lakes several hundred miles west of Churchill.

It was expected that the computed melt for Brochet and Ennadai Lakes would not be very satisfactory. The results of the analysis (Figure 6) indicate, however, that this simplified formula gives a reasonable estimate of the rate of melting of ice covers on sheltered lakes. Its accuracy will depend on how closely the necessary assumptions, such as that concerning the albedo, fit the actual situation at a particular lake. The accuracy of the formula should be greater the closer the radiation observation site is to the lake.

It should be emphasized that estimating the rate of ice melt with equation (4) is only part of the problem of estimating the date of break-up. For exposed lakes and rivers the influence of wind and currents must also be considered. There is some indication that break-up of ice on a river (Stankiewicz, 1947) or of ice on a large exposed lake (Richards, 1963) will occur after a specific amount of melting, and that equation (4) or some modification of it can be used to estimate the date when a river or large lake will clear. Studies would have to be made, however, at individual sites for a number of years before the value of such formulae for estimating break-up dates could be established.

Conclusions

The relation between freeze-up, break-up, and weather conditions was investigated for five winter seasons at a small fresh-water lake sheltered from wind action where heat exchange with the atmosphere is the dominant factor affecting ice formation at freeze-up and melting at break-up. The following conclusions were obtained:

1. Freeze-up occurs in two stages. During the first stage there is gradual cooling of water to a depth of 15 to 20 ft. below the surface until all the water is isothermal at a temperature of about 39 °F. During the second stage, from the time the water is isothermal at 39 °F until sheet ice forms, only a thin layer of surface water is cooled and thus surface temperature is much more sensitive to fluctuations in the weather.

2. During the first stage of cooling, air temperature can be correlated with surface water temperature. The rate of change of water temperature dT_W/dt can be equated with $K(T_W - T_A)$ where K is a constant and $T_W - T_A$ is air-water temperature difference. Using this relationship the date when a sheltered lake becomes isothermal at 39 °F can be calculated from air temperature records.

3. During the second stage of cooling accumulated degree-days (daily air temperature minus 32 °F) can be correlated with the time taken for a solid ice cover to form. Using this correlation the time taken for 4 to 6 in. of ice to form can be calculated from air temperature records.

4. Observations taken on the small fresh-water lake show that the rate at which ice melts during the spring break-up period can be estimated from the heat exchange formula

$$Qm = 0.5\,\Sigma \mathrm{Rsw} + 10\,\Sigma \Delta T$$

where Qm = the heat available for melting lake ice,
ΣRsw = the heat available from solar radiation accumulated over the break up period,
$\Sigma \Delta T$ = accumulated degree days over the break-up period.

The formula can be used to estimate the time taken for ice to melt from a lake, provided wind or currents do not acclerate the ice break-up process.

5. The usefulness of the relationships developed will depend not only on the accuracy of radiation and air temperature forecasts, but also on the length of period in the future for which these forecasts can be made. The real challenge of predicting freeze-up and break-up dates is thus a part of the more general challenge of improving the accuracy of short-term and long-term (30-day) weather forecasts.

References

Bilello, M. A., "Formation, growth and decay of sea-ice in the Canadian arctic archipelago," *Arctic*, Vol. 14, no. 1 (1961), pp. 2-25.

Bilello, M. A. "Method for predicting river and lake ice formation," *J. Applied Meteorology*, Vol. 3, no. 38 (1964), pp. 38-44.

Burbidge, F. F., and Launder, J. R., "A preliminary investigation into break-up and freeze-up conditions in Canada," Ottawa, Dept. of Transport, Meteorological Branch, Circular 2939 (1957).

Cote, J., *Etude sur la débâcle*, Province of Quebec, Meteorological Bureau, Bull. 17 (1950).

MacKay, D. K., *River ice conditions in the Nelson river drainage system*, Ottawa, Dept. Mines and Technical Surveys, Geographical Branch Paper 34 (1962).

MacKay, D. K., *Trends and factors affecting break-up and freeze-up dates in the Nelson river drainage system,* Ottawa, Dept. Mines and Technical Surveys, Geographical Branch Paper 35 (1962).

MacKay, J. R., "Freeze-up and break-up of the lower Mackenzie river, Northwest Territories," G. O. Raasch, ed., *Geology of the Arctic* (Toronto, University of Toronto Press, 1961), pp. 1119-34.

Meteorological Branch, Dept. of Transport, *Break-up and freeze-up dates of rivers and lakes in Canada*, Circ. 3156 (1959).

Ragotzkie, R. A., *Compilation of freezing and thawing dates for lakes in north central United States and Canada*, Dept. of Meteorology, University of Wisconsin, Tech. Rept. 3 (1960).

Richards, T. L., "Meteorological factors affecting ice cover on the Great Lakes," University of Michigan, Great Lakes Research Division Publication 10 (1963), pp. 204-15.

Rodhe, B., "On the relation between air temperature and ice formation in the Baltic," *Geografiska Annaler*, Vol. 34, nos. 3-4, (1952), pp. 175-202.

Smith, M. W., "Limnology and trout angling in Charlotte County lakes," *Journal of Fisheries*, New Brunswick Research Board (1951).

Stankiewicz, M. J., "Break-up can be foretold," *Pulp and Paper Magazine,* Vol. 48 (1947), pp. 118-20.

Williams, G. P., "Heat transfer coefficient for natural water surfaces," *Proc. Int. Assoc. Sci. Hydrology*, IUGG, Publication 62 (1963), pp. 203-12.

Williams, G. P., "Probability charts for predicting ice thickness," *Eng. J.,* Vol. 46, no. 6 (1963), pp. 31-35.

Williams, G. P., and Gold, L. W., "The use of dust to advance the break-up of ice on lakes and rivers. Canada," *Proceedings of Eastern Snow Conference* (1963), pp. 31-56. Reprinted as NRC 7725.

16

Instrumentation for Study of Ice Formation on the Niagara River

H. F. Cork

R. C. Chapil

Introduction

The Meteorological Branch of the Department of Transport and the Hydro-Electric Power Commission of Ontario are co-operating in a study of ice formation on the Niagara River as an International Hydrologic Decade Project. A study of the energy balance of the river above the Falls for conditions of potential and actual ice formation and dissipation will be made. The objective is to establish the relation of the timing and amount of ice formation in the river to meteorological conditions and the temperature of water entering the river from Lake Erie.

The Niagara River ice study is being undertaken over a river reach of sixteen miles, starting approximately one half mile southwest of the International Peace Bridge at Fort Erie and ending at the Ontario Hydro control dam one mile upstream from the Horseshoe Falls.

The Problem

Large quantities of ice are manufactured by Lake Erie and the upper Niagara River. The ice which moves down the river must be carried

---

Reprinted from International Hydrological Decade Progress Report No. 1 (1966), Dept. of Transport, Meteorological Branch Circular 4496, Tec. 627.

past the power intakes, over the Falls and past the power houses without jamming.

In the past, severe jams have occurred even with the full river flow available for passing ice.[1] The problem is now much more severe because of large diversions being made for power purposes. Under present regulations by the International Niagara River Board of Control, 50,000 cfs (scenic water) must go over the Falls in the winter. This is about 30 per cent of winter flow; the remaining 70 per cent may be diverted for power production.

It is desirable to carry the ice past the power installations using only the scenic water, so that all the remaining water will be available for power. When the amount of ice to be moved is very large, the scenic flow may not be adequate and diversions for power must be reduced to augment the amount of water available for transporting ice.

There are three important sources of ice in the Upper Niagara River, indicated by the terms frazil ice, anchor ice and lake ice.

When a small lake or slowly flowing river is cooled in the fall and early winter, convective currents are set up, and at one stage the whole body of water becomes isothermal with depth at 39°F, the temperature of maximum density for water. With further cooling, the colder water stays on top and an ice layer forms, the insulating effects of which reduces further ice formation.

However, under conditions of rapid flow (greater than about 1½ feet per second) mixing occurs and water temperatures throughout the whole depth of the river may be reduced to 32°F. Super-cooling of a few hundredths of a degree takes place, and then a dynamic form of ice begins to appear.

At first this will consist of separate circular disks,[2] usually less than 0.05 mm. thick and less than 2 mm. in diameter. The concentration may be very high, approaching one million per cubic meter, giving the water a milky appearance. Next, needle-like structures grow out from the disks, and the resulting irregular particles rapidly clump together in spongy masses which resemble cinders if they are lifted from the river and the water allowed to drain out. This form of ice is usually called "frazil" from the French "fraisil" for "cinders".

Conditions necessary for the formation of frazil include:

(1) a section of the river with sufficiently rapid flow that turbulence prevents the formation of an insulating layer of ice at the top.

---

[1] D. M. Foulds, "Niagara Power Versus the Ice Machine," (CEP Montreal, The Hydro-Electric Power Commission of Ontario, 1964).

[2] G. P. Williams, 'Frazil Ice: A Review of its Properties with a Selected Bibliography," *Eng. Journal*, Vol 42, no. 11 (1959), pp. 55-60.

(2) water entering the section of rapids at a temperature close to freezing, and
(3) severe cooling conditions with air temperature near 0°F or lower, sufficient wind to prevent formation of a heat island over the river, and little or no heat gain by radiant exchange.

Frazil ice will be "manufactured" continuously as long as suitable cooling conditions continue. The amount formed may be very large. At night with water temperature 32°F, and air temperature 0°F, a section of rapids 1000 feet long by 100 feet wide can produce frazil at the rate of 50 tons per hour.

Occasionally, further down river, pans of frozen slush may jam and form an ice bridge across the river. Extensive deposition of the frazil on the bottom surface of the bridge may occur, partly blocking the flow, resulting in a large increase in stage with perhaps extensive flooding.

At night, under strong cooling conditions "anchor" ice may form on the bottom of shallow rapids, raising the stage of the river and perhaps restricting flow out of the lake. If the sun provides sufficient radiation on a day following extensive anchor ice formation, large quantities will break loose and float to the top, often carrying small boulders and mud from the bottom. The surface of the river may be nearly covered in a short time with this type of ice. The cycle of formation and release of anchor ice causes important fluctuations in water available for power production.

Jams which have caused serious loss of power in the past resulted from complex situations which differed considerably from one occasion to the next. It is hoped that a detailed study of meteorological parameters may lead to better prediction of the quantity of ice formation, reducing, on the one hand, risk of ice jams which could lower power production for several months, and avoiding, on the other hand, use of unnecessarily large quantities of power water to move ice past the power installations. It should also improve predictions of flow fluctuations under anchor ice conditions.

Instrumentation

Radiation

The C.S.I.R.O. Net Pyrradiometer (Funk) used in (this study measures the vertical flux of solar plus terrestrial radiation and is mounted approximately five feet above the surface of the Niagara River (Figure 1). The essential parts of the net pyrradiometer are a pair of black horizontal surfaces, one facing upward and the other downward, with thermopiles for measuring the differences in temperature between these surfaces. This instrument is linked to a Honeywell strip chart recorder with a millivolt output of minus ten to plus forty.

One net pyrradiometer was installed on Pier No. 2 at the Niagara River Control Dam on March 9, 1966 (Figure 1). A framework of channel iron was bolted to the concrete pier, with a hinged twelve foot aluminum boom extending over the water. A system of two universals, counter-balanced by weights, provided a simple, self-leveling mounting for the net pyrradiometer. A mechanical reel was constructed making it possible to raise and lower the boom to service the sensor and adjust its height above water.

A shielded signal wire is connected to a Honeywell strip chart recorder located in the International Niagara River Control Center operated by Ontario Hydro. Honeywell recorders equipped with Libascope integrators and magnetic tape pulse recorders have been ordered, and it is anticipated these will be available by fall, 1966. With use of the magnetic tape recording system an electric pulse will be recorded with each make and break of a micro-switch making it possible to have an integrated value of radiation in each five-minute period, which can be tabulated and analyzed directly by a computer. The radiation data at present, are hand-scaled and the results expressed in Langleys per hour.

Figure 1. Net pyrradiometer at the Niagara River control dam.

The second pyrradiometer will be installed during the fall of 1966 nearer Lake Erie over a part of the river where no fixed shore ice forms during the winter. The operation of the pyrradiometer will be from December to April inclusive, which is the critical period of ice formation and dissipation on the Niagara River.

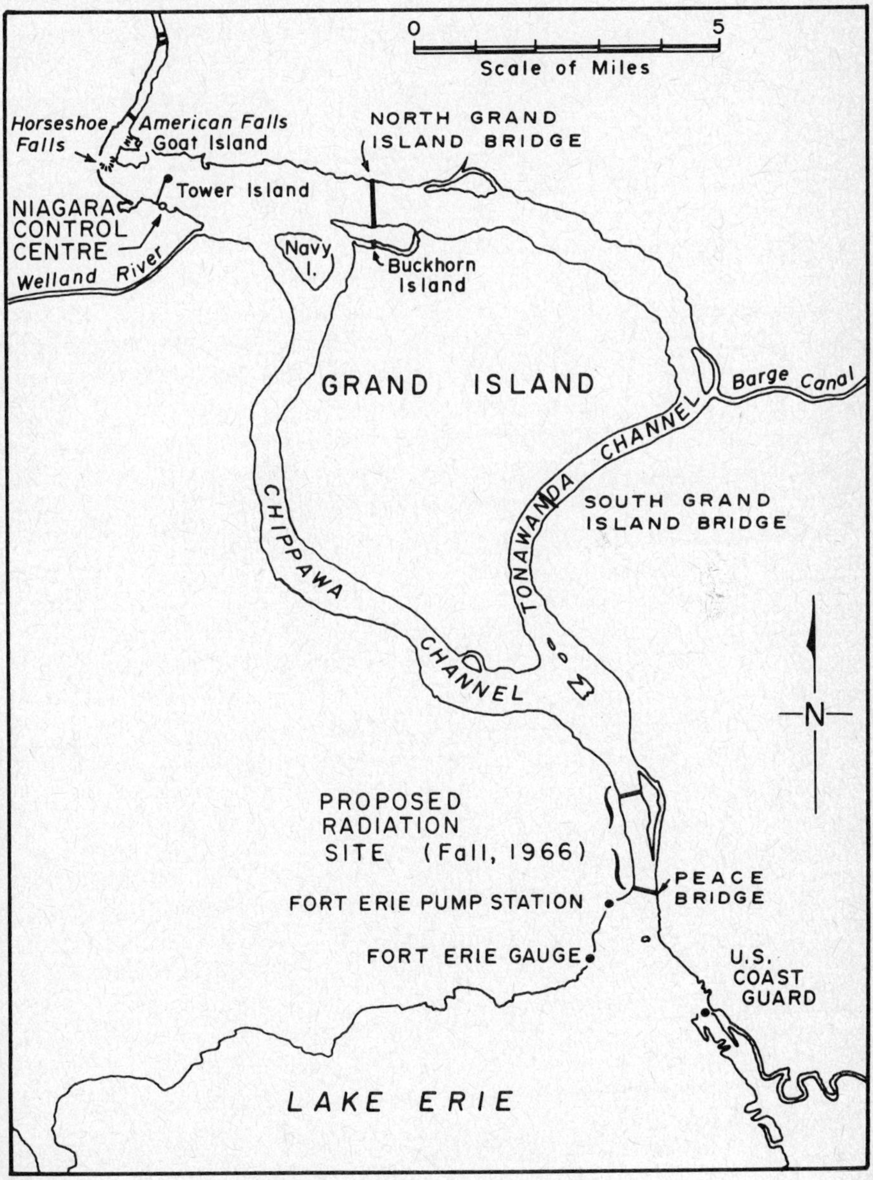

Figure 2. Niagara River – I.H.D. instrumentation sites.

The Net Pyrradiometer (Funk) located at the Niagara Control Dam is the first all-wave radiometer to be installed over a natural body of water by the Meteorological Branch.

Wind

A Bendix-Friez recording anemometer is located on the Niagara Control Dam, approximately mid-way between Tower Island and the Niagara Control Center (Figure 2).

Due to poor exposure for westerly winds at Fort Erie, the United States Weather Bureau has been asked to install a recording anemometer on the eastern side of the river at the Buffalo Coast Guard Station.

Air Temperature

Three Lambrecht thermohygrographs, providing week-long records of temperature and humidity, were installed (Figure 3).

One thermohygrograph, at the Fort Erie Pump Station, is expected to give a representative air temperature and humidity for the southern end of the Niagara River. Two thermohygrographs at the control dam near the Falls, one on the rampway to Tower Island and one on the grass area next to the Niagara Control Centre, will provide comparative temperature records. Maximum and minimum thermometers are located at all three sites with observations being taken daily at 0800 and 1600 L.S.T.

Using thermohygrographs in conjunction with twice daily read maximum and minimum thermometers, a correction is applied to the points corresponding to these temperatures, and an average correction is applied to the recorded thermograph chart. Comparative temperature reports can then be assessed. Thermohygrograph charts are used as a means of detecting mistakes in thermometer readings, as a back-up for missing data, and to provide a continuous temperature record for all hours of the day.

The daily maximum and minimum temperatures at Niagara Control Dam are also used for obtaining a mean daily temperature for net pyrradiometer evaluation.

Water Temperatures

Water temperatures are now measured at the entrance to the filtration plant at the Buffalo City intake, and at the control dam at Niagara Falls.

In the near future Ontario Hydro plans to install additional temperature recorders with greater sensitivity near the freezing point in positions significant for ice formation and control. Also, an infra-red radiation thermometer will be used by the Meteorological Branch for surface water temperature observations during several intensive observation periods when ice is forming.

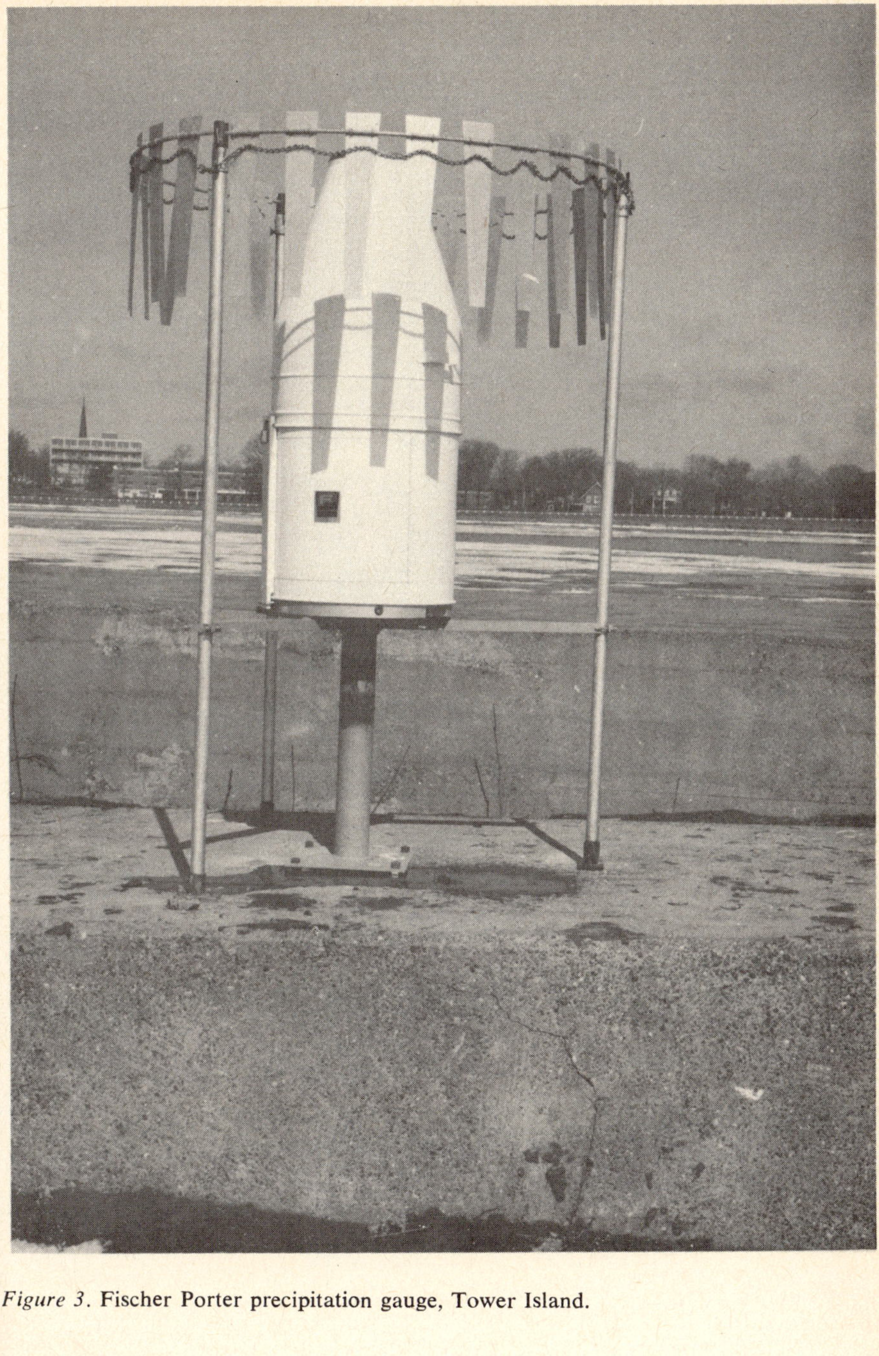

Figure 3. Fischer Porter precipitation gauge, Tower Island.

Precipitation

Two precipitation stations have been established, one at Fort Erie Pump Station giving a record of precipitation where Lake Erie empties into Niagara River, and another on Tower Island near the control dam. Using the Fischer-Porter paper tape output precipitation gauge recorder with a five-minute punch-out a record of rain or snow, with intensities of accumulated amounts for five-minute periods will make it possible to assess energy added to or taken from the river by channel precipitation.

The Fischer-Porter precipitation gauge consists of a collection bucket for receiving and storing precipitation, a weighing device for measurement, a dial indicator to show the amount of precipitation in the bucket, and a recording mechanism. The gauge reads in increments of one-tenth of an inch of precipitation and the bucket has a capacity of twenty inches.

The precipitation gauge and Alter wind shield at Tower Island were located on a concrete block 75 feet long, 65 inches wide, and projecting 24 inches above the ground. A 24-inch metal stand was bolted to the concrete block with the precipitation gauge and Alter shield bolted on top (Figure 3). This gauge was put into operation March 4, 1966. The second gauge was also mounted on a metal stand and was installed on the lawn at Fort Erie Pump Station on March 31.

The Meteorological Branch has purchased a Fischer-Porter Series 50TL1550 automatic translator, consisting of a tape reader and Model 1559 Time and Date Console. This equipment, when connected to a 047 IBM tape-to-card punch will convert the paper punched tape output from Fischer-Porter precipitation gauges into punched cards, which can be used to produce, by computer, precipitation summaries of various kinds.

Data from the Fischer-Porter precipitation gauge is checked for time, date, and mechanical breakdown before being processed by automatic translator and computer. Standard MSC rain gauges are located at Tower Island and Fort Erie and can be used for back-up data should malfunction in the recording precipitation gauge occur.

17

Evaporation from Snow Covers in Eastern Canada

G. P. Williams

Hydrologists charged with the responsibility of estimating the amount of runoff from melting snow have long been interested in estimating the amount of evaporation from a snow cover. Although there have been a number of investigations on snow evaporation, it has been difficult to find an analysis that gives an order-of-magnitude for evaporation from the comparatively shallow snow covers that are subject to the weather conditions typical of Eastern Canada. The purpose of this paper is to review the literature and experiments on evaporation from snow covers in other countries and to make a preliminary evaluation of some snow evaporation measurements taken at Ottawa during the 1956-57 season.

The theory of evaporation from snow covers will not be discussed in this report; Diamond has given a good analysis of the theory which would be of value to engineers.[1] Sverdrup[2] developed an equation for predicting the rate of evaporation and melt from a snow cover which, from a theoretical viewpoint, is still a classic. Light[3] used Sverdrup's equation for heat and water flux to develop an equation for predicting theoretical rates of snow evaporation.

The basic theory of evaporation from a snow cover is the same as evaporation from a water surface. Good résumés of this theory, particularly the turbulent-exchange theory on which some evaporation formulas are based, are given by Thornthwaite[4] and E. R. Anderson and others.[5]

Reprinted from Division of Building Research, Research Paper 73 (January, 1959). Originally published in Proceedings, Eastern Snow Conference (1958).

The main problem in evaporation studies is to obtain reliable measurements under field conditions. Evaporation rates from open water, bare soil or grass have been estimated by measuring the rate of evaporation from some standard pan or tank and from these measurements obtaining a coefficient which could be used to estimate evaporation from the surface area in question.[6,7] Evaporation rates from snow covers have usually been measured by filling pans with snow and placing them into the snow cover so that the surface of the pan is flush with the snow surface. The pans are usually painted white or are made of plastic to minimize radiation effects. The rate of evaporation from the pan is assumed to be representative of that from the larger snow surface area.

De Quervain[8] found that radiation did affect the evaporation rate if the pans were too shallow. Pans also interfere with the possible upward flow of vapour through the snow cover and there is some question as to what effect this has on the rate of evaporation from the snow surface. It is also impossible to obtain reliable results from pans during periods of precipitation or blowing snow. Some field measurements of evaporation were made in the Fraser Experimental Forest using a volumetric method and the pan method.[9] Because of difficulties of measurement they were unable to obtain good comparisons between the two methods. It is apparent that the use of pans to measure snow evaporation has several drawbacks which make it difficult to obtain continuous, reliable records. Nevertheless, their use appears to be the only way that the order-of-magnitude of snow evaporation rates can be ascertained.

Later in this report the pan evaporation measurements of several workers in different countries will be compared. The coefficients calculated from these studies for the various snow evaporation formulas will be compared with the evaporation coefficient calculated from the snow evaporation measurements taken at Ottawa during the 1956-57 winter.

Field Measurements at the Ottawa Site

Snow evaporation was measured at Ottawa by placing circular pans (4 cm deep with 355 sq cm area) filled with snow, flush with the snow surface and weighing the pans at different time intervals. The weight loss or gain during the time period was assumed to be due to evaporation or condensation from the snow surface during the period of observation. The light aluminum pans were painted white to reduce the absorption of radiation from the sun.

The observation site was in an exposed location near an area where standard meteorological observations are made. The standard meteorological records obtained at this station were used in all calculations.

There were many days when it was impossible to obtain reliable measurements because of weather conditions. Whenever there was blowing snow, precipitation, or melting snow, observations were not taken because of uncertainty in the measurements. Furthermore, during the 1956-57 winter, the duration of the snow cover was below normal in the Ottawa area and the opportunities for observation were limited.

Analysis of Results

For each day during which evaporation was measured, the evaporation rate was calculated from the following equation:

$$Q = 0.00011\, p\, U_a\, (e - e_s)$$

Where Q = evaporation rate, $gm/cm^2/min$
p = mean density of air, gm/cm^3
U_a = mean wind speed, cm/sec (at height a cm)
e = vapour pressure of air in Stevenson screen, (millibars)
e_s = saturated vapour pressure over ice at the mean temperature of the snow surface, (millibars).

The form of this equation was developed from the more complicated Sverdrup equation[2] by assuming constant values for surface roughness and air pressure.

The mean wind speed was obtained by averaging the hourly wind speed values during the period of observation. The vapour pressure of the air was calculated from the mean relative humidity value recorded at 8.30 a.m. and at 4.30 p.m. The vapour pressure at the snow surface was calculated by assuming that the mean snow surface temperature is equal to the mean air temperature recorded in the Stevenson screen.

Figure 1 shows the relationship between the calculated evaporation rate and the measured evaporation rates at the Ottawa site. The measured evaporation rate is the average rate from the three evaporation pans. Table I lists the results used to calculate the values shown in Figure 1.

Calculated Evaporation Rates During March

The evaporation observations listed in Table I were all made during the colder winter months. During the spring melt period, which occurred in March in 1957, no evaporation measurements were made. Whenever melting conditions occur, water is present with the snow in the pan and it is doubtful whether evaporation rates from this mixture are the same as evaporation rates from a wet snow surface where the free water can drain downward. The order-of-magnitude of evaporation rates during the spring melt period was estimated using equation 1 and the appropriate meteorological data. Whenever the snow was melting the vapour pressure was assumed to be equal to the saturated vapour pressure over ice at 32°F. Condensation was treated as negative evaporation.

The average calculated evaporation loss from the snow cover during the month of March for the years 1953-57 inclusive are plotted on Figure

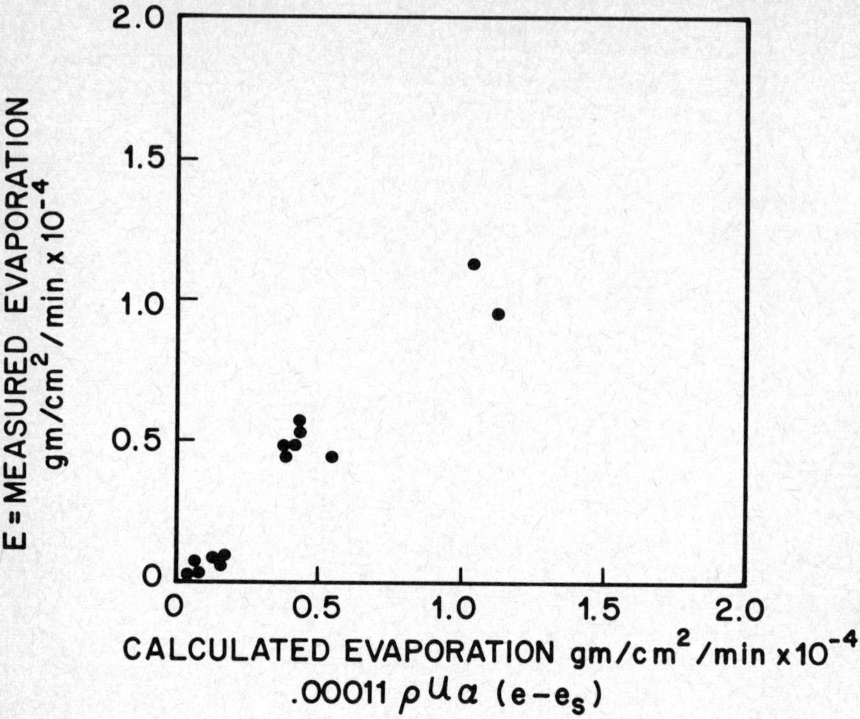

Figure 1. Relationship between calculated evaporation rate and measured evaporation rate, Ottawa, 1956-57.

2. This total monthly evaporation loss was obtained by accumulating the calculated rates for a 10-hour day from 8 a.m. to 6 p.m. Because of the uncertainty of relative humidity values during the night, the evaporation rates were not calculated for this period. It is generally assumed that evaporation losses will be less during the night when there is no heat supplied by radiation and when the relative humidity values are higher than during the daylight hours. Croft[10] found that evaporation losses from a snow cover during the daylight hours were three to four times as great as losses during the night period.

Thus Figure 2 only indicates total evaporation losses during part of the day and is only intended to indicate that average losses from a snow cover are generally under one inch of water during a month of spring runoff, and are generally much less than the precipitation during that month.

Comparison of Evaporation Coefficients

The general form of equation 1 is the same as that used by other investigators of snow evaporation. Research workers at the Central Sierra Snow Laboratory[11] have used the following form: Q_e equals K_e $(e_a - e_s)$

TABLE I. LIST OF DATA USED TO CALCULATE VALUES SHOWN IN FIGURE 1

Date	*Wind Velocity Ua cm/sec*	*Vapour Pressure Difference* $e\text{-}e_s$ *mb*	*Density of air gm/cm*³ (*x* 10^{-3})	*Calculated Evaporation Rate gm/cm*²*/min* (*x* 10^{-4})	**Measured Evaporation Rate gm/cm*²*/min* (*x* 10^{-4})
Dec. 6/56	134.1	1.08	1.30	0.208	0.15
Jan. 3/57	268.2	0.61	1.36	0.244	0.122
	246.9	0.61	1.36	0.225	0.147
Jan. 17/57	219.5	0.21	1.40	0.071	0.02
Jan. 18/57	179.8	0.32	1.38	0.089	0.08
	179.8	0.32	1.38	0.089	0.10
Jan. 24/57	460.3	0.57	1.36	0.392	0.45
	460.3	0.57	1.36	0.392	0.46
Jan. 28/57	268.2	1.36	1.33	0.540	0.45
Feb. 11/57	292.6	0.95	1.35	0.415	0.53
Feb. 12/57	499.9	0.57	1.36	0.43	0.56
Feb. 19/57	524.3	1.50	1.32	1.14	0.95
Feb. 20/57	597.4	1.18	1.33	1.03	1.14
Feb. 22/57	170.7	1.61	1.32	0.40	0.47

*Average of 3 pans—450 sq cm in area.

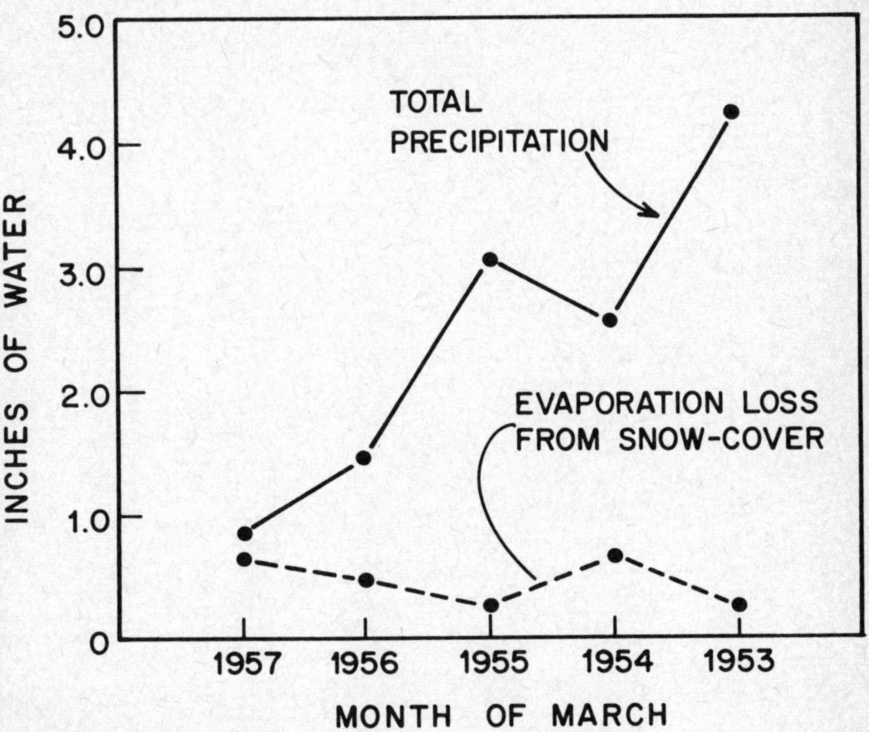

Figure 2. Calculated evaporation rates from snow-cover, Ottawa.

V_b; where Q_e is the amount of evaporation or condensation, e_a and e_s are the vapour pressure of the air and the saturated vapour pressure over ice at the mean snow surface temperature respectively, V_b is the wind speed, and K_e is a constant. In addition, they calculated a K_e' equal to 8.5 K_e $(Z_aZ_b)^{1/6}$, where the subscripts a and b identify the levels of measurement of the wind speed and air vapour pressure respectively, and K_e' is the value of the coefficient when the wind speed and vapour pressures are measured one foot above the snow surface.

They have compared the evaporation or condensation constant K_e' for several workers as follows:

Central Sierra Snow Laboratory Snow Investigations	(1954)	0.0540
Sverdrup	(1936)	0.0674
de Quervain	(1951)	0.0770

By using the same units of vapour pressure in millibars, mean wind speed in miles per hour, evaporation or condensation in inches, and heights of measurement in feet, it was possible to calculate a K_e' from the Ottawa measurements given in Table I. The mean K_e' thus obtained was 0.046 which compares reasonably well with the 0.054 value obtained by the investigators at the Central Sierra Snow Laboratory. The fact that these coefficients are in the same general range strengthens the confidence in the use of this form of equation to determine the order-of-magnitude of snow evaporation rates.

Comparison of Measured Evaporation Rates

There have been a surprising number of measurements of snow evaporation rates but, because of the difficulties of measurement, most of the records are sporadic and all the results must be qualified because of these difficulties. Although it is difficult to compare evaporation rates because of differences in sites and weather conditions, it was considered that if the reported mean monthly rates and reported maximum daily rates were tabulated, a check on the order-or-magnitude of evaporation losses would be obtained. In addition, the references provide a useful bibliography on snow evaporation. This tabulation of references and reported evaporation rates are listed in Table II.

All the reported measurements appear to be within the same general range. The mean monthly rates compare favourably with the monthly rates calculated for the Ottawa site.

Discussion of Results

It is reasonable to conclude that the order-of-magnitude of snow evaporation from large snow covered areas can now be determined using present techniques, but it is difficult to determine the degree of accuracy

TABLE II. EVAPORATION RATES FROM SNOW DURING THE SPRING MELT PERIOD

Investigator	*Maximum Daily (in. of water/day)*
de Quervain (10)	0.032
Ångström (12)	0.051
Kehrlein (13)	0.050
Shul'ts (14)	0.020
Rudovits (15)	0.019
	Mean Monthly (in. of water/month)
Guy (16)	0.31
Poliakov (17)	1.18
Chomicz (18)	1.15
Kaitera (19)	0.48
Fitzgerald (20)	0.60

of these calculated evaporation losses. Kaitera's experiments indicate that evaporation rate varies a great deal with exposure.[19] Evaporation losses from a snow-covered slope facing north will be less than from a similar slope facing south under comparable conditions. Evaporation losses from snow cover in sheltered sites protected by forest cover will be much less than evaporation from snow cover in exposed sites. Evaporation losses must be calculated for representative sites which are typical of conditions over the area in question.

Any formula used to calculate snow evaporation is limited by the basic accuracy with which the variables of wind speed and vapour pressure can be measured. The measurement of accurate vapour pressure differences by standard meteorological instruments leaves much to be desired, particularly at low temperatures. The measurement of wind speeds is not too difficult for one site, but it is questionable how representative such observations are of the average conditions required for the calculation of evaporation losses over a large area. Attempts to estimate evaporation rates from a large area must take into account these factors.

One condition not covered by any evaporation formula is the effect on evaporation when the wind is strong enough to disturb the snow surface. In exposed areas where the wind can move large quantities of snow, this factor might result in evaporation rates considerably higher than a general evaporation equation would indicate. This factor usually is not important when the snow cover is melting as wet snow is cohesive and is not generally moved even by strong winds.

The only time the order-of-magnitude of snow evaporation amounts might be significant compared to snow melt is when there is a shallow snow cover and the snow-melt period is extended over several weeks. In this case, the evaporation from small pools of water and from exposed saturated soil near the snow cover might be of importance. When the snow-melt period is short, such as during the Chinook conditions reported by Hoover,[21] the evaporation from the snow should be much less than the snow melt. As Diamond[1] points out, the theory of evaporation indicates that in these cases a much larger amount of snow must be melted than evaporated.

The order-of-magnitude of snow evaporation amounts might be of significance from a heat balance point-of-view for certain special investigations. From a hydrological viewpoint, evaporation losses from a large watershed area are likely to be less than the errors in obtaining samples of the total water content of the snow cover. For this reason, and because of the great difficulties in obtaining representative measurements of snow evaporation rates, it appears that hydrologists will have to be satisfied with calculated evaporation rates which can give an indication of the order-of-magnitude of snow evaporation losses.

References

1 M. Diamond, "Evaporation or melt of snow cover," *U. S. Snow, Ice and Permafrost Research Establishment, Research Paper no. 6* (November, 1953).

2 H. U. Sverdrup, "The eddy conductivity of the air over a smooth snow field," *Geofysiske Publikasjoner*, Vol. 11, no. 7 (1936), 69 pp.

3 P. Light, "Analysis of high rates of snow melting," *Trans. Amer. Geophys. Union, Pt. I* (1941), pp. 195-205.

4 C. W. Thornthwaite and B. Holzman, "Measurement of evaporation from land and water surfaces," U.S. Dept. Agric., *Tech. Bulletin no. 817* (May, 1942), 143 pp.

5 E. R. Anderson and others, "A review of evaporation theory and development of instrumentation," U.S. Navy Electronic Lab., San Diego, California, *Aero U.S. Navy N.E.L. Report* (1950).

6 M. Kohler, J. J. Nordenson and W. E. Fox, "Evaporation from pans and lakes," *U.S. Dept. of Commerce, Research Paper no. 38* (May, 1955).

7 Arthur A. Young, "Evaporation investigations in Southern California," Soil Conservation Service, Div. of Irrigation, Pomona Calif. (April, 1945).

8 M. de Quervain, "Evaporation from the snowpack," U.S. Army Corps of Engineers, South Pacific Division, Research Note no. 8, Translation (October 16, 1952). 22 pp.

9 "Report of the Committee on Evaporation and Transpiration, 1945-46," *American Geophysical Union, Trans. Amer. Geophys. Union*, Vol. 27 (October, 1946), pp. 720-25.

10 A. R. Croft, "Evaporation from snow," *Bulletin of the American Meteorological Society*, Vol. 25 (1944), pp. 334-37.

11 "Snow hydrology," Summary report of the snow investigations, North Pacific Division, U.S. Army, Corp of Engineers, Portland, Oregon (June 30, 1956), 437 pp.

12 Anders Ångström, "Measurements of evaporation in Sweden," Union Géodésique Géophys. Intern. Assoc. Hydrologie Sci. Commission de Potamologie, 1, Quest. 1, Rapport 6, Washington, D.C. September, 1939), 5 pp.

13 Oliver Kehrlein and others, "High Sierra snow ablation observations," *Proc. 21st Western Snow Conference 1953* (September, 1953), pp. 47-50.

14 V. L. Shult's "Study of the heat balance of a snow field," (Opyt izucheniià teplovogo balansa snezhnika; Text in Russian). *Biùlletin' Akademii Nauk SSSR (Tashkent)*, no. 11 (1946), pp. 13-16.

15 L. Rudovits, "Snow evaporation," (Isparenie snega; Text in Russian), *Meteorologisheskii Vestnik*, Vol. 34 (1924), pp. 50-52.
16 L. T. Guy, snow studies on Bogong High Plains by the State Electricity Commission of Victoria, Australia, International Union Geodesy and Geophysics, International Association of Scientific Hydrology, Edinburgh (September 1936).
17 B. V. Poliàkov, "Total evaporation" (Summarnoe isparenie; Text in Russian), Gidrologicheskii Analiz i Raschety, by B. V. Poliàkov, Leningrad, Meteoizdat (1946), pp. 126-45.
18 K. Chomicz, "The importance of snow-cover studies for water management" (O przystosowanie badan nad pokrywa sniezna do potrzeb gospodarki wodnej; Text in Polish), *Gospodarka Wodna*, Vol. 13 (March, 1953) pp. 108-11.
19 P. Kaitera, "The evaporation from snow," International Union of Geodesy and Geophysics, International Association of Scientific Hydrology, Washington, D.C. (September, 1939).
20 D. Fitzgerald, "Evaporation," *Trans. Am. Soc. Civil Engrs.*, Vol. 15 (1886), pp. 581-646.
21 O. H. Hoover, "Effects of Chinook (Foehn) winds on snow cover and runoff," Int. Union Géodésique Géophys., Intern. Assoc. Hydrologie Sci. Assemblée Générale d'Oslo 192-28 août 1948, Vol. 2, pp. 86-100.

18

Annual Heat and Mass Transfer at an Ottawa site

L. W. Gold

D. W. Boyd

Heat and moisture transfers at the ground surface undergo an annual cycle as a consequence of the annual variation in solar radiation. Observations indicated that net radiation and heat flow from the ground at an Ottawa site were closely approximated by a simple harmonic function with period of 1 year, and that the amplitude and phase of this function did not change appreciably from year to year. This suggested that useful approximations of the other components of the heat balance could be obtained for the site with similar functions. Although it was not possible to measure these components continuously, it was considered that the amplitude and phase of the simple harmonic approximation could be deduced for the moisture transfer from estimates of the average maximum and minimum values, and the times of the year at which these extremes occur. The simple harmonic approximation to the convective exchange could then be obtained as the difference in the heat balance equation. It was realized that the approximations might at times be in considerable relative error, particularly during the period of minimum exchange, but it was considered that they would provide a useful picture of the annual exchange, a reference framework for more accurate measurements and probably as accurate an indication

Reprinted from *Can. J. of Earth Sciences,* Vol. 2 (1965), pp. 1-10.

of heat and moisture loss over a long period of time as could be obtained with present-day instruments. It was with this in mind, as well as the intention of obtaining additional insight into the relationships between climate and ground thermal regime, that the study recorded in this paper was undertaken. The results presented are based on observations made over the 2-year period January 1, 1961, to December 31, 1962.

Instrumentation

The observation area was some 60 meters square and covered with field grass allowed to grow to a height of between 15 and 20 cm before being cut. There were a number of one- and two-storey buildings in the vicinity, the nearest about 45 meters north and south of the measurement area. The site was open to the west, and buildings were at least 120 meters away to the east. There was a single row of trees about 30 meters to the southwest.

Air temperature about 120 cm above the soil surface was measured in a Stevenson screen with a ventilated thermocouple immersed in oil in a glass tube about 0.75 cm in diameter and 20 cm long. Wind speed was measured 2 meters above the soil surface with a type 45 anemometer made by the Meteorological Service of Canada. Dew point at the 2-meter level was measured continuously on the north side of a small instrument hut with a "Dewcel" manufactured by the Foxboro Co. Ltd. Evaporation observations were made with a class A pan supplied by the Meteorological Service of Canada. The net radiation observations were made with a radiometer (designed by the Commonwealth Scientific and Industrial Research Organization of Australia) (Funk 1959) mounted about 2 meters above the soil surface. In November of 1961 observations were begun on net solar radiation using two Eppley pyrheliometers mounted horizontally about 2 meters above the soil surface. One instrument faced upward, the other downward.

Ground temperatures have been observed at the site since 1957. A description of the installation is given in a paper by Pearce and Gold (1959). The ground heat flow component was estimated from the measured temperatures, the necessary soil thermal properties having been determined in the earlier study (1959).

Calculation of Components

Net Radiation

Monthly averages of the measured daily net radiation are presented in Figure 1. Shown as well is the Fourier approximation given by

$$q_r = R_0 + R_1 \cos (wt + \delta)$$

where q_r is the daily net radiation, positive for heat supplied to the surface,

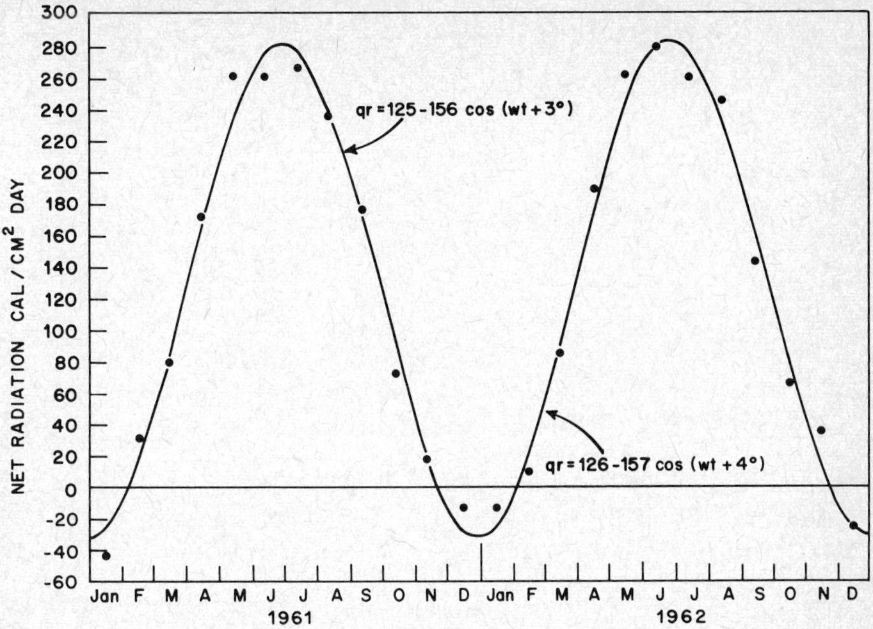

Figure 1. Dependence of net radiation on time of year. The plotted points are monthly averages of daily net radiation.

w is equal to 360 deg. per year,
t is the time in years measured from the beginning of January,
δ is the phase angle in degrees,
R_0 is the annual average of the daily net radiation,
R_1 is the amplitude of the Fourier component with period 1 year.

The monthly average values calculated from the Fourier approximation are within ± 20 cal/cm² day of the observed values. This difference is less than 10% of the observed daily average for each month in mid-summer, but almost 100% of the observed average in mid-winter.

These figures indicate the significance of the winter exchange relative to the larger summer exchange. The relative winter error is not considered significant in this present study because it is very difficult to measure net radiation in the field with an accuracy better than about ± 30 cal/cm² day using radiometers available at present (Boyd 1962).

Ground Heat Flow

The Fourier approximation of the heat flow from the ground is given by

(1) $q_g = \sqrt{(w/\lambda)}kT_1 \exp(-z\sqrt{(w/2\lambda)}) \cos(wt + \delta - z\sqrt{(w/2\lambda)})$

where q_g = heat flow from ground, positive for heat flow to the surface,

λ = thermal diffusivity of soil,
k = thermal conductivity of soil,
z = depth of observation below surface,
δ = phase angle of the heat flux at the surface,
T_1 = amplitude at the surface of Fourier component of the ground temperature with a period of 1 year.

Using the soil thermal properties given by Pearce and Gold (1959) ($\lambda = 10.8$ cm² h, $k = 8.45$ cal/cm °C h) and calculating T_1 from ground temperature observations, the following approximations were obtained for ground heat flow at the surface:

Period	Fourier approximation
1 January 1960 – 31 December 1960	$18.9 \cos(wt + 20°)$ cal/cm² day
1 January 1961 – 31 December 1961	$17.5 \cos(wt + 24°)$ cal/cm² day

For the purpose of the present discussion, the Fourier approximation for the 2-year period was taken as $18 \cos(wt + 22°)$ cal/cm² day.

Mass Transfer

The heat flux associated with the average mass transfer, including evaporation, transpiration, and sublimation, often is estimated for periods of 1 day or longer with semiempirical formulae of the form

$$q_e = A(U_a + B)(e_a - e_s), \tag{2}$$

where q_e is the heat flux associated with the average mass transfer, positive for vapor transfer to the surface,
U_a is the average wind speed at the measuring height a,
e_a is the average water vapor pressure at the same height,
e_s is the saturation vapor pressure at the average surface temperature,
A and B are constants.

Earlier observations indicated that at Ottawa the average daily snow surface temperature is usually between 0 and 1 C deg colder than the average daily air temperature, and that in summer the average daily soil surface temperature is about 1 to 3 C deg warmer than the average daily air temperature (Gold 1963). The average daily air temperature was used, therefore, in calculating the saturation vapor pressure at the surface for periods of snow cover. The saturation vapor pressure at the surface in summer was assumed to be that for a temperature equal to the average daily temperature of the soil surface at the base of the grass stems. Average daily values for e_a were calculated from observed average hourly values. Average monthly values of $e_a - e_s$ as well as Fourier approximations calculated from 5-day averages of $e_a - e_s$, are plotted in Figure 2.

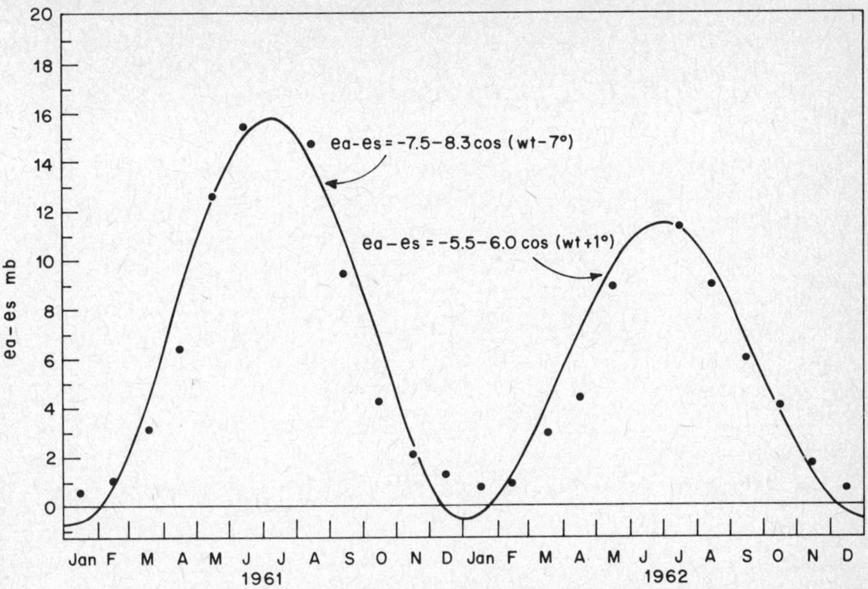

Figure 2. Time dependence of the monthly averages of the vapor pressure difference between the 2-metre level and the ground surface. The vapor pressure at the surface is assumed to be the saturation value at the average surface temperature.

Penman (1948, 1956) gave, for a saturated surface, the following values for the constants in equation (2):

$$A = 9.75 \times 10^{-7} \text{ cal/cm}^3 \text{ mb}$$
$$B = 32.5 \times 10^4 \text{ cm/h}$$

for U_a measured in cm/h at a height of 2 meters above the surface; e_a and e_s in mb. The average wind speed at the site in July (the time of maximum evapotranspiration) was 53.4×10^4 cm/h in 1961, and 59.9×10^4 cm/h in 1962. The average daily value for $e_a - e_s$ at the time of maximum was about $-$ 18.6 mb in 1961, and about $-$ 13.3 mb in 1962. Assuming $e_a - e_s$ to be $-$ 16 mb and the average wind speed 56×10^4 cm/h, equation (2) gives

$$q_e = -13.8 \text{ cal/cm}^2 \text{ h}$$
$$= -330 \text{ cal/cm}^2 \text{ day.}$$

The measured average evaporation rate from the class A pan was observed to be 0.49 cm water per day in July 1961 and 0.55 cm in July 1962, equivalent to about 290 cal/cm² day and 320 cal/cm² day, respectively. Converting class A pan observations to lake evaporation gives about 0.38 cm in 1961 and 0.42 cm in 1962, equivalent to about 225 and 250 cal/cm^2 day, respectively. Since the average surface temperature of a lake is probably cooler than the average surface temperature of a freely evapotranspirating grass surface subject to the same weather, the daily moisture loss from such a surface should be between that for a lake and that for the class A pan. The class A pan observations, therefore, indicate a value for evapotranspiration somewhat lower than that obtained from equation (2) using constants established by Penman. The value of e_s used in equation (2), however, was based on average temperatures measured at the base of the grass stems, and these temperatures would be expected to be higher than the average temperature for the transpirating surfaces. This would result in an absolute value for $e_a - e_s$ in equation (2) that is too large and, therefore, an estimate of the evapotranspiration loss that is too high. It was assumed for the purpose of the present calculations, therefore, that the average moisture loss from a freely evapotranspirating surface during July at Ottawa was equivalent to 280 cal/cm^2 day, and recognized that this estimate may be in error by about $\pm$ 10%.

If the soil moisture content is not at field capacity, evapotranspiration will be less than that from a saturated surface. Penman found that, for the grass surfaces upon which he made his observations, evapotranspiration was between 0.60 and 0.80 of that calculated from equation (2), depending on the amount of water stored in the soil (Penman 1956). The value of 0.60 was obtained under winter conditions in southeast England.

During July at Ottawa it is normal for evapotranspiration to exceed rainfall by almost 6 cm of water, but conditions vary a great deal from year to year. In July 1961 the soil moisture content remained nearly at field capacity for 3 weeks (according to calculations based on the Thornthwaite method (Thornthwaite and Mather 1957)) but was rapidly reduced during the last week to about 4.5 cm below field capacity. In July 1962 the soil moisture content was about 8 cm below field capacity at the beginning of the month and remained between 8 and 12 cm throughout the month. These calculations are in agreement with observations made by the Soil Mechanics Section of the Division of Building Research, National Research Council, on the soil water pressure measured with a piezometer at the 3-meter depth at a site about 12 meters east of the row of trees. In July 1961 the soil water pressure at that depth varied between 253 and 300 cm of water, and in July of 1962, between 195 and 241 cm of water.

The observations on soil moisture and the Thornthwaite calculations indicated a good supply of soil water. It was assumed, therefore, based on Penman's observations, that average evapotranspiration during July for the 2 years was 0.80 of the assumed evaporation from a saturated surface (280 cal/cm^2 day), that is, about 225 cal/cm^2 day.

The ground at the observation site is fairly level, usually with no surface runoff in summer. Because the underlying soil is clay, it was considered that there was no net loss of water from the site due to ground water flow. Measured rainfall, therefore, should give the lower limit to the evapotranspiration loss. In July of 1961 the average rainfall was 0.38 cm/day, equivalent to about 225 cal/cm^2 day. In July 1962 the average rainfall was 0.33 cm/day, equivalent to about 195 cal/cm^2 day. These values are in good agreement with the estimate obtained from equation (2) and the class A pan observations.

For sublimation in winter, reasonable values for the constants in equation (2) are

$$A = 1.5 \times 10^{-6} \text{ cal/cm}^3 \text{ mb},$$
$$B = 0 \text{ (Williams 1959).}$$

The average wind speed at the site in January is about 65×10^4 cm/h (4 m.p.h.). Figure 2 indicates that the monthly average value for $e_a - e_s$ at the time of minimum average sublimation loss in winter is about -0.67 mb. Substituting these values into equation (2) gives

$$q_e = -0.65 \text{ cal/cm}^2 \text{ h},$$
$$= -15.6 \text{ cal/cm}^2 \text{ day.}$$

On the basis of the preceding discussion, it was assumed that the Fourier approximation of the heat loss by evaporation and sublimation had a maximum of 225 cal/cm^2 day in summer, a minimum of 15 cal/cm^2 day in winter, and an annual average value of 120 cal/cm^2 day. From the observations on $e_a - e_s$, it was assumed that the phase angle was -2 deg, with time measured from the beginning of January, giving maximum evapotranspiration in early July and minimum sublimation in early January. In choosing the phase angle, greater weight was given to the observations for 1962 than to those for 1961. The equation obtained is given below:

$$q_e = -120 + 105 \cos (wt - 2) \text{ cal/cm}^2 \text{ day.}$$

Figure 2 indicates that for a saturated surface the Fourier component with period of 1 year is only a fair approximation of the observations of $e_a - e_s$. The reason for this, in part, is the non-linear dependence on temperature of the saturation vapor pressure of water. The drying of the soil from evapotranspiration will reduce the actual difference between the average vapor pressure of the air and at the surface during the summer months. This reduction would probably cause the Fourier component of period of 1 year to be a better approximation of the evapotranspiration loss than is indicated in Figure 2.

Convective Loss

No observations were available from which estimates could be made of the average convective loss at the time of maximum in summer and minimum in winter. The expression for the convective loss was obtained by putting it equal to the sum of the other three components of the heat balance. The phase angle obtained (7 deg) is about 8 deg greater than that given by observations on the difference between the average daily air and surface temperatures. The approximations for all four components of the heat balance are given below and plotted in Figure 3.

Fourier approximation of components of heat and mass transfer at grass surface at Ottawa in cal/cm^2 day ($t = 0$, January 1)

$$q_r = 125 - 157 \cos (wt + 3°),$$
$$q_g = 18 \cos (wt + 22°),$$
$$q_e = -120 + 105 \cos (wt - 2°),$$
$$q_e = -5 + 35 \cos (wt + 7°).$$

Incoming solar $= 329 - 208 \cos (wt + 13°)$.
Net solar $= 252 - 211 \cos (wt + 4°)$.
Net long wave $= -127 + 54 \cos (wt + 8°)$.

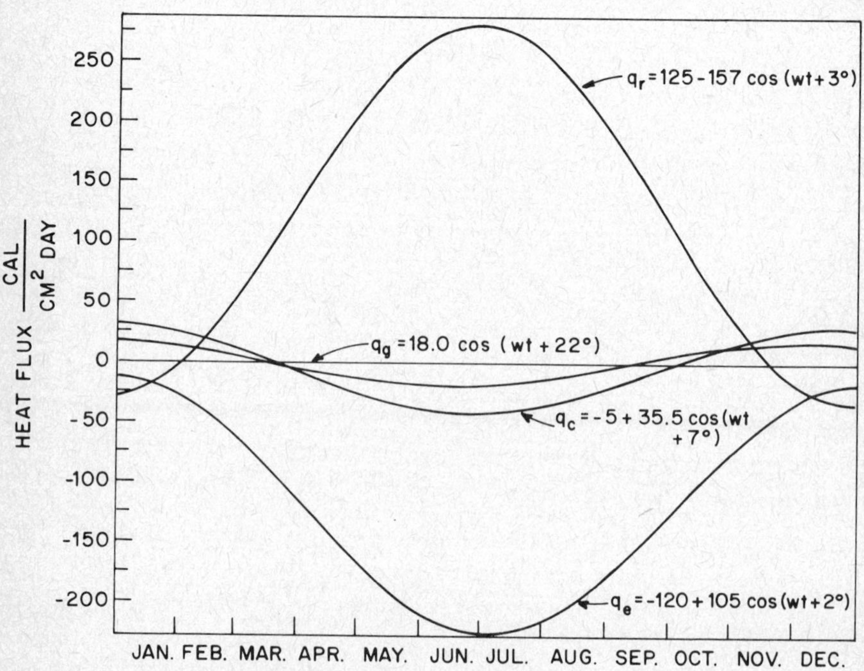

Figure 3. Fourier component of period 1 for the net radiation, convection, mass transfer, and ground heat flow at the Ottawa site.

Solar Radiation and Albedo

Incoming solar radiation on a horizontal surface was measured near the site with an Eppley pyrheliometer supplied by the Meteorological Service of Canada. Average monthly values for the daily radiation for the 2-year period are plotted in Figure 4. Shown also is the Fourier component with a period of 1 year. Average monthly values for the albedo obtained with the two Eppley instruments mounted back to back are as follows.

	Jan.	Feb.	Mar.	Apr.	May	June	July	Aug.	Sept.	Oct.	Nov.	Dec.
Albedo	65	61	55	10	13	16	15	16	17	15	21	20
Percentage			10									65

A comparison with other radiometers indicated that these albedos are accurate to within 5 albedo units. The two albedo values for the months of December and March are for the periods with and without snow cover. Of interest is the marked increase in albedo when the ground becomes covered with snow and the gradual increase for the grass cover during the summer. A low value for the albedo in summer and a relatively high value in winter are characteristic of most terrains in Canada.

The albedos above were used to calculate for each month the average daily net solar radiation from the incoming solar radiation observations. Average daily net long-wave radiation for each month was obtained by

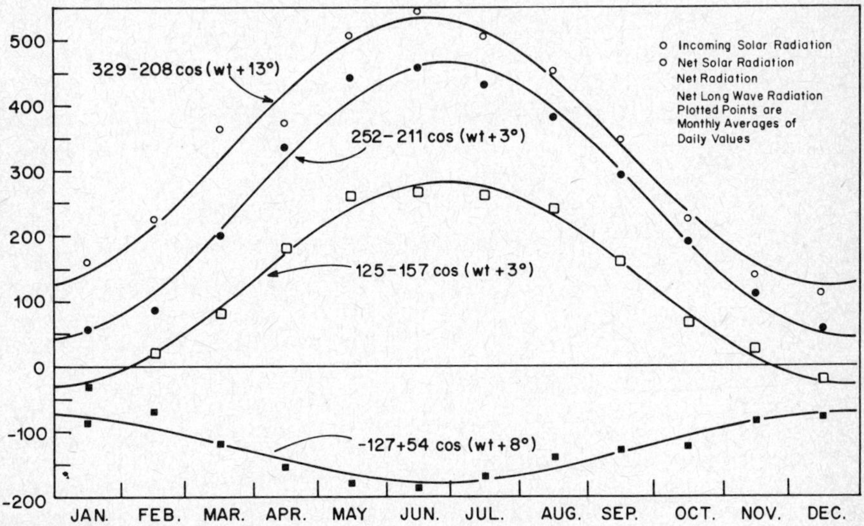

Figure 4. Time dependence of the incoming solar, net solar, net long-wave, and net radiation.

subtracting net solar radiation from the net radiation. The monthly averages for the daily net solar and long-wave radiation are plotted in Figure 4. Given also are the Fourier components with period 1 year.

Discussion

One of the more difficult aspects of establishing the heat balance at ground surfaces is to determine the probable accuracy of observations and calculations. The Fourier approximation for the net radiation given in Figure 3 agrees within ± 20 cal/cm^2 day with the observed daily values for each month for the 2-year period. A check of the calibration of the net radiometer in 1963 gave a calibration factor for long-wave radiation within 2% of that used for the period of observations.

It is considered that the Fourier approximation for the ground heat flow agrees within ± 3 cal/cm^2 day with the average daily values for each month for the 2-year period. The approximation for evapotranspiration and sublimation is considered to be within ± 25 cal/cm^2 day of the average daily values for each month during the period from early winter to the middle of summer, but may be in error by more than this amount in the fall and late summer. Little can be said about the accuracy of the approximation for the convective component, except to acknowledge that it probably has the largest relative error.

A check was made on the evapotranspiration approximation by integrating the equation in Figure 3 over the period 1 April to 30 September. The loss was calculated to be equivalent to 57 cm of water. This is in good agreement with the observed rainfall of between 45 and 50 cm as there was a net loss of soil water during that period for both years. The heat associated with the approximation for each of the other components was obtained by integration for the above period as well. The values obtained are given in Table I, expressed both in cal/cm^2 and as the height of a column of water 1 cm^2 in cross-sectional area that could be evaporated at 0 °C by the heat associated with each component. From this

TABLE I. NET HEAT ASSOCIATED WITH EACH FOURIER COMPONENT DURING THE PERIOD APRIL 1 TO SEPTEMBER 30

Component	*Heat, cal/cm^2*	*Equiv. ht. of column of water evaporated, cm*	*Percentage of net solar radiation*
Incoming solar	8.38×10^4	140.8	
Net solar	7.05×10^4	118.5	
Net long wave	-2.95×10^4	49.6	41.8
q_r	4.10×10^4	68.9	
q_e	-3.41×10^4	57.3	48.4
q_c	-0.51×10^4	8.5	7.0
q_g	-0.20×10^4	3.3	2.8

table it may be seen that for the Ottawa site about 48% of the net solar heat gain in summer was dissipated by evapotranspiration, 42% by long-wave radiation, and 7% by convection and that only 3% was stored in the ground. Because the components are almost in phase, the above figures indicate also the approximate average partitioning of heat amongst the components throughout the summer. The heat used for plant growth, probably less than 1% of the net solar radiation, was not considered in the study.

The components given in this paper are expressed as average daily values, but it should be kept in mind that each component undergoes a daily fluctuation as well. For example, unless there is marked advection, most of the evapotranspiration will occur during the daytime in association with the absorption of solar radiation. During the daytime convection will dissipate heat, but at night, if there is a significant long-wave radiation loss, convection will probably carry heat from the atmosphere to the ground surface. The partitioning of the heat over long periods as determined in the present analysis, therefore, gives no indication of the partitioning for each instant.

The calculated heat flow rates are all about the same absolute size in winter and, except for the ground heat flow, about equal to the probable error in the observations or calculations. This demonstrates one of the difficulties that is encountered in trying to determine the heat balance at the ground surface in winter for periods of 1 day or longer. It is of interest that the calculations indicate a net convective heat gain to the surface in winter.

References

Boyd, D. W. "Comparison of CSIRO and Suomi radiometers," Symposium on the Heat Exchange at Snow and Ice Surfaces, Associate Committee on Soil and Snow Mechanics, Ottawa, Tech. Man. no. 78, (1962).

Funk, J. P., "Improved polythene-shielded net radiometer," *J. Sci. Instr.*, Vol. 36, no. 6 (1959), p. 267.

Gold, L. W., "The influence of snow cover on the average annual ground temperature at Ottawa, Canada." Presented to the Intern. Union Geodesy Geophys., Intern. Assoc. Sci. Hydrol., General Assembly, Berkeley, Calif. Pubn. no. 61 (August, 1963), pp. 82-91.

Pearce, D. C. and Gold L. W., "Observations of ground temperature and heat flow at Ottawa, Canada," *J. Geophys. Res.*, Vol. 64, no. 9 (1959), p. 1293.

Penman, H. L., "Natural Evaporation from open water, bare soil and grass," *Proc. Roy. Soc. London*, Ser. A. Vol. 193 (1948), p. 120.

Penman, H. L., "Evaporation: an introductory survey," *Netherlands J. Agric. Sci.*, Vol. 4, no. 1 (1956), p. 9.

Thornthwaite, C. W. and Mather, J. R., "Instructions and tables for computing potential evapotranspiration and the water balance," *Publ. Climatology*, Vol. 10, no. 3 (Centreton, N.J., Drexel Instit. Technology, Laboratory of Climatology, 1957).

Williams, G. P., "Evaporation from snow covers in eastern Canada," Div. Bldg. Res. Natl. Res. Council Canada, N.R.C. no. 5003 (1959).

19

Some Observations on the Influence of Climatic and Terrain Features on Permafrost at Norman Wells, N.W.T., Canada

R. J. E. Brown

Field observations were carried out during the summers of 1959 and 1960 at Norman Wells to obtain quantitative values of the influence of some climatic and terrain features on permafrost. Evaporation (including potential evapotranspiration) and net radiation were measured at the ground surface. Depth of thaw measurements were carried out and ground temperatures were measured in both the seasonally thawed and perennially frozen ground.

Norman Wells (65° 18′ N., 126° 49′ W.) is located on the right bank of the Mackenzie River, 240 ft above sea level, 90 miles south of the Arctic Circle and about 500 miles north of the southern limit of permafrost (Figure 1). The depth to permafrost at Norman Wells varies from 1 to 2 ft in undisturbed moss-peat areas to a maximum of 15 ft and more in areas disturbed by construction. Permafrost occurs everywhere under the ground surface and is about 200 ft thick.

Reprinted from *Can. J. of Earth Sciences,* Vol. 2 (1965), pp. 15-31.

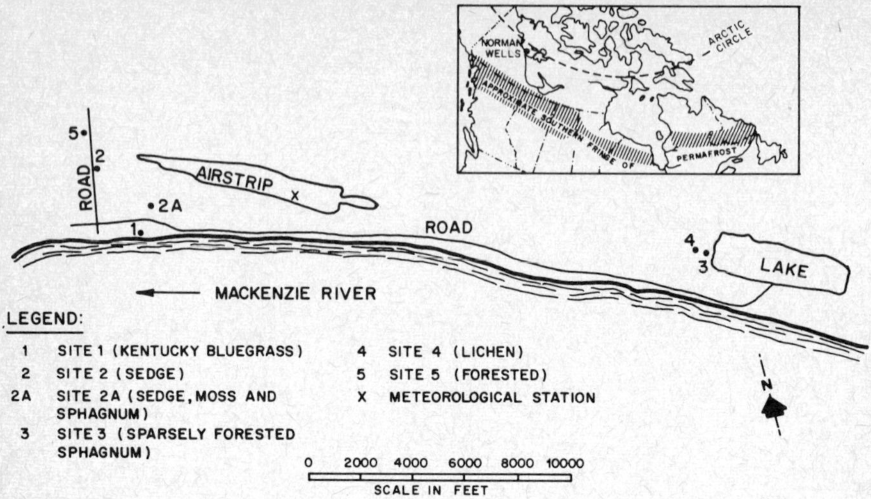

Figure 1. Location of test sites at Norman Wells, N.W.T.

Description of Observation Sites

Observations were taken at five sites in the Norman Wells region. Site No. 1 was a Thornthwaite potential evapotranspiration site; the other four were chosen to include the various types of vegetation indigenous to the Norman Wells region.

Site 1

Vegetation: Kentucky bluegrass, 2 in. high (Figure 2).

Soil: 0 in. to 4 in. Black decomposed organic matter and silt.
Below 4 in. Brown clayey silt, scattered stones, organic matter.

Site 2

Vegetation: Grass-like sedge, 1 ft high with scattered dwarf willow and ground birch (Figure 3).

Soil: 0 in. to 7 in. Black decomposed organic matter (peat).
7 in. to 1 ft 8 in. Black organic silt.
Below 1 ft 8 in. Clayey sandy silt.

Site 3

Vegetation: Scattered spruce and tamarack up to 6 ft high, dwarf willow and ground birch, 2 ft high with ground cover of Labrador tea and *Sphagnum*.

Soil: 0 in. to 5 in. *Sphagnum* (scattered patches exceeded 12 in.).
5 in. to 1 ft 11 in. Black decomposed organic matter (peat).
Below 1 ft 11 in. Grey clayey silt.

Figure 2. Site (Thornthwaite installation) at Norman Wells, N.W.T. Grass-covered pans with tensiometers in foreground. Supply tanks and overflow housing in background. August 1956.

Figure 3. Site 2 (sedge area).

Figure 4. Site 4 (lichen area) showing installation of thermocouple cable in permafrost.

Site 4

Vegetation: Scattered spruce, tamarack, and birch up to 10 ft high, Labrador tea, *Sphagnum*, and lichen (*Cladonia alpestris, Cetraria nivalis, Cetraria cucullata*) (Figure 4).

Soil:	0 in. to 2 in.	Lichen.
	2 in. to 4 in.	Black decomposed organic matter (peat).
	4 in. to 2 ft. 0 in.	Brown decomposed organic matter (peat).
	Below 2 ft 0 in.	Grey clayey silt.

Site 5

Vegetation: Dense spruce and tamarack up to 20 ft high, willow and alder up to 4 ft high with ground cover of Labrador tea, *Sphagnum* and other mosses (*Thuidium abietinum, Pleurozium* sp.) (Figure 5).

Soil:	0 in. to 4 in.	*Sphagnum* and other moss.
	4 in. to 7 in.	Black decomposed organic matter (peat).
	Below 7 in.	Brown clayey silt.

The Kentucky bluegrass at site 1 is not native to Norman Wells but is a cover commonly used at Thornthwaite installations in temperate non-permafrost regions (Thornthwaite 1948). At this site, the depth to permafrost is between 5 and 10 ft. The depth to permafrost at site 2 is about 6 ft and at the other sites it is less than 5 ft.

Figure 5. Site 5 (forested area). 29 August, 1959.

Apparatus and Observation Methods

Evaporation (Including Evapotranspiration)

The purpose of the Thornthwaite installation at site 1 was to test the validity of the Thornthwaite world climatic classification system, which is based on potential evapotranspiration (PE), for northern areas that have long days during the vegetative growing season (Sanderson 1954). PE was measured at the other four sites to observe quantitative differences in various vegetation types which grow naturally in the permafrost region and to compare these with values obtained from the Thornthwaite installation.

At all sites a smaller evapotranspirometer was used; it consisted of half a 45-gal oil drum about 56 cm in diameter. On the bottom of each drum was welded a threaded outlet from which a pipe led to an overflow container about 3 ft away. Measurements were taken daily from 15 to 31 August 1959 and 30 June to 17 September 1960. After the overflow was recorded, a measured quantity of water in excess of the anticipated PE was sprinkled uniformly over the surface.

Air temperatures and rainfall were measured daily at each site. At sites 2 to 5, some perenially frozen soil had to be excavated to accommodate the oil drums and overflow apparatus but the low soil temperatures did not interfere with the evapotranspiration or overflow.

The latest evaporability of the air was measured with an Alundum disc atmometer, loaned by the Canadian Department of Agriculture. This atmometer was installed in the southwest corner of site 1. In 1959, daily readings were taken from the beginning of August until early September. In 1960, daily readings were taken from 11 June until 17 September. The results of both summers were compared with observations from the Thornthwaite installation.

In 1960, a class A evaporation pan was supplied by the Meteorological Branch, Canada Department of Transport, to measure the evaporation from the surface of water in an open pan. This installation was located 15 ft west of the Thornthwaite installation. Daily readings were taken from 11 June to 17 September and the results compared with the observations from the Thornthwaite installation.

Net Radiation

In 1959, weekly readings of net radiation were taken at each site with an economic net radiometer (Suomi and Kuhn 1958) to obtain some estimate of the relative importance of the various ground plant species on the heat exchange between the atmosphere and the ground.

Ground Thermal Regime

In the summers of 1957 to 1959 inclusive, weekly depth of thaw measurements were made at each site to determine quantitative variations under different types of vegetative cover and to assess the relative importance of the individual species comprising the vegetative cover.

Ground temperatures were measured weekly in the thawed layer above the permafrost and in the permafrost at each site to observe variations in the thermal regime under the different types of vegetation. Thermocouples of 20-gauge copper-constantan duplex wire were used to measure the temperatures. In the thawed layer, thermocouple were placed at intervals varying from 3 to 12 in. to the following depths: site 1, 70 in.; site 2, 60 in.; site 3, 22 in.; site 5, 14 in. Weekly temperature readings were also taken on two special thermocouple strings installed in thick *Sphagnum* (site 3) and lichen (site 4). Thermocouples on these strings were spaced at 1-in. intervals from the ground surface to the 12-in. depth to obtain temperature gradients through the living plant cover to determine whether it was uniform throughout or whether it varied with depth, and to determine whether there was any difference in the gradient between *Sphagnum* and lichen. In the permafrost, thermocouple cables were placed at intervals varying from 1 to 1.5 ft to the following depths: site 2, 19.5 ft; site 3, 18.5 ft; site 4, 11.5 ft.

Results and Discussion

Evaporation (Including Evapotranspiration)

The observed daily PE at site 1 (Table I) was compared with Thornthwaite's (1948) formula in which he derived PE from the length of day and a heat index related to the number of degrees of the mean daily temperature above 32°F. The amount of precipitation was not considered because he assumed that the amount of water evaporated and transpired is determined by temperature and length of day. Using the formula, the computed PE for the 1959 and 1960 periods was 4.47 and 24.75 cm respectively compared with the observed PE of 3.28 and 19.41 cm respectively. In 1960 at site 1, the observed PE was 22.90 cm at the oil drum, which is slightly higher than the Thornthwaite installation.

Measurements of the latent evaporability (LE) of the air obtained from the Alundum disc atmometer were converted to PE using the formula suggested by the Canadian Department of Agriculture (Robertson 1954):

$$\text{LE} \times 0.0034 \times 2.54 = \text{PE (cm)}. \tag{1}$$

Applying this formula, the computed PE for the 1959 and 1960 periods were 4.19 and 19.47 cm respectively compared with the PE of 3.28 and 19.41 cm respectively observed at site 1.

The evaporation of water in 1960 from the class A evaporation pan was compared with the PE observed at site 1. The former was considerably higher, being 25.63 cm compared with 19.41 cm for the latter (Table I).

TABLE I. EVAPORATION (INCLUDING EVAPOTRANSPIRATION) AT NORMAN WELLS TEST SITES

	Site 1, 1960	*Site 2, 1960*	*Site 3, 1960*	*Site 4, 1960*	*Site 5, 1960*	*D.O.T. Station 1960*
Elevation above sea level, ft.	318	339	300 ft. from site 5 at same elevation	348	344	340
Mean daily air temperature, °F	54.6	53.2	Not measured	53.6	54.2	55.2
Mean daily max. air temperature, °F	64.3	64.7	Not measured	65.0	66.9	63.8
Mean daily min. air temperature, °F	44.7	41.7	Not measured	42.2	41.5	46.5
Rainfall, cm	17.89	18.63	19.30	19.10	18.89	19.06

P.E., cm	1959	1960					
Thornthwaite installation	3.28	19.41					
Oil drum	—	22.90	12.51	13.37	10.75	11.76	
Class A pan	—	25.63					
Atmometer	4.19	19.47					
Thornthwaite (computed)	4.47	24.75					
Penman	4.51	19.32					

NOTE: 1959 period = 15 to 31 August inclusive (17 days).
1960 period = 30 June to 17 September inclusive (80 days).

In addition to being compared with Thornthwaite's formula, the PE observed at site 1 was compared also with the formula developed by Penman (1956).

Penman's mass transfer formula is:

$$E_0 = 0.35\,(e_0 - e_d)\,(0.5 + (U_2/100)), \quad (2)$$

where

E_0 = evaporation, cm;

$e_0 - e_d$ = vapor pressure difference between surface and air, mm of Hg;

U_2 = mean wind speed, miles per day.

It was assumed that the mean temperature of a saturated surface is approximately equal to the mean air temperature. Wind velocity and air temperature records were obtained from the Department of Transport's Meteorological Station at Norman Wells. The wind factor was adjusted to the 2-meter level by assuming that the wind velocity decreases with height according to the one-seventh power law. The Penman totals for the 1959 and 1960 periods were 4.51 cm and 19.32 cm respectively compared with the observed PE of 3.28 and 19.41 cm respectively. A comparison was made between the calculated Penman formula and observed PE for 10-day and monthly totals, resulting in good correlation (Figure 6).

In 1960, the evaporation observed on the Alundum disc atmometer and computed by the Penman formula (Table I) gave the closest values to the PE measured at the Thornthwaite installation (80-day observation period). None of the values measured in 1959 was close to the PE measured on the Thornthwaite installation possibly because the observation period was so short (only 17 days).

The use of the Thornthwaite formula to compute PE at high latitudes has been criticized because the length of day factors and the heat index have been computed only north to 50° (Norman Wells lies at 65° 18′). Farther north the days are longer during the evapotranspiration period but the amount of incoming solar radiation is less because of the lower angle of the sun. It is not known how well these factors balance out.

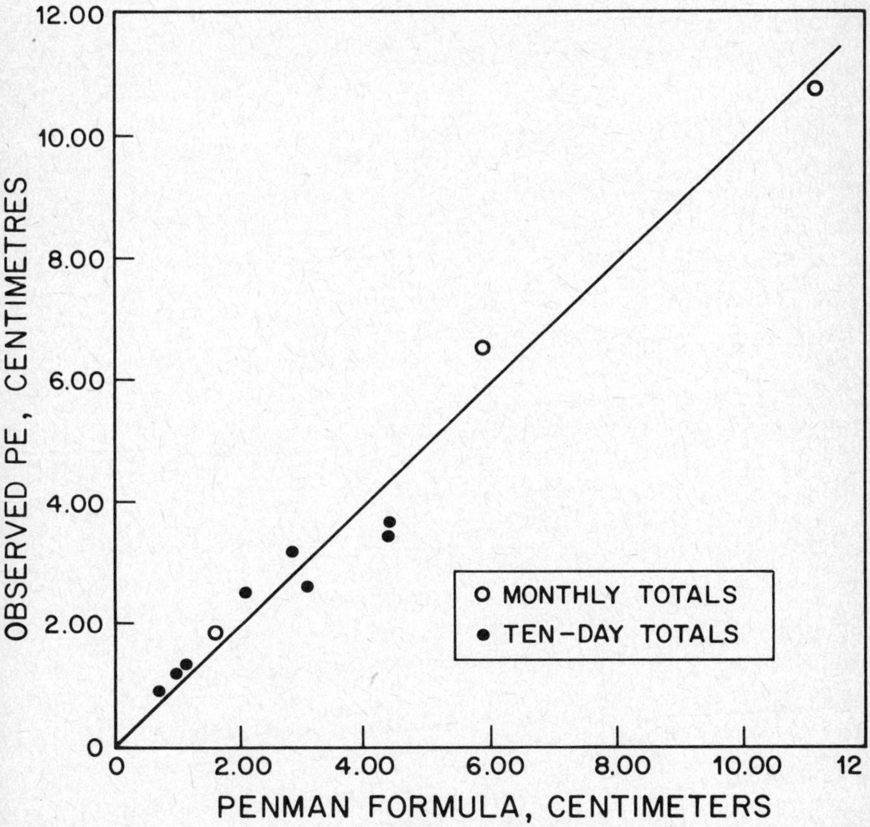

Figure 6. Comparison of observed potential evapotranspiration from Kentucky bluegrass at site 1 (Thornthwaite installation) with Penman evaporation formula, for summer 1960.

The disregard by the Thornthwaite and latent evaporation formulae for different plant types would seem to be an objection particularly for northern areas where mosses and lichens grow widely. The mechanisms of moisture movement in these non-vascular plants are quite different from vascular plants. *Sphagnum* and other mosses are strongly hygroscopic and can lose moisture rapidly and in large quantities. Lichens, however, have very dry surfaces at times, even when lower layers near the soil are very wet. It is not certain whether the transfer of water vapor from the wet basal layer to the atmosphere above the lichen takes place by evaporation or merely by diffusion exhange. The area of vegetative surface per unit ground area of mosses and lichens may be very different from that for the higher plants. For these reasons, the quantity of moisture transfer (and resulting heat transfer) between the permafrost and the atmosphere may be quite different from temperate areas where mosses and lichens are not predominant.

PE measurements in 1960 at sites 2 to 5, inclusive, are presented in Table I. (Results from 1959 are not presented because they are so scattered probably because of the very short length of the observation period.) The PE at these four sites is considerably less than at site 1. Because of the uncertain accuracy of the oil drum method of measuring PE, it is impossible to give quantitative differences among sedge, moss, and lichen. Nevertheless PE from these plant types was significantly less than from Kentucky bluegrass.

Mean daily air temperatures were recorded at four of the five sites but there was no apparent correlation with PE. Site 1 had the highest PE; its mean temperature for the 80-day period of 54.6°F was higher than that of site 5 (which was shaded) and site 2 but lower than sites 3 and 4 near a small lake. The higher degree of exposure to wind of site 1 would contribute to its having a higher PE, but sites 3 and 4 were also fairly exposed. These last two sites had PE values similar to site 5, which was sheltered.

It has been stated that meteorological factors play a prominent role in evapotranspiration rates where soil moisture is not the limiting factor. Nevertheless, it is possible that the physiological characteristics and radiation and thermal properties of plant materials such as moss and lichen, which maintain a high permafrost table, are significant factors in determining the contribution of evapotranspiration to the energy exchange of permafrost. One discrepancy that arises is the fact that the sedge does not maintain a high permafrost table, but has PE rates comparable to those for moss and lichen. This may be caused by its lower insulating qualities which permit a greater depth of thaw during the summer.

Net Radiation

During the summer of 1959, weekly net radiation measurements of various vegetative cover types in the Norman Wells area were taken using an economic net radiometer. To obtain net radiation with approximately 3 to 5% accuracy, the short expression for radiation in Langleys per minute was used:

$$R_n = 1.25\sigma (T_t^4 - T_b^4) + 0.0025 (T_t - T_b) \quad (3)$$

where

R_n is net radiation, Langleys per minute;
σ is the Stefan-Bolz constant;
T_t = temperature of top face, °C;
T_b = temperature of bottom face, °C.

This expression considers incoming radiation positive and outgoing radiation negative.

The calculated net radiation results and air temperatures at the time of reading are tabulated in Table II. This table also includes the exact

TABLE II. NET RADIATION VALUES OBTAINED FROM VARIOUS TYPES OF VEGETATION AT NORMAN WELLS, N.W.T., USING ECONOMIC NET RADIOMETER, 1959

Date	*Site*	*Net radiation g-cal/cm² min.*	*Time P.S.T.*	*Air temp., °F*	*Sky conditions*
Aug. 28	Site 1	0.34863	1155	57.3	Clear
"	Site 2	0.44000	1408	59.1	Clear
"	Site 3	0.38050	1447	60.0	Clear
"	Site 4	0.04525	1025	54.3	Cloudy bright
"	Site 5	0.55762	1522	60.0	Clear
Sept. 7	Site 1	0.12213	1343	38.4	Cloudy bright
"	Site 2	0.12525	1509	38.5	Overcast
"	Site 3	0.11250	1415	38.3	Overcast
"	Site 4	0.02288	0953	37.1	Cloudy bright
"	Site 5	0.24675	1059	37.3	Cloudy bright
Sept. 11	Site 1	0.27875	1351	42.9	Clear
"	Site 2	0.17400	1500	43.0	Clear
"	Site 3	0.12400	1532	43.0	Clear
Sept. 12	Site 4	0.09638	0913	37.4	Scattered cloud
Sept. 11	Site 5	0.06013	1603	43.0	Clear
Sept. 23	Site 1	0.05988	1604	43.3	Cloudy bright
Sept. 24	Site 2	0.08763	0935	34.8	Cloudy bright
"	Site 3	0.18175	1051	36.8	Cloudy bright
Sept. 23	Site 4	0.16550	1022	43.6	Cloudy bright
Sept. 24	Site 5	0.18713	1413	40.1	Cloudy bright
Sept. 30	Site 1	0.11425	0950	36.8	Overcast
"	Site 2	0.02275	1516	38.2	Overcast
"	Site 3	0.03413	1540	38.0	Overcast
"	Site 4	0.02288	1049	38.3	Overcast
"	Site 5	0.00563	1631	38.0	Overcast

times at which observations were taken and the percentage of sunlight recorded during the hour in which each reading was made. No continuous measurements of incoming solar radiation were made with which the observed net radiation values could be compared. It was thus impossible to compare the differences of net radiation through different types of vegetative cover with the incoming solar radiation at the time of observation.

Despite the questionable validity of the observations, a general qualitative pattern is evident. For example, it was anticipated that the net radiation values for lichen would be the lowest because of this plant's light color and consequent high reflective properties. This was true on clear days but not on partly cloudy or overcast days when differences in net radiation from one plant type to another were small. On the other hand, net radiation values for the sedge, which is also a light-colored plant, were much higher than for lichen on clear days and higher than the darker-colored plants such as grass and moss. Net radiation values at site 1 were the highest on some days but not on others. Net radiation through *Sphagnum* and other mosses was highest on clear days but not always on overcast days.

From these crude measurements, it appears that net radiation through moss is higher than through lichen. Nevertheless, these plants maintain the permafrost table at about the same level in a given area and near surface ground temperatures are similar under both plant types. Consequently, if lichen rejects a higher proportion of the net radiative flux than moss this may be compensated by the moss rejecting a higher proportion than the lichen of some other component of the energy exchange.

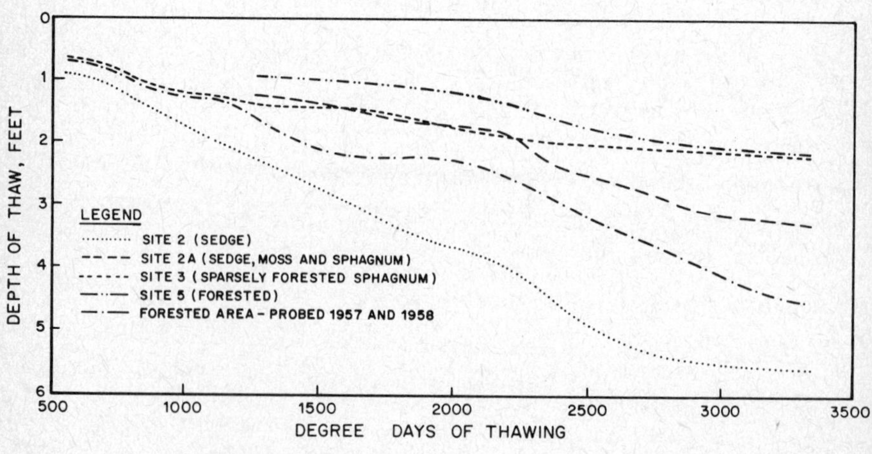

Figure 7. Average depth of thaw vs. cumulative degree days of thawing at Norman Wells, N.W.T., 1957, 1958, 1959.

Ground Thermal Regime

Depth of thaw measurements in various types of vegetative cover were plotted against Fahrenheit degree days for the summers of 1957, 1958, and 1959. The use of degree days gives a more realistic comparison than calendar dates because of the variation in air temperature from one summer to another. For example, the depth of thaw in 1959 was considerably less than in the two previous summers because of the lower air temperatures. Nevertheless, a comparison of depth of thaw with degree days of thawing indicated a similar relationship for the three summers of observations. Average values of depth of thaw vs. degree days of thawing were obtained for each vegetation cover type and plotted (Figure 7).

The greatest depth of thaw occurred at site 2, sedge and no moss cover overlying 7 in. of peat. The next greatest depth of thaw was observed in a forested area, probed only in 1957 and 1958. This site had a moss cover 4 in. thick overlying 3 in. of peat. Actually the 1958 depth of thaw was much greater than in 1957, a condition which may be explained by the inadvertent disturbance of the site in 1957 resulting in an unnaturally deep thaw the next year. A treeless area having a ground cover of sedge (*Scorpidium* sp.) and 3-in. thick cover of *Sphagnum* and other mosses (Figure 1, site 2A) overlying 6 in. of peat experienced less depth of thaw than the previously mentioned area. The shallowest depth of thaw was observed at site 3, sparsely forested area that had a 5 in. to > 12 in. *Sphagnum* cover overlying about 18 in. of peat and site 5, forested area with 4 to 12 in. of *Sphagnum* and other mosses overlying several inches of peat.

The depth of thaw is greatly influenced by the thickness of moss and peat. The degree the depth of thaw is influenced by the shade of shrubs and trees is uncertain. This is probably a relatively minor factor compared with the influence of moss and peat. This is supported by the air temperature observations obtained at the various sites during the summer of 1960. The mean daily air temperature at the sedge area (site 2) for the period of 30 June to 17 September 1960, which includes a major portion of the thawing period, was 53.2°F. The mean daily air temperature for the same period at the forested area (site 5) was 54.2°F although the depth of thaw was several feet less. The predominating influence of the moss and peat is demonstrated by comparing maximum depth of thaw at locations with widely different thawing indices. At Aklavik, N.W.T. (68° N), the mean thawing index is about 2 300; at Thompson, Manitoba (56° N), it is about 3 100. Nevertheless, the maximum depth of thaw in areas of thick moss and peat accumulation at each settlement is similar.

Precipitation also influences the depth of thaw and soil temperatures. No observations were made at Norman Wells and there is little available information in the literature. Russian observations (Shvetsov and Zaporozhtseva 1963) indicate that the situation is complex and that two factors should be considered. First, the amount of moisture in the soil immediately before it freezes in the autumn determines the ice content and depth of thaw the following summer. Second, the moisture content of the soil surface and the infiltration of atmospheric water influence the heat transfer to the frozen soil during the thaw period.

Weekly ground temperatures taken during the summers of 1959 and 1960 in the thawed layer above the permafrost showed that significant differences existed under different types of vegetation. The ground temperatures were the highest at site 1 and progressively less at site 2; in the treeless sedge area with a cover of *Sphagnum* and other mosses (site 2A); and at site 3. In August, site 3 had lower temperatures than site 5 and site 4 had higher temperatures. There appeared to be a general decrease in temperature with increased moss cover and peat thickness. Although the depth of thaw did not appear to be significantly less in areas shaded by trees, these areas experienced slightly lower temperatures in the thawed layer than treeless areas.

Ground temperatures were compared with degree days of thawing. The cooler nature of the 1959 summer was confirmed by the fact that the cumulative total of degree days of thawing on 30 September totalled 2 513 compared with 2 954 on 23 September 1960. A plot of ground temperatures vs. degree days of thawing did not produce similar patterns, however, for the two summers. For any given degree day total, the ground temperatures were much higher in the warmer summer of 1960. For example, in 1959, the temperature at the 12-in. depth at site 1 on 28 August was 46.6°F, at which time the cumulative total of degree days of thawing was 2 232. Almost the same degree day total was reached on 5 August 1960 (2 202), at which time the temperature at the 12-in. depth was 54.5°F.

Mosses and lichen maintain a high permafrost table. In summer, the air temperature frequently rises above 80°F but the temperature beneath these plants remains close to 32°F. At site 3, the *Sphagnum* exceeded 12 in. thickness at the location of the 12-in. thermocouple string so that its full length was in the *Sphagnum*. At site 4, the lichen was only 2 in. thick so that the 12-in. thermocouple string extended 10 in. below the lichen.

In 1959 the temperature gradients for the two strings were higher in the top 8 in. and lower in the bottom 4 in. This means that the gradient was uniform throughout the lichen and the top few inches of the underlying peat, in contrast to the *Sphagnum* where the gradient changed in the *Sphagnum* itself.

In 1960 the same large differences between surface temperatures and temperatures at the lower end of the thermocouple strings were evident.

The chief difference between the two summers was that temperature profiles in 1960 showed changes in gradient at about the 4- or 5-in. depth in contrast to the 8-in. depth in 1959. In all cases, the temperature observed at the surface of the *Sphagnum* and lichen was several degrees higher than the screen air temperature observed at the time of the ground temperature readings.

In 1959, the temperatures observed at the 12-in. depth were the same in both the *Sphagnum* and lichen, but in 1960 the temperature at the bottom of the thermocouple string in the *Sphagnum* was consistently about 5°F higher than in the lichen. It is difficult to know whether temperatures at the 12-in. depth actually varied by this amount between the two plant types or whether one of the thermocouple strings was faulty. It is possible that the thermocouple string in the moss was defective because the 12-in. thermocouple gave unreliable readings and the 11-in. thermocouple was used instead. The readings on the lichen were probably closer to reality but even they were about 4°F higher than temperatures observed at the 12-in. depth in the wooded area.

Taking all the weekly readings from 28 August to 30 September 1959, which was the period when the depth of thaw reached its maximum and the differences caused by variations in vegetative cover were most pronounced, the average temperature at the 1-ft. depth was calculated. It was noted in the hourly readings taken in site 1 (mentioned below) that over a 24-hour period the temperature fluctuation at the 1-ft. depth was 1.5°F. All the ground temperature readings at the various sites were taken between 9.00 a.m. and 4.00 p.m. during which period the temperature fluctuation at the 1-ft. depth at site 1 was 0.8°F. (This would probably be as large a variation as would ever occur during the summer because it was measured through a day which was clear with a large air temperature fluctuation.) The mean air temperature for this period was 41.2°F. The mean ground temperatures were as follows:

Site 1	(Kentucky bluegrass)	40.0°F	Site 3 (thick *Sphagnum*)	32.5°F
Site 2	(sedge)	36.5°F	Site 4 (lichen)	32.6°F
Site 2A	(sedge, moss *Sphagnum*)	35.0°F	Site 5 (forested)	32.8°F
Site 3	(sparsely forested *Sphagnum*)	33.7°F		

The lichen and *Sphagnum* had approximately the same average temperature over this period although the lichen was only 2 in. thick in contrast to the 12 in. of *Sphagnum,* both overlying similar thicknesses of peat.

Means of the weekly temperatures at the various depths for different time periods were calculated for 1960. Mean ground temperatures and ground temperature fluctuations for the period of 15 July to 16 September are plotted in Figure 8 showing them to be much higher at sites 1 and 2 than under the *Sphagnum* and other mosses at site 5. Differences between the mean air and mean ground temperatures at the 12-in. depth at the various sites for the period of 28 August to 30 September 1959 were compared with the differences for the period of 15 July to 16 September 1960. The differences were:

Site 1 (Kentucky bluegrass)	1.2°F (1959),	3.3°F (1960)
Site 2 (sedge)	4.7°F (1959),	5.6°F (1960)
Site 3 (thick *Sphagnum*)	8.7°F (1959), about	12°F (1960)
Site 4 (lichen)	8.6°F (1959), about	17°F (1960)
Site 5 (forested)	8.4°F (1959),	19.8°F (1960)

This appears to confirm the thermal resistance of the moss and lichen because the much higher air temperature means in 1960 did not cause ground temperatures that were much higher than those recorded in 1959.

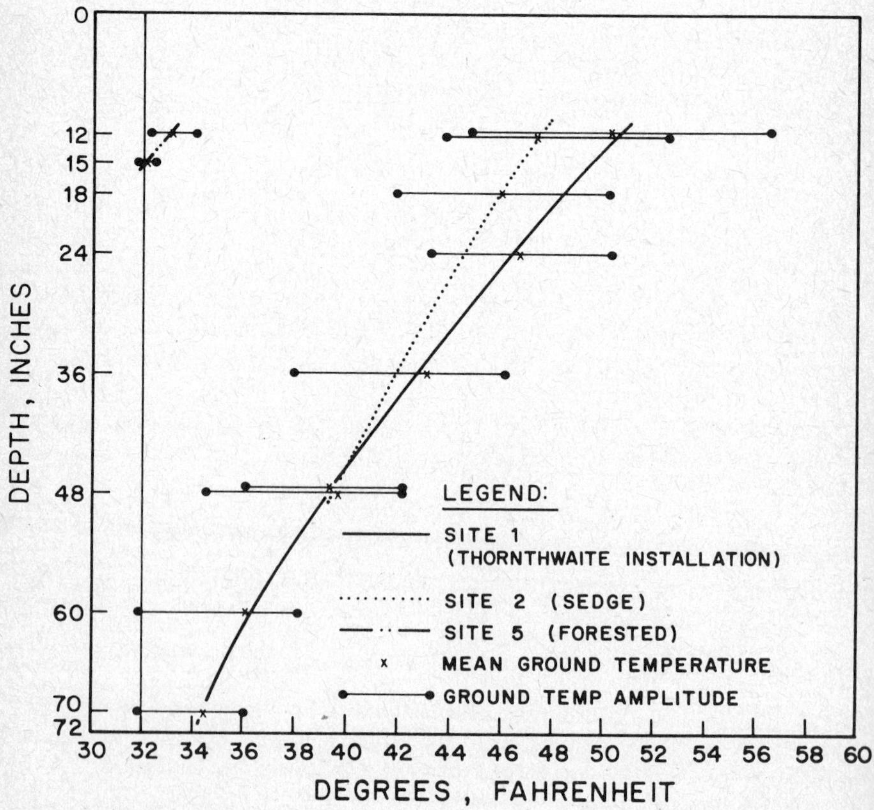

Figure 8. Mean ground temperatures and ground temperature fluctuations (15 July to 16 September, 1960).

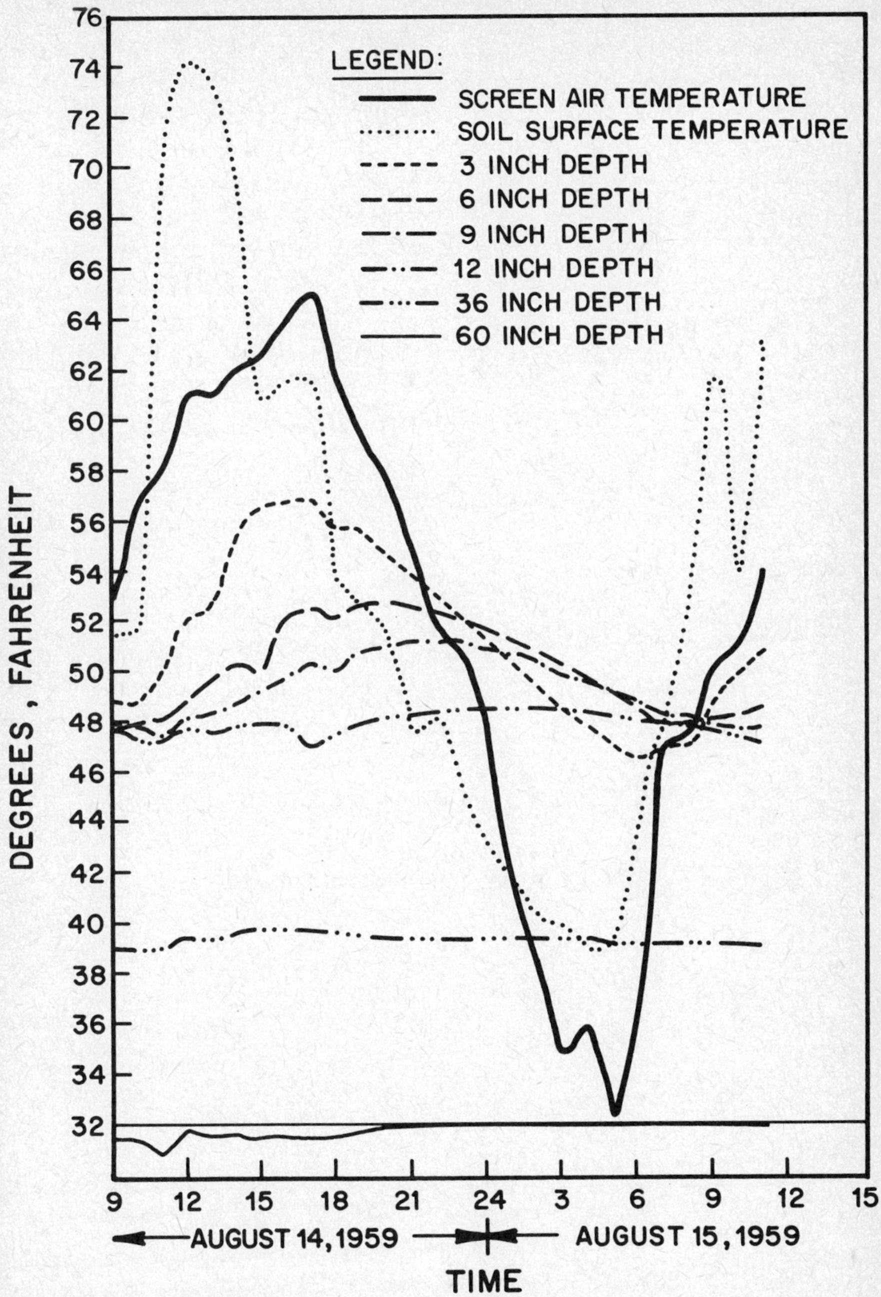

Figure 9. Ground temperatures at site 1 (Thornthwaite installation) (hourly readings from 0900 on 14 August, 1959 to 1100 on 15 August, 1959).

In 1959, a series of 27 consecutive hourly observations of screen air temperature, ground surface temperature, and temperatures at depths of 3, 6, 9, 12, 36, and 60 in. were taken at site 1 from 9.00 a.m. 14 August to 11.00 a.m. 15 August. A plot of these observations showed that the largest temperature fluctuations occurred in the air and at the soil surface; fluctuations decreased progressively with soil depth. The lag in maximum and minimum temperatures also increased with depth (Figure 9).

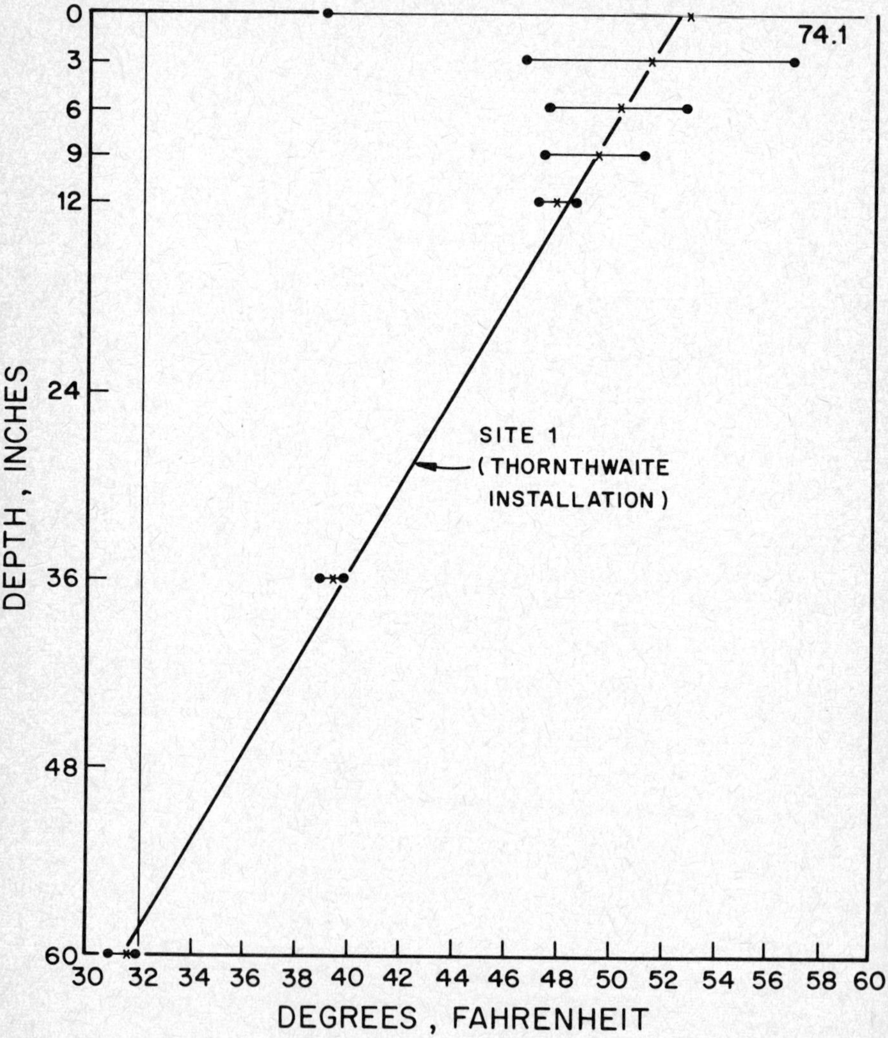

Figure 10. Mean temperature of 24-hour period vs. depth at site 1 (Thornthwaite installation). (Hourly readings 0900, 14 August, 1959, to 0900, 15 August, 1959). Mean air temperature for this period +51.7°F with fluctuations from 65.0°F to 32.5°F.

Mean temperatures for the 24-hour period from 9.00 a.m. 14 August to 9.00 a.m. 15 August were calculated for the 3-, 6-, 9-, 12-, 36-, and 60-in. depths, the soil surface, and the screen air temperature (Figure 10). A plot of the means on a graph indicates a linear decrease with depth from the surface to the frost table at the 60-in. depth and a decrease of ampli-

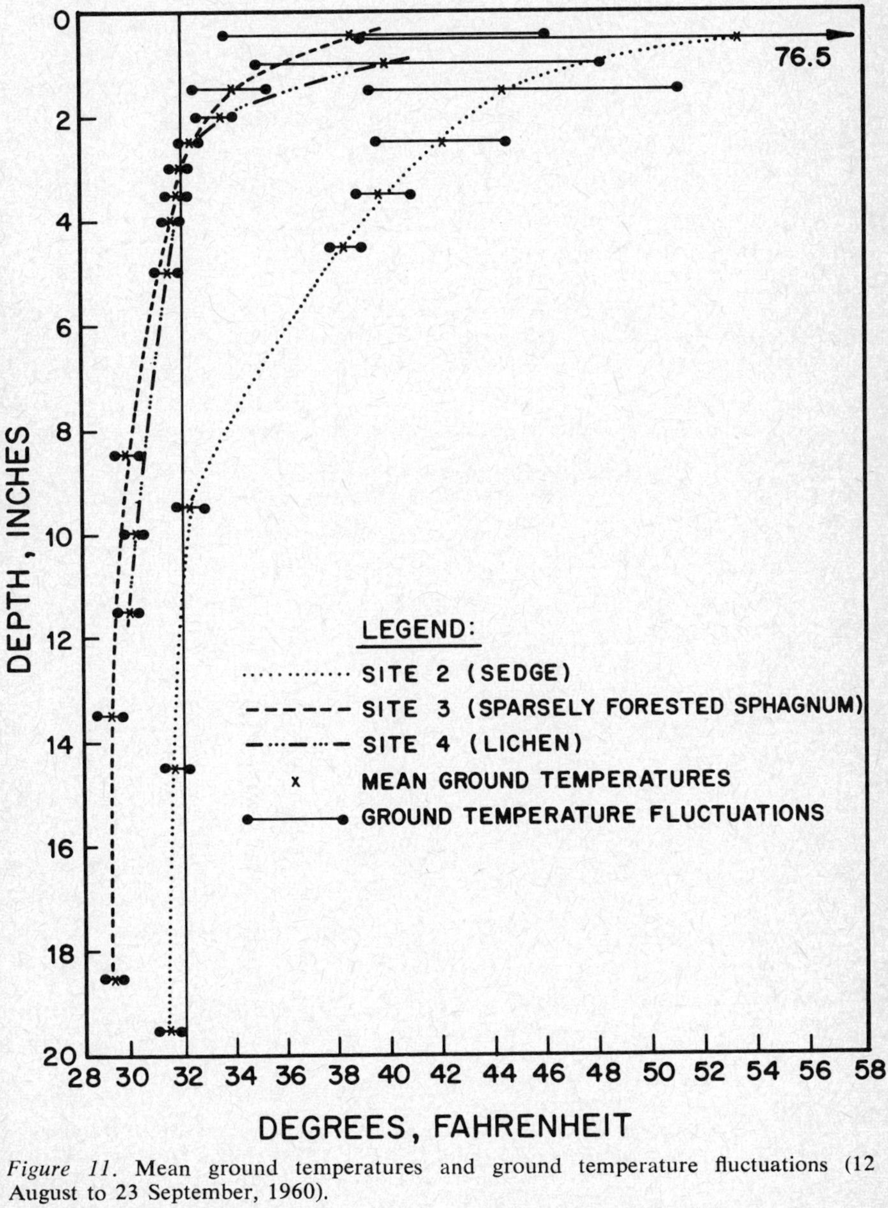

Figure 11. Mean ground temperatures and ground temperature fluctuations (12 August to 23 September, 1960).

tude. The highest mean temperature for the period occurred at the soil surface (52.9°F) and the mean air temperature in the screen (51.7°F) was almost the same as the ground temperature at the 3-in. depth (51.5°F).

Weekly readings were taken in 1960 on the three thermocouple strings in permafrost from the end of July to 23 September. The means and fluctuations of the seven weekly sets of ground temperature readings taken in August and September 1960 are shown in Figure 11. The lower summer ground temperatures under the moss and lichen as compared with the sedge are immediately evident. In addition, the temperature fluctuation in the top few feet is much greater under the sedge than under the moss or lichen.

Conclusions

1. The PE values computed from the evaporation recorded on the Alundum disc atmometer and from the Penman mass transfer formula compared favorably with the PE measured at the Thornthwaite installation during the 80-day period in 1960. There was not the same good correspondence in 1959 possibly because of the short observation period, only 17 days. The PE measured at the oil drum evapotranspirometer, and the PE computed from the Thornthwaite formula and from the evaporation recorded at the class A pan, were considerably higher than the PE measured at the Thornthwaite installation in 1959 and 1960.

2. The measurement of PE by adding water to the surface of a small plot of vegetation and recording the overflow is questionable because the daily addition of water actually changes the climate of the vegetation under test. In addition, the results do not give a true representation of the natural moisture regime.

3. The use of very small test plots such as oil drums for measuring PE is questionable because the edge effect is so pronounced that it probably changes the evapotranspiration.

4. Despite the difficulties of measuring exaporation (including PE), it appears that the grass-like sedge, moss, and lichen at Norman Wells gave significantly lower PE results than the Kentucky bluegrass.

5. The economic net radiometer is a crude instrument capable only of spot measurements. A continuous recording net radiometer is required to measure the thermal energy entering the ground through the vegetative cover. Nevertheless, from the observations taken with the economic net radiometer, it appears that values are somewhat higher for *Sphagnum* and other mosses than for lichen. Because the thermal regime of the ground under moss and lichen appears to be similar, as manifested by similar depth of thaw and ground temperature regime, differences between the two

plant types must occur in some other component of the energy regime which was not studied.

6. The depth of thaw under moss and lichen growing on peat was less than in areas supporting other types of plant growth.

7. Ground temperature observations in the thawed layer and in the perennially frozen ground showed that temperatures were lower in areas covered by moss and lichen than in areas supporting other types of plant growth.

The influence of climatic and terrain features is so complex and the features so closely interrelated that it is difficult to isolate each one or measure them accurately. The measurements at Norman Wells, described above, were carried out to provide some preliminary quantitative information on the relative importance of these features in the permafrost environment. More sophisticated instrumentation and a more comprehensive study program is required to obtain a complete picture of the permafrost regime and the role of the various features influencing it.

References

Penman, H. L., "Evaporation: An introductory survey," *Netherlands J. Agr. Sci.*, Vol. 4, no. 1 (1956).

Robertson, G. W., "Latent evaporation: its concept, measurement and application", Paper presented 29 January 1954 at the regular weekly seminar of the Field Husbandry, Soils Agr. Eng. Div., Exptl. Farms Serv., Dept. Agr., Ottawa.

Sanderson, M., "Measuring potential evapotranspiration at Norman Wells, Northwest Territories, Canada, 1949, 1950," *The measurement of potential evapotranspiration* edited by J. R. Mather, Seabrook, N.J., (Publications in Climatology, The Johns Hopkins University, Laboratory of Climatology, 1954), Vol. VII, no. 1, pp. 84-90.

Shvetsov, P. F. and Zaporozhteseva, I. V., Povtoryayemost' i inzhenerno-geokriologicheskoye znacheniye dvukh-trekhletnikh povysheniy temperatury gruntov v subarktike (The recurrent nature and permafrost engineering significance of 2-3 year soil temperature increases in the subarctic). Problemy Severa (Problems of the North), no. 7 (1963), pp. 22-45.

Suomi, V. E. and Kuhn, P. M., "An economical net radiometer," *Tellus,* Vol. 10, no. 1 (1958), p. 160.

Thornthwaite, C. W., "An approach toward a rational classification of climate," *Geographical Rev.,* Vol. XXXVIII, no. 1 (1948), p. 55.

20

Notes on Late-Quaternary Climatic Changes in Canada

J. Terasmae

Introduction

In recent years research projects in the earth and biological sciences undertaken in all parts of Canada have significantly widened the range and improved our knowledge of geologic and biological events of late-Quaternary age. The scope of many of these studies has included climatic changes as a framework on which to base the interpretation of results. Advance and retreat of continental glaciers, sea-level fluctuations, migrations of plants and animals, formation of soil profiles and drainage patterns, changes in sedimentary sequence and sequences of fossils all depend in some measure on changes in climate. Knowing the present-day relationships, past climatic environments can be reconstructed with some degree of certainty.

In other parts of the world enough detailed studies have been made to enable compilation of fairly accurate climate curves and maps showing, for example, the composition of "fossil forests" at certain times in the post-Glacial episode. Although similar patterns are beginning to appear, much more careful work is required before such compilations can be made in Canada, except for local areas.

Reprinted from New York Academy of Science, *Annals,* Vol. 95 (1961), pp. 658-75.

It is with this sobering thought in mind that I wish to emphasize the word notes, in the title of this article. Nevertheless, there is a reasonably large number of these notes available, and some lines of approach to paleoclimatic problems of the late-Quaternary time in Canada will be outlined in the following.

Studies pertaining to Quaternary research in Canada have passed through an encouraging development in the past 10 to 15 years, and it is the opinion of the writer, based on numerous discussions with his colleagues in various fields of science, that this has been a phase of basic studies and reconnaissance characterized by conservatism in interpretation and critical observation and gathering of field data. Only recently have many preliminary studies been completed and reports published on the obtained results.

Scientists have at times ventured guesses at paleoclimates as inferred from their studies, but the bravest lot of gamblers, by far, have been the palynologists who have had definite ideas on paleoclimates for a long time. Perhaps fortuitously such paleoclimates usually have been described in general terms, such as "cool, moist" and "warm, dry," as compared with the present climate in the region where the studies were made. Such designations of climate may be quite adequate for some purposes, but they are insufficient and even misleading for more specific discussions. For example, warm, cool, dry, and moist are relative terms, and absolute figures, even approximate, are much more descriptive and definite and can be compared with available meteorological data.

References to the literature are intended as examples only, and I make no claims for adequate coverage of available published reports. Several of the references listed, however, contain useful lists on bibliography in their specific fields of study.

Sources of Information

Meteorological observations. Records on observations of weather in Canada are available for more than 100 years in only rare cases; generally reliable meteorologic observations extend back 30 to 50 years at a number of localities. Meteorologic stations in more outlying areas are rather recent. Scattered notes on weather can be found in written records much farther back in time, but these are of less interest and value because of their discontinuous nature (Koeppe, 1931).

Geologically speaking and also from the paleoclimatologic point of view, the written records on climate in Canada are recent. Nevertheless, *such records form the only sound basis for interpretation of past climates* (Canada Year Book, 1960; Chapman, 1952; Chapman, 1938, 1953; Hare, 1950, 1952*a* and *b*; Kendrew and Currie, 1955; Kendrew and Kerr, 1955; Rae, 1951; Shenfeld and Slater, 1960; Thomas, 1953). Paleoclimatologic studies would certainly benefit if meteorologic records from more localities in Canada were available. It is hoped that the excellent work carried on by the Meteorological Branch, Air Services, of the Canada Department of Transport can be extended considerably in the future.

Glacial geology and related studies. Geological studies of Quaternary deposits made across southern Canada have established in some detail the stratigraphy and correlation of glacial, interglacial, and interstadial deposits (Antevs, 1925, 1931; Armstrong, 1957; Chapman and Putnam, 1951; Coleman, 1933, 1941; Craig, 1956, Deane, 1950; Dreimanis, 1960*a* and *b*; Elson, 1955, 1957, 1959, 1960; Fyles, 1956; Gadd, 1960; Glacial Map of Canada, 1958; Gravenor, 1957; Henderson, 1959 *a*, *b*, and *c*; Hughes, 1956; Johnston, 1917; Karlstrom, 1956; Karrow, 1957, 1959; Kupsch, 1960; Laverdiere and Courtemanche, 1959; Lee, 1955, 1957; MacClintock and Terasmae, 1960; Mirynech, 1961; Prest, 1957; Stalker, 1960; Terasmae and Hughes, 1960*a*; Watt, 1953). Owing to the fact that the regime of glaciers is ultimately dependent on climate, the studies in glacial geology have provided much valuable information of paleoclimatologic importance.

Geologic studies have shown that the beginning of the last, Wisconsin, glacial stage consisted of several advances and retreats of the glacier similar in magnitude to the final retreats and advances of the Wisconsin ice. Thus it can be inferred from geologic studies that climatic oscillations leading to the Wisconsin glaciation began with a trend towards lower annual mean temperatures in southeastern Canada. Average total precipitation may have remained unchanged, but the annual snowfall in relation to rain must have increased. Lower annual mean temperatures could have been caused by lower summer temperatures; higher winter precipitation, particularly as snow, could have resulted from generally milder winters with fewer extremely cold periods. Such conditions would increase the total snowfall and reduce the melting period during summer. Further evidence for this statement will be presented later.

The main Wisconsin ice advance covered southern Canada, and thus no records of climatic conditions were preserved during the height of glaciation. However, retreat of the Wisconsin ice, interrupted by halts and readvances, left abundant evidence of climatic conditions, only a part of which has been studied. Rapid retreat of the ice margin about 11,000 to 7000 years ago is clearly indicated by abundant evidence of melt-water erosion and deposition. The magnitude of such phenomena require large volumes of water and often torrential currents. Formation of large ice-dammed lakes such as Glacial lakes Agassiz, Barlow-Ojibway and Iroquois, and particularly the great volumes of sediments deposited in them required correspondingly huge sources of supply that could be derived only from the melting of the continental glaciers.

The steep gradiant of tilting of the beaches formed around the shores of some of the ice-dammed lakes indicate rapid differential uplift and release from the load of the ice. All these data indicate that at the close of the Wisconsin glaciation a significant and quite abrupt increase in temperature is required for an almost "catastrophical" melting of the continental glaciers.

Geologic studies made in southern Canada stimulated the extension of similar studies northward through the provinces and into the Northwest Territories (Antevs, 1925; Bird, 1959; Cook, 1960; Craig, 1960; Craig and Fyles, 1960; Geol. Survey of Canada Bull. 54, 1959; Glacial Map of Canada, 1958; Hare, 1959; Henderson, 1959*b*; Hughes, 1956; Ignatius, 1956; Ives, 1959*a, b,* 1960; Laveriere and Courtemanche, 1959; Lee, 1959, 1960; MacKay, 1958; Nicholson, 1960; Prest, 1957; Terasmae and Hughes, 1960*b*). Much important paleoclimatological information has been gleaned from these investigations. First, the last remnants of ice melted from areas and west of Hudson Bay about 7000 to 6000 years ago. It has been postulated that annual mean temperatures then may have been about 5°F. warmer than at present (Ives, 1959*a*). However, following the hypsithermal interval in central Labrador, a cooling of climate with resulting greater accumulation of snow is indicated by formation of the nivation hollows (Henderson, 1959*b*). Further evidence for such a cooling of climate is found in advances of alpine glaciers in western Canada and Alaska (Goldthwait, 1960; Karlstrom, 1960). The history of the Tyrrell Sea (the post-Glacial marine inundation in the Hudson Bay region) has been outlined by Lee (1960), and this study is also pertinent to climatic interpretations in that region.

Studies of permafrost, its regime, and southern limit have proved to have paleoclimatologic implications (Cook, 1960).

Certain biologic, geographic and archaeological studies in the Canadian Arctic have in several ways contributed to climatologic considerations (Bird, 1959; MacKay, 1958; Meldgaard, 1960; Nicholson, 1960; Porsild, 1955, 1957, 1958; Terasmae and Craig, 1958; Terasmae, 1959).

Studies of modern glaciers in the Arctic and the Cordillera have yielded climatologic information extending from the present to some considerable antiquity (Hattersley-Smith, 1958, 1959; Mathews, 1951).

Oceanographic studies. Evidence of former higher and lower sea levels has been recognized for many years, but only recently have such phenomena been related more closely to changes in climate (Fairbridge, 1958, 1960; Shepard and Suess, 1956). Along the coasts of Canada, however, studies of the past sea-level fluctuations have been complicated by isostatic and other crustal movements that have been of varying magnitude in different regions.

Much more useful paleoclimatological information has been gleaned from studies of fossils in marine sediments (Wagner, 1959) coupled with radiocarbon or other methods of dating of marine fossils (Elson, 1960). Stratigraphic, paleontologic and sedimentologic studies of Quaternary marine sediments in Canada, now in progress, are already yielding promising data also on paleoclimatic problems. Such studies in the Arctic will probably eliminate much speculation inherent in current theories on the Ice Ages.

Paleontologic studies. Investigations of Quaternary vertebrate fossils (Hibbard, 1949) in Canada have been made in some areas widely scattered in both space and time (for earlier studies see Baker, 1920), and there is much need for more paleontologic studies. Taxonomic, genetic, and ecologic studies of present animal life form the necessary basis for interpretation of fossils (Rand, 1948, 1954). To give examples of some recent contributions, the archaeological studies by R. S. MacNeish in the Yukon (Firth River area) have supplied much useful information on fossil faunas and paleoclimate. Other rich fossil localities are known from the Yukon, Alaska, where, at Dawson, a geologic and paleontologic study is in progress by O. L. Hughes of the Geological Survey of Canada. Many contributions have been made by scientists of the Royal Ontario Museum and the National Museum of Canada, for example.

Lists on Quaternary invertebrate faunas exist from many areas. Such lists of species are certainly useful; but from the climatological as well as the paleoecologic point of view, data on the relative abundance of different species and their distribution in time, and also information on representative size and weight of individuals is much desired and would add considerably to the usefulness of such studies. This approach on molluscan faunas of fresh-water origin, presently pursued by La Roque (1960), Reynolds (1959), and Elson (1960), has clearly shown the advantages of a wider scope used in studies of marine molluscs of the Champlain Sea episode in the St. Lawrence lowlands. Elson's studies (1960) have shown that the Champlain Sea episode, lasting from 11,500 to 9500 years B.P. (Terasmae, 1960), can be divided into four phases:

(1) *Pecten* phase. Probable deep water phase.

(2) *Hiatella* phase. Subarctic sea with summer temperatures from 0°C. to 5°C. and salinity from 28 to 23‰. Probable age from 11,300 to 10,600 years B.P.

(3) *Mya arenaria* phase. Summer temperatures increasing from 5°C. to perhaps 12°C. (locally 15° to 20°C.). Salinity probably decreased from 25‰ or 20‰ as the water freshened. Probable age 10,600 to 10,000 years B.P.

(4) The *Lampsilis* lake. Brackish to fresh water environment. End of the Champlain Sea episode.

Botanical studies. Under this heading two major points deserve mention. First, many species of plants have not had their present distribution ranges in the past; and, second, the migration and existence of plants within their specific ranges are largely governed by ecologic factors that in turn are intimately related to climate. For example, a change in climate affects the ecologic site and this either favors or limits the existence of certain plant species at that site. Depending on such changes, the plant communities change in composition. Migration of plants may occur in a direction of more favorable environments.

It is here that studies of taxonomy, ecology, genetics, history, and geography of modern plants yield pertinent information on paleoclimatology (Baldwin, 1958; Danserau, 1953; Deevey, 1949; Dillon, 1956; Dore and Gillett, 1955; Grayson, 1956; Griggs, 1937; Halliday, 1937; Hare, 1959; Horton and Bedell, 1960; Hustich, 1949; Löve and Löve, 1957; Löve, D., 1959; Löve, A., 1959; Native Trees of Canada, 1956; Polunin, 1940; Porsild, 1955, 1957, 1958; Raup, 1941; Savile, 1956; Scoggan, 1957). Extensive botanical studies are in progress by J. A. Calder in British Columbia and the Yukon by D. B. O. Savile in the Arctic Archipelago, and by J. C. Ritchie in Manitoba, to mention a few.

Discontinuous ranges of distribution of some species may indicate that such ranges were continuous in the past, and thus that climatic conditions may have prevailed that favored that particular species. An advancing or retreating tree line, for example, may indicate that climatic conditions are either improving or deteriorating (Griggs, 1937).

Certain soil profiles are associated with specific types of plant communities and, where such association does not exist, a climatic change may have caused a change in vegetation, whereas the soil profile, being more resistant to change, remains unaltered for some time. A situation where forest is invading prairie or vice versa may illustrate this point.

Study of soils. The different physiographic regions and phytogeographic regions have their distinctive climatic conditions anad are characterized by differences in the soil profile that has developed. Particularly in areas that were covered or strongly influenced by Pleistocene glaciations, and where migrations of plants have taken place, changes in the soil profile also have occurred. Such changes are easier to detect in some areas than in others. In border line areas between distinct physiographic and phytogeographic regions these changes can be observed quite readily. An interesting study has been made by Sharp (1942) in the Yukon where he noticed patterned ground on the slopes in the St. Elias Range. These features occur largely above timber line and below the permanent frost line. Near and below the timber line these features are stable, as indicated by invasion by vegetation and some modification by weathering and erosion. Similar features near the snow line are active. In the intermediate area some stone stripes rest on earthy subsoil containing abundant plant roots. Sharp believes that this relation indicates a rejuvenation of soil structure development caused by the accession of cooler climate following a mild period during which frost action has been less severe.

Another interesting example has been observed by Hills (1959, and personal communication) from the Clay Belt of northern Ontario where soil drainage conditions, following the drainage of Glacial Lake Barlow-Ojibway about 6000 to 7000 years ago, were better then than later, partly owing to greater slope towards James Bay in early post-Glacial time. It is possible that during the drier and warmer hypsithermal interval, about 7500 to 5000 years ago, even fine-grained silty clay could have been reasonably well drained and a corresponding soil profile developed. Later, a cooling of climate or an increase in precipitation, or both, caused the development of the glei layer that is now characteristic of the poorly drained soils of the Clay Belt. This glei layer seems to have been developed in, and superimposed on, an earlier, different soil profile.

Although such plant-soil dissociations and superimposed soil profiles or soil profiles out of harmony with their physiographic regions can be good indicators of climatic changes, it is often difficult to determine the age of such changes from the soil profiles, except in a relative sense.

Palynology and paleobotany. Palynological and paleobotanical studies help to reconstruct the past development, migration, and composition of vegetation. Paleobotanical studies of Canadian Quaternary deposits extend back about 100 years and several reports include results from these investigations (Baker, 1920; Coleman, 1941; Hinde, 1878). The main drawback in paleobotanical studies, from the paleoclimatologic point of view, is the lack of quantitative data. In most cases, only a list of plant species is is given.

Palynological studies, however, have overcome this difficulty. Furthermore, plant microfossils can be found in a much greater variety of sediments, and sufficient numbers of pollen and spores, to allow statistical treatment of the results, can be recovered from small samples. These are important advantages that have been used with success in stratigraphic and paleoecoligic work.

Palynological studies made of Quaternary deposits in Canada are rather widely scattered. Somewhat better coverage is available for Quebec and Ontario (some results still unpublished), where numerous studies have been made in recent years (Auer, 1930; Grayson, 1956; Ignatius, 1956;

Potzger, 1953*b*; Potzger and Courtemanche, 1956; Terasmae, 1958, Wenner, 1947).

In British Columbia palynological studies have been made by Hansen (1949, 1950) and Heusser (1955), and by myself in scattered localities in that province (Terasmae and Fyles, 1959).

Palynological studies have been planned by J. C. Ritchie in Manitoba and Mrs. M. Steeves in Saskatchewan where the writer has made isolated studies of late-glacial deposits.

Palynological investigations (unpublished) made by the writer in New Brunswick, Nova Scotia, Prince Edward Island and Newfoundland have yielded much useful information, pertinent to paleoclimatologic speculation. A detailed study has been made in Cape Breton island by Livingstone and Livingstone (1958), and Ogden (1960) has studied pollen stratigraphy in Kings County, Nova Scotia.

I have made palynological studies of Quaternary beds in several areas in the Northwest Territories (Terasmae and Craig, 1958; Terasmae, 1959) but this work is still in progress.

Palynological studies that I have made in the Arctic islands are still in the preliminary stage and hence unpublished. However, these studies have shown that continuous sequences of sediments extend beyond the range of the radio-carbon dating in some areas indicating that such areas were not glaciated at least during the last glaciation. Livingstone's studies in Alaska (1955) have been very helpful for reference anad comparison.

Climatic Changes

Interpretation of climatic changes (Deevey, 1949; Ewing and Donn, 1956; Flint, 1947; Landsberg, 1949) beyond the range of written records is a fascinating field for speculation. The factors to be considered are numerous and the pitfalls are many, some of which are known: others not. The reliability and accuracy of paleoclimatologic studies largely depends on the alertness, experience, and ability of observation and interpretation of the scientist. Certain kinds of evidence are very obvious: for example, the existence of dry lake basins in arid regions constitute evidence of former climates with more moisture. Changes in abundance of certain species within a plant association are much more difficult to interpret from a climatologic point of view and require intimate knowledge of ecologic tolerance and requirements of the species concerned. An excellent example of such a study is the report on white and red pine by Horton and Bedell (1960). Experience has shown that interpretation of past climates should be based on as many criteria as possible and this calls for team work where scientists from several fields join their forces for mutual benefit.

The plant-animal-climate relationship is very complex, and great caution should be exercised in reaching paleoclimatologic conclusions. In such conclusions adequate attention must be given also to edaphic factors and the history of the species with its genetic potentialities and deficiencies. The geological history of the region considered is of equal importance.

Southern Ontario and Quebec. In this region evidence of late-Quaternary climates extends back about 100,000 years beginning with the interglacial Don beds at Toronto (Coleman, 1932; Hinde, 1878; Terasmae, 1960; Watt, 1953). Although an unknown length of time is not represented by deposition in the beginning of that interglacial interval (probably Sangamon), evidence found in deposits of the remaining part of that interval indicates that climatic conditions were closely similar to the present at Toronto, and toward the end of the interglacial interval a deterioration of climate occurred. It is inferred from palynological studies that the annual mean temperature was lowered by about 15°F. in the closing phase of that interglacial interval, bringing about a change from mixed hardwood forest to a boreal forest. A further lowering of the annual mean temperature by another 15° to 20°F. would have introduced tundra conditions, but satisfactory evidence for this is lacking. At present it seems more likely that the forest in the Toronto area may have been eventually overridden by the main Wisconsin ice. However, the Wisconsin ice did not advance to the limit of its maximum extent in one grand event. Halts and retreats of the ice margin occurred, as indicated by nonglacial deposits (Dreimanis, 1960*b*; Terasmae, 1960). It is probable that an increase of some 10°—15°F. in the annual mean temperature occurred in southern Ontario during these advance interstadial intervals of the Wisconsin glaciation.

In all the previous reasoning it was assumed that the annual precipitation did not decrease; rather it was the ratio of rain and snow that decreased. At present the annual precipitation at Toronto consists of 25.5 inches of rain and 54.6 inches of snow. In the boreal forest region (at Kapuskasing) this ratio is about 18.4 inches rain and 95.8 inches snow and in the central areas of Labrador (at Knob Lake) the ratio is 14.7 inches of rain and 128.6 inches of snow. In all these areas there is not too much difference in the total precipitation, measured as water. If this trend is followed and accompanied by a slight decrease in the annual mean temperature, accumulation of glacier ice will inevitably result, first in areas such as central Labrador-Ungava.

Some areas in southwestern Ontario became ice-free some time before 12,000 years ago and, possibly, 13,000 to 14,000 years B.P. as shown by radiocarbon dates. With all probability tundra conditions existed at least locally, and palynological studies I have made lend some support to this statement. From the time of deglaciation until some 10,000 years B.P. boreal forest covered southwestern Ontario. It seems reasonable to assume that climatic conditions were somewhat similar to these in the present boreal forest region of Ontario. January mean temperatures may have ranged from 0°F. to 10°F. Extreme low temperatures in the winter may have been frequently —50°F. to —60°F. Late frosts may have occurred in June, and summer temperatures reached 90°F. to 100°F. July mean temperature probably ranged in the 60s. Summers were short and frosts may have occurred in September and October. Precipitation during summer was higher than in the winter. Average total precipitation ranged from 20 to 30 inches, and winter snowfall averaged from 60 to 100 inches.

These climatic conditions quite effectively eliminated most hardwood species from the forest composition.

TABLE I.

	Pine area	*Extremes in pine range*
Annual average of daily mean temperature	40° F	32 to 44° F.
January average of daily minimum temperature	−3° F.	−15 to 18° F.
July average of daily maximum temperature	79° F.	64 to 81° F.
Average total precipitation	30.2″	20.3 to 53.1″

The episode from 10,000 to 8000 years ago is characterized by a decrease in spruce and a corresponding increase in jack pine. This change could have been caused by a slight decrease in summer precipitation or an increase in summer temperature resulting in a longer, drier summer, whereas extreme cold spells in the winter still may have been frequent. It is possible that jack pine was favored under such conditions.

About 8000 to 7000 years ago the jack pine was replaced first by red pine and then by both white pine and red pine. Horton and Bedell (1960, p. 17) have described the climate within the range of white and red pine (Table I). For the area with greatest pine concentration they state on page 17, ". . . the data indicate a moderate temperature, but somewhat low precipitation. Summers are warm but somewhat droughty. The abundance of the pine in the various parts of the range is controlled to a large extent by climate as it favors or discourages competing species and the incidence of fires." Furthermore, on p. 53: "Generalizing, it is apparent that the pines prefer dry conditions. Stands are abundant and relatively stable only in the climatically drier sections, and in moister sections the pine species are largely restricted to dry sites by strong competition from tolerant hardwoods or, to the north, boreal mixedwoods." These statements seem to be well substantiated by palynological studies.

Hardwood species were present in southern Ontario all through the pine period lasting from 7000 to 5000 years ago. Then a marked decrease in pine occurred, with a corresponding increase of hemlock and beech.

The study by Horton and Bedell supports the interpretation, based on palynological studies, that a change in climate at the close of the pine period involved more moisture during summer. The average annual mean temperature probably remained unchanged and about 5°F. higher than the present. Mixed hardwood forest was favored under these conditions.

About 2500 to 3000 years ago a cooling of climate took place. This change again favored the boreal species.

During the hemlock-beech period fluctuations in climate are indicated by a definitely bimodial maximum in the hemlock graph in southeastern Canada. More detailed palynological studies, coupled with radiocarbon dating, are planned to investigate these more recent climatic oscillations.

Based on the palynological evidence, coupled with other available data, an attempt has been made to construct a curve showing changes in the annual mean temperature in southeastern Canada in the last 14,000 years. This generalized curve is admittedly tentative and may have to be modified as more information accumulates (Figure 1).

The maritime provinces. A development of climate essentially similar to that of southern Ontario and Quebec in late- and post-Glacial time has

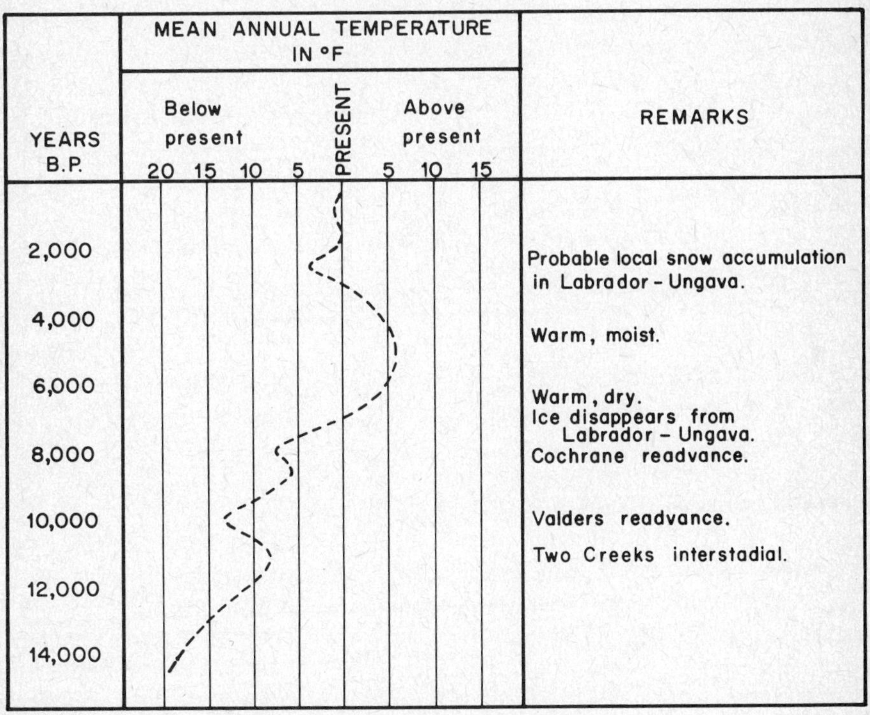

Figure 1. Tentative paleotemperature curve for southeastern Canada, based on palynological studies and other available data.

been established for the Maritime Provinces, based on about 10 pollen diagrams worked out by myself in that region and additional studies by Auer (1930), Livingstone and Livingstone (1958), and Ogden (1960).

Certain differences are characteristic of pollen diagrams from Newfoundland. First, one is dealing with fewer indicator species of tree pollen, commonly used by palynologists elsewhere and, second, climatic changes have been moderated by the influence of the Atlantic Ocean. However, the palynological studies made so far in Newfoundland are too few to permit a detailed discussion on paleoclimates.

Northern Ontario and Quebec. Palynological studies made in these areas are not numerous. The time covered extends back about 7000 years from the present (Grayson, 1956; Ignatius, 1956; Potzger and Courtemanche, 1956; Terasmae anad Hughes, 1960; Wenner, 1947). I have found evidence in the basal parts of pollen diagrams for a climate probably warmer than the present. This opinion is in agreement with other studies from the region, and it seems logical to expect that the remnants of the continental ice melted during the hypsithermal interval. The annual mean temperature then was about 5°F. higher than the present; summer and, possibly, also winter precipitation was probably lower. The hypsithermal interval came to a close when the average annual mean temperature gradually dropped some 10°F. and was accompanied by heavier accumulation of snow. The climate then was more severe than it is now and may have resulted in expansion of the permafrost area. Supporting evidence has been found in the James Bay lowland (Baldwin and Porsild, personal communication).

Present climatic conditions in Labrador-Ungava and northern Quebec deserve special mention because this region probably was the area of initial accumulation of snow gradually leading to formation of the continental ice sheet.

The vast area east of Hudson and James Bays extending to Hudson Strait in the north, to the coast of Labrador in the east, and to the Gulf of St. Lawrence in the south is largely a highland plateau with a monotonous surface strewn with lakes and muskeg. Winters in this region are long and cold. At Schefferville (Knob Lake) the annual mean temperature is about 24°F., with January mean ca. —12°F. Summers are cool, but warm days may occur, with temperatures reaching 90°F. The frost-free season varies from 40 to 80 days. A large section of this region has permanently frozen subsoil, chiefly north of latitude 55°N. Precipitation varies considerably with topography and elevation, averaging from about 40 inches in the interior to 15 inches to the west and north. Snowfall is heavy, more than 100 inches in the interior, and total accumulation of snow on the ground at the end of winter exceeds four feet.

It is quite possible that a slight lowering of average summer temperatures, with precipitation remaining unchanged, will allow accumulation of perennial snowbanks at suitable sites. If such lower mean temperatures, about 5°F. to 10°F. below the present, persist over a number of years continuous snowfields may form over large areas in Labrador-Ungava. Thus it seems plausible that even relatively small changes in temperature and, perhaps, in precipitation can initiate climatic conditions which, if prevailing over a period of time, may well become the beginning of another glaciation.

Western plains region. Palynological studies in this region are entirely too few to allow reconstruction of a climatic sequence for the postglacial time. However, the available evidence indicates that the retreat of ice was followed, perhaps after a short tundra episode, by the boreal forest and not directly by a prairie vegetation. In late-Glacial time, 12,000 to 10,000 years ago, some areas in southern Saskatchewan were covered by boreal forest (Kupsch, 1960; my own unpublished studies).

Palynological studies in the extensive prairie region will encounter certain specific difficulties because of the largely treeless vegetation and interpretation of the pollen and spore record.

West Coast and interior British Columbia. Studies made thus far have indicated that a paleoclimatologic sequence can be worked out for the West Coast (Hensen, 1950; Heusser, 1955; Terasmae and Fyles, 1959). Precipitation seems to have been adequate or excessive at all times during the post-Glacial period, and temperature is the controlling factor although even this is moderated by the influence of the Pacific Ocean. Hansen has demonstrated climatic changes in post-Glacial time on the West Coast, and it appears that a similar climatic sequence to that of eastern North America also applies here. A post-Glacial warmer and drier period on the West Coast seems to be the correlative of the hypsithermal interval.

Conditions in the interior British Columbia are entirely different, and precipitation is the limiting factor in the valleys and on the plateaux. Further difficulties are introduced by the markedly different vegetation on the north- and south-facing slopes of the mountains. It is not clear how a climatic change will influence the sum total of pollen and of spores derived from the north and south slopes and from the valley bottom in any particular locality.

Arctic Canada. Preliminary studies have shown that much more information pertaining to late-Quaternary climates can be found in the Arctic than had been expected previously. Until a few years ago the Canadian Arctic was still largely unexplored. The recent economical and political interest in the far north has opened up an entirely new episode for the Arctic. This increased activity has enabled the scientist to gain access to remote islands in the Arctic Archipelago and to northern coastal areas of the mainland. Large scale geologic reconnaissance by the Geological Survey of Canada (1959) has provided the necessary basic coverage, and a number of separate research teams have carried on field work in the Arctic. Most of these studies, however, are still in progress and only preliminary reports have been issued.

Meteorologically the Canadian Arctic is still almost unexplored. Weather records from the stations in the far north are available for the last few years only; hence they may or may not be representative of a longer period of time. This is unfortunate, for interpretation of paleoclimates is very much dependent on the present plant-animal-soil-climate relationships. This lack of reference data on present climate is felt more in the Arctic than elsewhere because marked differences between physiographic regions and from season to season have been noticed. Furthermore, plants and animals in the Arctic Archipelago are existing in environments where a small change in climate means the difference between survival and extinction.

The Arctic is a land of contrast, both real and imaginary. The view that the Arctic is a land of snow and ice is true in a sense, yet it is a true rock desert at the same time. Annual average precipitation is extremely low; only 2 to 3 inches in northern Arctic Archipelago and less than 10 inches in southern parts of the Archipelago. The Arctic climate region is that part of Canada north of the July 50°F. isotherm, which runs from MacKenzie delta to Churchill, Manitoba, around the southern Hudson Bay and down the Labrador coast. This isotherm corresponds in general to the northern limit of tree growth. The extremely long nights in winter and days in summer are an important factor in climatic control, and this latitudinal control, involving also incoming radiation coupled with areal distribution of land and water, determines in a general way the climatic limits within the Arctic region. In addition to the MacKenzie Basin, the Arctic climatic region includes the Barrens and the Arctic Archipelago. The Barrens comprise the part of Northwest Territories that stretches northward to the Arctic Ocean from the northern limit of tree growth and lies between the Hudson Bay and the MacKenzie delta area. Countless lakes, swamps, muskeg, and rock outcrop characterize this low-lying region. Summers are very short and cool. The average frost-free period is 40 to more than 60 days although freezing temperatures can occur during any month of the year. July and August are the only months with an average temperature above 40°F. Winters are extremely long and cold. Precipitation averages from 7 to 8 inches in the north to about 12 inches in the southern part of the Barrens. Snowfall contributes about one half of the total precipitation. Clear skies characterize the winter months along the Arctic coast, but late summer and early fall are cloudy, accompanied by the precipitation maximum at that time. Generally strong winds comprise another important climatic factor in the Barrens.

It is interesting to note that although the mean annual temperatures are lower in the Arctic Archipelago than anywhere else in Canada, ranging from 15°F. to 0°F. on Melville and Ellesmere Islands, the extremes are not as severe as they would be in continental areas of the same latitude. Summer temperatures are kept uniformly cool by the ice-filled polar waters, with surface temperature near 30°F., which prevent the air in contact with them from warming. On the other hand, winter temperatures are probably slightly moderated by radiation from the water beneath the ice cover, which is usually less than 6 feet in thickness.

The occurrence of occasional mild periods during the Arctic winter deserves mention. The intense cyclonic activity in the Davis Strait area can bring the relatively warm air from the Atlantic over the eastern Arctic. According to weather records one such warm spell occurred in January 1958, when the mean temperatures for the north were some 15°F. above normal in northern Ellesmere Island. These mild spells usually do not extend west of Cornwallis and Somerset Islands.

It is possible, as a suggestion, that such abnormal circulation conditions may contribute somewhat to the moisture regime on the large eastern islands of the Arctic Archipelago in order to maintain small ice caps on Baffin, Devon, and Ellesmere Islands, which lie in a high mountain range with peaks rising above 10,000 feet.

A very interesting and important problem in Quaternary research has been the existence of unglaciated areas in the far north. Views both for and against the existence of such areas have been presented, and it is interesting to note that the term "unglaciated" may have a different meaning depending on whether it is used in a geologic or biological sense. In the first case, unglaciated areas are those that were not covered by glaciers and where there is no evidence of erosion or deposition by ice. Biologically the term generally means that plant and animal life could have existed in such areas. Thus it is possible that an area shown as unglaciated on a geologic map was glaciated in a biological sense. A low summer temperature and a frequently overcast sky with generally high winds, coupled with cold winters and low precipitation can produce conditions where a growth period for plants is almost nonexistent, and snow covers the more sheltered sites, whereas exposed sites may not support any plant life.

Certain areas may have escaped all or at least some of the glaciations, and records, coupled with those from interglacial deposits, can be discovered there of past plant and animal life. Interglacial pollen and spore-bearing deposits have been discovered already on several of the Arctic islands. Studies of these deposits are in progress and as yet unpublished. Observations in the field (J. G. Fyles, personal communication) and palynological studies have shown that in interglacial intervals trees were growing on Banks Island, and appreciable thickness of peat has accumulated in severel areas on the island in late-Quaternay time.

On the mainland, palynological studies have shown that the distribution ranges of some plant species extended farther north about 5000 to 6000 years ago, indicating climatic conditions more favorable than the present (Terasmae and Craig, 1958). Alluvial and lacustrine sediments in the Thelon Valley, Northwest Territories, in which plant fossils were found, dated at 5400 ± 230 years B.P. (L-428), have been frost-heaved, and they form a pingo indicating a deterioration of climate later than about 5000 years ago.

When similar geologic observations become more numerous and many more palynological and paleobotanical studies, supported by radiocarbon dating are made, it will perhaps be possible to compile a climatic sequence covering most of the Quaternary epoch for the Canadian Arctic. It will be possible to achieve this only when information from all sources is compiled and after such data have been critically examined and interpreted by scientists in both the physical and biological fields.

Time Equivalence of Late-Quaternary Climatic Changes

Time equivalence of climatic changes in the Quaternary epoch is the fundamental requirement on which all correlations using palynological and paleontologic evidence are based. Recent studies seem to indicate that this is true probably on a world-wide basis, although more studies are required to substantiate the statement. At least within the limits of any one continent the climatic changes appear to satisfy the requirement of time equivalence.

However, terms such as late-Glacial, post-Glacial and Recent are necessarily relative depending on the geographical location. It is obvious that the late-Glacial episode becomes younger in a south to north direction following the retreat of the ice margin. This presents some difficulties in correlating, for example, the late-Glacial deposits of Wisconsin age. When the plant communities followed and colonized the area uncovered by the retreating ice margin it is possible that similar fossil assemblages, but of different age, were to be found in two localities separated by some distance in a north-south direction. Radiocarbon age determinations have helped to overcome these problems.

The problem of migration rates for plants and animals is of major importance in paleoclimatological interpretations (Savile, 1956). For example, in Canada many species must have covered hundreds and even thousands of miles in late- and post-Glacial time in order to occupy their present distribution ranges. Studies made of fossils in late-Quaternary beds seem to indicate that these migrations occurred rapidly and, perhaps, in a haphazard manner for several species which, in some measure, may account for the present discontinuous ranges. However, the demonstrated ability of many species to migrate over long distances rapidly may make less convincing the argument for the existence of refugia, that is, isolated areas in which plants could survive the glaciation. Such refugia certainly did exist in Alaska and in Peary Land (Northern Greenland) in the far north. Further studies of plant and animal fossils in such areas will certainly contribute to our knowledge of paleoclimates of late-Quaternary time.

References

Antevs, E., "Retreat of the last ice-sheet in Eastern Canada," *Geol. Survey Can. Mem. 146* (1925).
Antevs, E., "Late-glacial correlations and ice recession in Manitoba," *Geol. Survey Can. Mem. 168* (1931).
Armstrong, J. E., "Surficial geology of New Westminster map-area, British Columbia," *Geol. Survey Can. Paper 57-5* (1957).
Auer, V., "Peat bogs in southeastern Canada," *Geol. Survey Can. Mem. 162* (1930).
Baker, F. C., "The life of the Pleistocene or glacial period," *Ill. Univ. Bull.*, Vol. 17, no. 41 (1920).
Baldwin, W. K. W., "Plants of the Clay Belt of northern Ontario and Quebec," *Natl. Museum Can. Bull. 156* (1958).
Bird, J. B., "Recent contributions to the physiography of northern Canada," *Z. Geomorph*, Vol. 3, no. 2 (1959), pp. 151-74.
Canada Year Book, "The climate of Canada," Met. Branch, Air Service, Dept. Transp. Can. (1960).
Chapman, J. D., "The climate of British Columbia," *Proc. Fifth Brit. Columbia Natl. Research Conf.* (1952).
Chapman, L. J., "The climate of southern Ontario," *Can. Geogr. J.*, Vol. 17, no. 3 (1938), pp. 136-41.
Chapman, L. J. and Putnam, D. F., *The Physiography of Southern Ontario* (Toronto, Univ. of Toronto Press, 1951).
Chapman, L. J., "The climate of northern Ontario," *Can. J. Agric. Sci.*, Vol. 33, no. 1 (1953), pp. 41-73.
Coleman, A. P., "The Pleistocene of the Toronto Region," *Ont. Dept. Mines Rept.*, Vol. 41, no. 7 (1933), pp. 1-55.
Coleman, A. P., *The Last Million Years* (Toronto, Univ. of Toronto Press, 1941).
Cook, F. A., "Selected bibliography on periglacial phenomena in Canada," Can. Dept. Min. Techn. Surveys, Geogr. Br. Bibliogr. Ser. 24.
Craig, B. G., "Surficial geology of the Drumheller area, Alberta, Canada," *Unpubl. Ph.D. Thesis,* Univ. Michigan, Ann Arbor, Mich. (1956).
Craig, B. G., "Surficial geology of North-Central District of MacKenzie, Northwest Territories," *Geol. Survey Can. Paper* 60 (1960), p. 18.
Craig, B. G. and Fyles, J. G., "Pleistocene geology of Arctic Canada", *Geol. Survey Can. Paper* 60 (1960), p. 10.
Danserau, P., "The postglacial pine period," *Roy. Soc. Can. Trans. 47,* Ser., Vol. 3, no. 5 (1953), pp. 23-38.
Deane, R. E., "Pleistocene geology of the Lake Simcoe district, Ontario," *Geol. Survey Can. Mem. 256* (1950).
Deevey, E. S., "Biogeography of the Pleistocene," *Bull. Geol. Soc. Am.* 60, no. 9 (1949), pp. 1315-1416.

Dillon, L. S. "Wisconsin climate and life zones in North America," *Science,* Vol. 123, no. 3188 (1956), pp. 167-76.
Dore, W. G. and Gillett, J. M., "Botanical survey of the St. Lawrence Seaway area in Ontario," *Can. Dept. Agric. Sci.* (1955).
Dreimanis, A., "Finite radiocarbon dates of the Port Talbot interstadial deposits in southern Ontario," *Science,* Vol. 131, no. 3415 (1960), pp. 1738-39.
Dreimanis, A., "Pre-classical Wisconsin in the eastern portion of the Great Lakes region, North America," *Proc. Intern. Geol. Congr. Norden,* Pt. IV, Sect. 4 (1960), pp. 109-19.
Elson, J. A., "Surficial geology of the Tiger Hills region, Manitoba," *Unpub. Ph.D. Thesis,* Yale Univ., New Haven, Conn. (1955).
Elson, J. A., "Littoral mollusks of the Champlain Sea," *Excursion guide,* Dept. Geol. Sci. McGill Univ., Montreal (1960).
Ewing, M. and Donn, W. L., "A theory of ice ages," *Science,* Vol. 123, no. 3207 (1956), pp. 1061-66.
Fairbridge, R. W., "Dating the latest movements of the Quaternary sea level," *Trans. N.Y. Acad. Sci. Ser. II,* Vol. 20, no. 6 (1958), pp. 471-82.
Fairbridge, R. W., "The changing level of the sea," *Scientific American,* Vol. 202, no. 5 (1960), pp. 70-79.
Flint, R. F., *Glacial Geology and the Pleistocene Epoch* (New York, Wiley & Sons, 1947).
Fyles, J. G., "Surficial geology of the Horne Lake and Parksville map-areas, Vancouver Island, British Columbia," *Dissert. Abstr.,* Vol. 16, no. 11 (1956), p. 2136.
Gadd, N. R., "Surficial geology of the Becancour map-area, Quebec," *Geol. Survey Can. Paper 59-8* (1960).
Geological Survey of Canada, "Helicopter operations of the Geological Survey of Canada," *Geol. Survey Can. Bull. 54* (1959).
Glacial Map of Canada, Geol. Assoc. Can. Toronto, Ont. (1958).
Goldthwait, R. P., "Dating the Little Ice Age in Glacier Bay, Alaska," *Inst. Polar Stud.* Ohio State Univ. (Mim. manuscr.) (1960).
Gravenor, C. P., "Surficial geology of the Lindsay-Peterborough area, Ontario," *Geol. Survey Can. Mem. 288* (1957).
Grayson, J. F., "Post-glacial history of vegetation and climate in the Labrador-Quebec region as determined by palynology," *Unpub. Ph.D. Thesis,* Univ. Michigan, Ann Arbor, Mich. (1956).
Griggs, R. F., "Timberlines as indicators of climatic trends," *Science,* Vol. 85, no. 2202 (1937), pp. 251-55.
Halliday, W. E. D., "A forest classification for Canada," Can. Dept. Min. Res., Forestry Br. Domin. Forest Serv. Bull. 89 (1937), pp. 1-50.
Hansen, H. P., "Postglacial forests in south central Alberta, Can.," *Am. J. Botany,* Vol. 36, no. 1 (1949), pp. 54-65.
Hansen, H. P., "Pollen analysis of three bogs on Vancouver Island, Canada," *J. Ecol.,* Vol. 38, no. 2 (1950).
Hare, F. K., "The climate of the eastern Canadian Arctic and Subarctic and its influence on accessibility," *Ph.D. Thesis,* Univ. Montreal, Montreal, Quebec (1950).
Hare, F. K., "The climate of the Island of Newfoundland: a geographical analysis," *Geogr. Bull.,* Vol. 2 (1952), pp. 36-88.
Hare, F. K., "Post-glacial climatic change in Eastern Canada," *Can. Br. Roy. Met. Soc. Papers 2,* no. 7 (1952).
Harlow, W. H. and Harrar, E. S., *Textbook on Dendrology* (New York, McGraw-Hill, 1941).
Hattersley-Smith, G., "Glaciological research in northern Ellesmere Island," *Can. Geographer,* Vol. 12 (1958), pp. 32-34.
Hattersley-Smith, G., "Research in the Lake Hazen region of northern Ellesmere Island in the International Geophysical Year," *Arctic Circular 12,* no. 1 (1959), pp. 2-12.

Henderson, E. P., "Surficial geology of Sturgeon Lake map-area, Alberta," *Geol. Survey Can. Mem. 303* (1959).
Henderson, E. P., "A glacial study of central Quebec-Labrador," *Geol. Survey Can. Bull. 50* (1959).
Henderson, E. P., "Surficial geology of St. John's, Newfoundland," *Geol. Survey Can. Map 35* (1959).
Heusser, C. J., "Pollen profiles from the Queen Charlotte Islands, British Columbia," Dept. Expl. and Field Research, Am. Geogr. Soc. New York (manuscript, 1955).
Hibbard, C. W., "Pleistocene vertebrate paleontology in North America," *Bull. Geol. Soc. Am.*, Vol. 60, no. 9 (1949), pp. 1417-28.
Hills, G. A., "A ready reference to the description of the land of Ontario and its productivity," Ont. Dept. Lands Forests, Div. Res. Prelim. Rept., (1959).
Hinde, G. J., "Glacial and interglacial strata of Scarborough Heights," *J. Can. Inst.* (Roy. Can. Inst.) (1878).
Horton, K. W. and Bedell, G. H. D., "White and red pine – ecology, silviculture and management," Can. Dept. North. Affairs Natl. Resources, *Forestry Branch Bull. 124* (1960).
Hughes, O. L., "Surficial geology of Smooth Rock Cochrane District, Ontario," *Geol. Survey Can. Paper 55-41* (1956).
Hustich, I., "On the forest geography of the Labrador peninsula. A preliminary synthesis," *Acta Geogr.*, Vol. 10 (1949).
Ignatius, H. G., "Late-Wisconsin stratigraphy in North-Central Quebec and Ontario, Canada," *Ph.D. Thesis,* Yale Univ. New Haven, Conn. (1956).
Ives, J. D., "The deglaciation of the Helluva Lake area, 50-70 miles northwest of Schefferville," *McGill Subarctic Research Papers,* Vol. 6 (1959), pp. 45-49.
Ives, J. D., "Glacial drainage channels as indicators of late-glacial conditions in Labrador-Ungave: a discussion," *Cahiers de Geogr. de Quebec,* Vol. 3, no. 5 (1959), pp. 57-72.
Ives, J. D., "The deglaciation of Labrador-Ungava – an outline," *Cahiers de Geogr. de Quebec,* Vol. 4, no. 8 (1960), pp. 323-43.
Johnston, W. A., "Pleistocene and recent deposits in the vicinity of Ottawa, with a description of the soils," *Geol. Survey Can. Mem. 101* (1917).
Karlstrom, T. N. V., "The problem of the Cochrane in late Pleistocene chronology," *U.S. Geol. Survey Bull. 1021-J* (1956).
Karlstrom, T. N. V., "The Cook Inlet, Alaska, glacial record and Quaternary classification," *U.S. Geol. Survey Prof. Paper,* no. 400-B (1960), pp. 330-32.
Karrow, P. F., "Pleistocene geology of the Grondines map-area, Quebec," *Ph.D. Thesis,* Univ. Ill., Urbana, Ill. (1957).
Karrow, P. F., "Pleistocene geology of the Hamilton map-area," *Ont. Dept. Min. Geol. Circular 8* (1959).
Kendrew, W. G. and Currie, B. W., *The Climate of Central Canada,* Ottawa, (Queen's Printer, 1955).
Kendrew, W. G. and Kerr, D., *The Climate of British Columbia and the Yukon Territory* (Ottawa, Queen's Printer, 1955).
Koeppe, C. E., *The Canadian Climate* (Bloomington, Ill., McKnight & McKnight, 1931).
Kupsch, W. O., "Radiocarbon-dated organic sediment near Herbert, Saskatchewan," *Am. J. Sci.,* Vol. 258, no. 4 (1960), pp. 282-92.
Landsberg, H., "Climatology of the Pleistocene," *Bull. Geol. Soc. Am.,* Vol. 60, no. 9 (1949), pp. 1437-42.
La Roque, A., "Quantitative methods in the study of non-marine Pleistocene mollusca," *Proc. Intern. Geol. Congr. Norden,* Pt. IV., Sect. 4 (1960), pp. 134-41.
Laveriere, C. and Courtemanche, A., "La geomorphologie glaciaire de la region du Mont Tremblant," *Revue Can. Geogr.,* Vol. 13, nos. 3-4 (1959), pp. 103-34.
Lee, H. A., "Surficial geology of Edmundston, Madawaska and Temiscouata Counties, New Brunswick and Quebec," *Geol. Survey Can. Paper 55-15* (1955).

Lee, H. A., "Surficial geology of Fredericton, York and Sunbury Counties, New Brunswick," *Geol. Survey Can. Paper 56-2* (1957).
Lee, H. A., "Surficial geology of southern District of Keewatin and the Keewatin ice divide, Northwest Territories," *Geol. Survey Can. Bull. 51* (1959).
Lee, H. A., "Late glacial and postglacial Hudson Bay Sea episode," *Science,* Vol. 131, no. 3413 (1960), pp. 1609-11.
Livingstone, D. A., "Some pollen profiles from Arctic Alaska," *Ecology,* Vol. 36, no. 4, pp. 587-600.
Livingstone, D. A. and Livingstone, B. G. R., "Late-glacial and postglacial vegetation from Gillis Lake in Richmond County, Cape Breton Island, Nova Scotia," *Am. J. Sci.,* Vol. 256 (1958), pp. 341-59.
Love, A., "Origin of the Arctic flora (in Problems of the Pleistocene and Arctic), McGill Univ., Museums Publ. 1, (1955), pp. 82-95.
Love, A. and Love, D., "Arctic polyploidy," *Proc. Genet. Soc. Can.,* Vol. 2 (1957), pp. 23-27.
Love, D., "The postglacial development of the flora of Manitoba: a discussion," *Can. J. Botany,* Vol. 37 (1959), pp. 547-85.
MacClintock, P. and Terasmae, J., "Glacial history of Covey Hill," *J. Geol.,* Vol. 68, no. 2 (1960), pp. 232-41.
MacKay, J. R., "The Anderson River map-area, N.W.T.," *Can. Dept. Min. Techn. Surveys,* Geogr. Branch, Mem. 5 (1958).
MacLean's Magazine, "An interim report on the next Ice Age," December 3, 1960 issue, Toronto.
Mathews, W. H., "Historic and pre-historic fluctuations of alpine glaciers in the Mount Garibaldi map-area, south-western British Columbia," *J. Geol.,* Vol. 59 (1951), pp. 357-80.
Meldegaard, J., "Origin and evolution of Eskimo cultures in the Eastern Arctic," *Can. Geogr. J.,* Vol. 60, no. 2 (1960), pp. 64-75.
Mirynech, E., "Surficial geology of the Trenton-Campbellford map-areas, Ontario," *Ph.D. Thesis,* Univ. Toronto, Toronto (1961).
Native Trees of Canada, Can. Dept. North. Affrs. Natl. Resources. Forestry Branch, Bull. 61 (1956).
Nicholson, N. L., "The Northwest Territories – geographical aspects," *Can. Geogr. J.,* Vol. 60, no. 1 (1960), pp. 2-27.
Ogden, J. G. III, "Recurrence surfaces and pollen stratigraphy of a postglacial raised bog, Kings County, Nova Scotia," *Am. J. Sci.,* Vol. 258 (1960), pp. 341-53.
Polunin, N., "Botany of the Canadian Eastern Arctic," Pt. I, *Natl. Mus. Can. Bull. 92* (1940).
Porsild, A. E., "The vascular plants of the western Canadian Arctic Archipelago," *Natl. Museum Can. Bull. 135* (1955).
Porsild, A. E., "Illustrated flora of the Canadian Arctic Archipelago," *Natl. Mus. Can. Bull. 146* (1957).
Porsild, A. E., "Geographical distribution of some elements in the flora of Canada," *Geogr. Bull.* Vol. 11 (1958), pp. 57-77.
Potzger, J. E., "History of forests in the Quetico-Superior country from fossil pollen studies," *J. Forestry,* Vol. 51 (1953), pp. 560-65.
Potzger, J. E., "Nineteen bogs from southern Quebec," *Can. J. Botany,* Vol. 31 (1953), pp. 383-401.
Potzger, J. E. and Courtemanche, A., "A series of bogs across Quebec, from the St. Lawrence Valley to James Bay," *Can. J. Botany,* Vol. 34, no. 4 (1956), pp. 473-500.
Prest, V. K., "Pleistocene geology and surficial deposits," Chapter 8 in *Geology and Economic Minerals of Canada,* Geol. Survey Can. Econ. Geol. Ser. 1 (1957).
Rae, R. W., "Climate of the Canadian Arctic Archipelago," *Can. Dept. Transp.,* Toronto (1951).
Rand, A. L., "Mammals of the Eastern Rockies and Western Plains of Canada," *Natl. Museum Can. Bull. 108* (1948).

Rand, A. L., "The ice age and mammal speciation in North America," *Arctic,* Vol. 7 (1954), pp. 31-35.

Raup, H. M., "Botanical problems in boreal America I," *Bot. Rev.,* Vol. 7, no. 3 (1941), pp. 147-208.

Reynolds, M. B., "Pleistocene molluscan faunas of the Humboldt deposit, Ross County Ohio," *Ohio J. Sci.,* Vol. 59 (1959), pp. 152-66.

Savile, D. B. O., "Known dispersal rates and migratory potentials as clues to the origin of the North American biota," *Am. Midland Natl.,* Vol. 56, no. 2 (1956), pp. 434-53.

Scoggan, H. J., "Flora of Manitoba," *Natl. Museum Can. Bull. 140* (1957).

Sharp, R. P., "Soil structures in the St. Elias Range, Yukon Territory," *J. Geomorphol,* Vol. 5 (1942), pp. 275-301.

Shenfeld, L. and Slater, D. F. A., "The climate of Toronto," *Met. Branch, Dept. Transp. Can. Circular 3352* (1960).

Shepard, F. G. and Suess, H. E., "Rate of postglacial rise of sea level," *Science,* Vol. 123, no. 3207 (1956), pp. 1082-1133.

Stalker, A. M., "Surficial geology of the Red Deer-Stettler map-area, Alberta," *Geol. Survey Can. Mem. 306* (1960).

Terasmae, J. and Craig, B. G., "Discovery of fossil *Ceratophyllum demersum* L. in Northwest Territories, Canada," *Can. J. Botany,* Vol. 36 (1958), pp. 567-69.

Terasmae, J., "Contributions to Canadian palynology, Pt. 1, The Use of palynological studies in Pleistocene stratigraphy; Pt. 2, Non-glacial deposits in the St. Lawrence Lowlands, Quebec; Pt. 3, Non-glacial deposits along Missinaibi River, Ontario," *Geol. Survey Can. Bull. 46* (1958).

Terasmae, J., "Palaeobotanical study of buried peat from the MacKenzie River delta area, Northwest Territories," *Can. J. Botany,* Vol. 37 (1959), pp. 715-17.

Terasmae, J. and Fyles, J. G., "Palaeobotanical study of late-glacial deposits from Vancouver Island, British Columbia," *Can. J. Botany,* Vol. 37 (1959), pp. 815-17.

Terasmae, J., "Contributions to Canadian palynology, 2; Part 1, A palynological study of postglacial deposits in the St. Lawrence lowlands; Part 2, A palynological study of the Pleistocene interglacial beds at Toronto, Ontario," *Geol. Survey Can. Bull. 56* (1960).

Terasmae, J. and Hughes, O. L., "Notes on glacial retreat in the North Bay area, Ontario," *Science,* Vol. 131, no. 3411 (1960), pp. 1444-46.

Terasmae, J. and Hughes, O. L., "A geological and palynological study of Pleistocene deposits in the James Bay lowlands, Ontario," *Geol. Survey Can. Bull. 62* (1960).

Thomas, M. K., "Climatological atlas of Canada," *Natl. Research Council,* Div. Bldg. Res. & Can. Dept. Transp. Met. Div., N.R.C. no. 3151 (1953).

Wagner, F. J. E., "Palaeoecology of the marine Pleistocene faunas of southwestern British Columbia," *Geol. Survey Can. Bull. 52* (1959).

Watt, A. K., "Glacial geology of the Toronto-Orangeville area, Ontario," *Guide Book,* Field Trip 3, Geol. Soc. Am. Meetings, Toronto, Ontario (1953).

Wenner, C.-G., "Pollen diagrams from Labrador," *Geogr. Ann.* Vol. 29 (1947), pp. 137-374.

21

Climatic Trends on the Canadian Prairies

M. K. Thomas

Synopsis

Our knowledge of the climates of the Prairie Provinces prior to the past 100 years is very limited. The first explorers were generally fur trade promoters and they travelled in the forested lands to the north of the true Prairies. Government sponsored scientific expeditions were sent into the region about 1850, but it is from the accounts of the Red River Settlement that we have the first records of favorable and unfavorable agricultural weather. Official meteorological observations began at Winnipeg in the 1870's and were extended across the Prairies within a decade.

By today's standards temperatures were relatively low on the Prairies before the turn of the century. There was a general increase of temperature until the 1930's, and by the late 1940's there was a marked depression in the temperature trend curves which culminated in the mid 1950's. There has been a recovery, and in Manitoba temperatures are now averaging as high as the previous maximum in the late 1940's. Precipitation was relatively abundant during the first few years of the century, but was lighter in the following three to four decades. Heavier precipitation was again

Reprinted from Symposium on Water and Climate, Report No. 2, Water Studies Institute of Saskatoon.

experienced in all provinces during the 1950's, and during the past few years there has been a tendency towards lower precipitation again in Alberta and Saskatchewan, but not in Manitoba.

Trend lines consisting of plots of successive median values have been constructed and since the quantitative effect of abnormal years is suppressed, these may be better trend indicators than the moving mean trend lines. Since it is not yet possible to forecast trends, the use of probability statistics is recommended for planning and design studies.

Introduction

When considering climatic change, we must first state the period of time in which we are interested. To geophysicists and geologists climatic change has to do with a span of time extending back millions of years; to historians or biologists five to ten thousand years; to most people only the decades of a life span. For meteorologists, interest stretches back for a few centuries—the period for which some instrumental data are available, although there are paleoclimatologists who, by using geophysical methods, are carrying out research on climatic change back many thousands of years. When such long periods of time are considered, there have been drastic climatic changes on earth. The Canadian Prairies have been buried under thousands of feet of ice, not once but several times. On the other hand, there have been long, hot, humid periods producing rank vegetation, giving rise to today's oil and gas fields. Changes in solar radiation, astronomical changes, changes in the transmissivity of the earth's atmosphere, and the resulting changes in physical geography, climate and vegetation are all interwoven into a vast complex which is only slowly being unraveled by scientists of many disciplines.

This paper, dealing with climatic change on the Canadian Prairies, is of necessity largely limited to the past century. The Europeans who first ventured into this region in the 17th, 18th and early 19th centuries were not scientific explorers. The first Europeans came into Hudson Bay in search of the Northwest Passage, but did not get inland. Next came men, who in pursuit of the fur trade, travelled inland to the interior plains, and early in the 18th century fur traders travelled into the West by the St. Lawrence and the Great Lakes. The letters, reports and books from the 18th century that are available refer generally to the park belt and the forested areas, since the true Prairies were not fur-bearing country. It was not until the first half of the 19th century that exploring parties were sent into the southern Prairies to report on the general nature of the country. These expeditions were generally government sponsored, were led by military men, and their reports contain some general information on weather and climate. After about 1850 the governments began sending out scientific parties to report on the geography, vegetation and climate for the planning of possible settlements and railroads. And with the building of the railroads and the telegraph lines, meteorological observing stations were established during the last quarter of the century.

For the purpose of this study, it is perhaps just as well that we are restricted to data for less than a century. Most of us are more concerned with what has happened in the past few decades, and what might be expected to happen in the next decade or so, than we are in conditions over vast ranges of thousands of years. Not that these long-range studies are unimportant. In fact, it is hoped that they will assist in leading to a better understanding of weather processes with the implications of better day-to-day and seasonal forecasts, and ultimately perhaps, to some measures of weather control. But to those concerned with the economics of agriculture, water supply and resources planning, the present and the next decades are very important. What information and data on climatic trends or fluctuations does the meteorologist have to offer?

The Pre-Meteorological Service Record

Excellent summaries of early writings on the weather and climate of the Prairies from the time of the first explorers to the beginning of official meteorological observations are summarized in a paper by Hope (1938), and quoted in detail in a book by Warkentin (1964). Explorers such as Umfreville, Thompson, Mackenzie and Richardson left journals and books which mention aspects of the Prairies' climate which are similar to today's conditions. The first book to give details of climatic variations was written by Ross, who in describing conditions in the Red River Settlement from 1812 to 1852, reported that the first good crop was harvested in 1817, but the climatic conditions were apparently favorable for grasshoppers during the first few years of the settlement, and many crops were ruined. Major floods were reported in 1826 and again in 1852. Until the mid 1830's there was an apparent normal assortment of dry, wet, cold and warm years. But from 1837 to 1848 it was particularly dry over the Prairies, and a wet period followed lasting until about 1857.

The first scientific exploring expeditions to report on climatic conditions were those of Palliser and Hind, both in the field just before 1860. Hector, of the British sponsored Pallister Expedition, wrote describing the climate, but because of the short period of the expedition was not able to add any information regarding fluctuations or variations in climate. Palliser, whose name has become associated with the relatively dry triangle in the southwestern portion of the Prairie Provinces where he dismissed the agricultural possibilities, expressed the need for more data—"With regard to the climate of the district of the Red River and the Assiniboine, we are in need of more complete and careful observations than have yet been obtained to justify our speaking with confidence on the subject." At much the same time the Canadian sponsored Hind-Dawson Expedition was in the field, and in his report Hind stressed the difference between the relatively humid regions of the Red River valley and the more arid regions further west. A later writer, Hargrave, in describing conditions in the Red River settlement after 1860, mentions the floods of 1861 and the subsequent hot, dry weather of the 1860's. The year 1868 appears to have been the last of the series of disastrous ones, and the 1870's proved to be as good for the Prairies as the 1860's had been bad. It was during this time that official meteorological observations began at Winnipeg, although there are some earlier unofficial records of temperature and precipitation available from Edmonton taken by Drummond from 1825 to 1827, and from the Winnipeg area taken by Gunn from 1855 to 1861.

The Meteorological Service of Canada was organized in 1870, and during the next 15 years the network spread across the Prairies from Winnipeg. By the turn of the century there were 37 climatological reporting stations in Manitoba and more than 50 in that part of the Northwest Territories which is now Alberta and Saskatchewan. It is interesting to note that on January 1, 1964, there were 733 official stations taking climatological observations each day in the three Prairie Provinces.

Previous Studies

The first Canadian meteorologist to write on the subject of climatic change in Canada was Sir Frederick Stupart (1917), the Director of the Service, who commented that winters at Winnipeg had been milder since 1900 than during the first three decades of observations. He noted the severity of the winters during the 1880's at Edmonton, and stated that there were evidences that 1897 ended a comparatively dry precipitation cycle. Papers by Connor (1933) and Thomson (1936) were attempts to correlate sun spots, weather and wheat yields. Thomson pointed out that since the correlation between weather and wheat yield is poor, sun spots would be of no value in predicting the yield of any particular year. And Connor, although he was able to show some correlation between sun spots and temperature and precipitation, stated that variations in the period and amplitude of sun spot fluctuations are so great that the number of sunspots present cannot be accurately forecast a year or so in advance. Hence, seasonal weather forecasts based on sun spot numbers would be unreliable. Johnson (1940) exploring the subject further, concluded that while the sun spot cycle probably bears some casual relation to precipitation and also to temperature and indirectly to crop yield, it nevertheless does not provide a basis for long-range weather forecasting owing to the confusing effects of other determining factors.

The first thorough study of temperature trends in Canada was conducted by Longley (1953). He wrote that the decade of the 1880's was cold not only in western Canada, but throughout the country and in many other parts of the northern hemisphere. There was then a fairly rapid increase in temperature in the Prairie Provinces amounting to 3 or 4° on an annual basis, over the next two to three decades after which the trend curve flattened out. The warmest decades were those ending in the 1930's, and by the late 1940's it was evident that western Canada was definitely cooler. Currie (1954) suggested that the moist conditions that prevailed on the Prairies during the 1870's, and the earlier part of this century, were exceptional, and that the precipitation over the period 1940 to 1954 was nearly typical of the Prairies. He also speculated that "a slow extension northward of the Prairies and the northern forests has now started, an extension that could continue for some hundreds or even thousands of years."

In 1956, Kendall and Thomas published a paper on the variability and trends of precipitation in the Prairie Provinces. They illustrated that precipitation is not as reliable in southeastern Alberta and southwestern Saskatchewan as it is in the rest of the region, and that the heavy precipitation during the first decade of the century was followed by a long period of generally subnormal precipitation. Their graphs indicated, however, that during the 1950's precipitation again increased to a relatively high level.

Analysis of Trend Data

Data and Methods

Although there are currently more than 700 official stations in the Prairie Provinces reporting daily observations of temperature and precipitation, less than a dozen have relatively complete homogeneous periods of record extending back to the 1880's. Even at these locations observational methods and instruments have changed over the years, and at each the actual observational site has been moved at least two or three times. Early meteorological observations were taken in forts or settlements by telegraph operators and by part-time observers, but in recent decades instruments and procedures have been standardized, and at principal stations observations are now taken by professional weather observers at airports. The homogeneity of early records is often in doubt, and even at present we are not too sure of the effect of busy airport traffic on the representativeness of the data. But by grouping the data from two or more locations in each province, an attempt has been made to improve both the homogeneity and the representativeness.

Perhaps the easiest method of obtaining trend lines to show fluctuations in the temperature and precipitation regimes, is to calculate and plot moving means. Averaged over ten years the random year-to-year fluctuation is subdued. For obvious simplicity periods of ten years were used, and in this study the ten-year values were credited to the final year in each decade. It should be stated at this point that moving means have been used here with full awareness of their limitations and the false impressions that can be gained from their use.

Another kind of trend line which has also been examined involves successive median values of temperature and precipitation. Again for simplicity sake, periods of eleven years were considered, and the annual values of temperature and precipitation were rounded off to the nearest whole degree or inch respectively. In using median values the effect of the markedly abnormal year is considerably reduced, and the resulting trend lines may be more representative than the ones based on mean values.

A more sophisticated and satisfactory method of studying cycles and trends is one using power spectrum analysis. This type of analysis, sometimes known as generalized harmonic analysis, is based on the premise that time series are not necessarily composed of a finite number of oscillations, but that they consist of virtually infinite numbers of small oscillations spanning a continuous distribution of wave lengths. It is becoming apparent that this statistical tool is the one at present that is the most uniformly efficient for identifying all possible alternatives to randomness in time series. To date, however, we have insufficient data available on punched cards to undertake such a study for stations in the Prairie Provinces. Preliminary work has been completed on data from Toronto, and the results confirm earlier clerical trend studies and also indicate that at least a few minor cycles are present. Steps are currently being undertaken to transfer long-term periods of record from stations across the country to punched cards to allow power spectrum analysis of data from many stations.

Before considering the present analysis, it might be well to examine some of the "raw data" on which the analyses have been made. Figure 1 shows the annual values of temperature and precipitation at Prince Albert from 1885 to 1964. Smooth trends from year to year are noticeably absent, the extreme years stand out, and from casual inspection it is impossible to select any recurring cycles.

Moving Means

Ten-year moving means of annual temperature at four Saskatchewan stations are shown in Figure 2. The data from each station are considered to be relatively homogeneous, although in each instance there has been a move from a town or from an experimental farm to the airport. This may, or may not be the reason for differences in the trend, for example, at North Battleford during the 1930's when a falling trend is shown compared to the trends at the other stations which were either constant or rising. To get away from homogeneity difficulties at individual stations, the data from these four Saskatchewan stations were combined in the

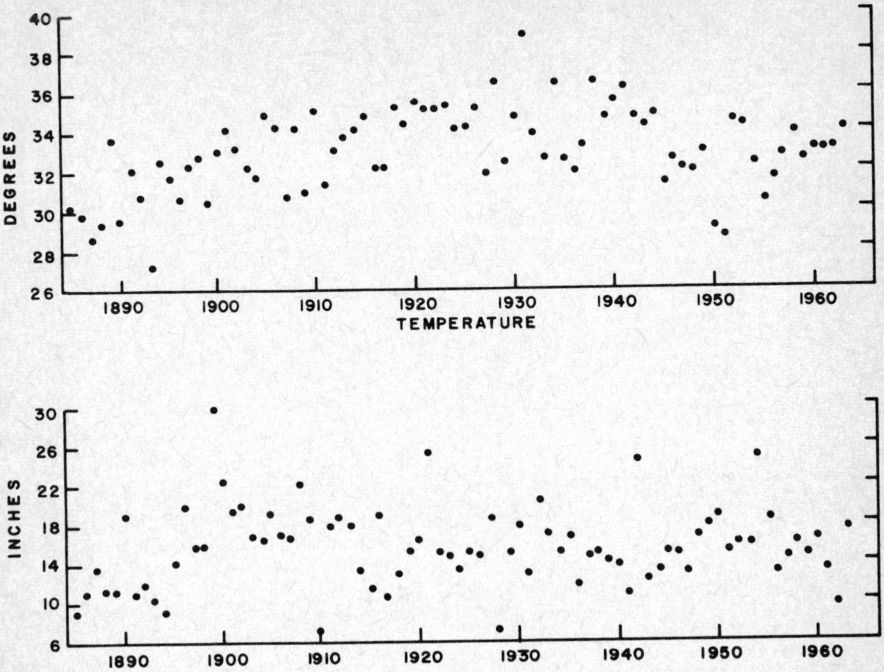

Figure 1. Annual values of temperature and precipitation at Prince Albert, Sask.

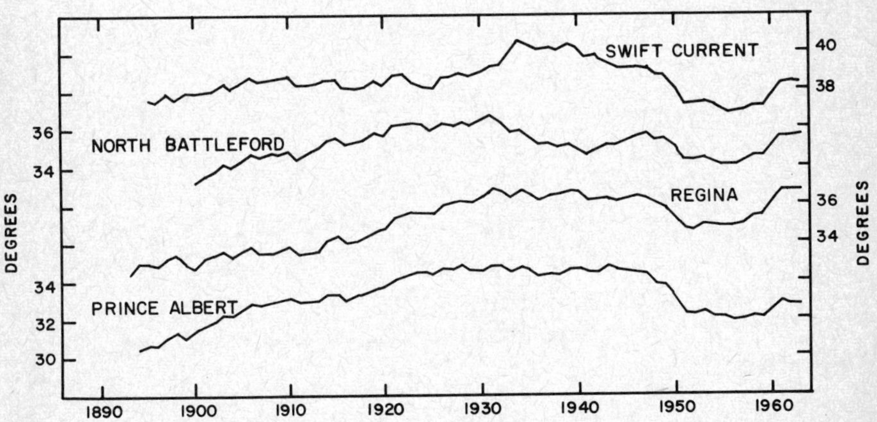

Figure 2. Ten year moving means of annual mean temperature at four Saskatchewan stations (°F).

provincial graph that follows. Similarly, Figure 3 shows ten-year moving means of annual precipitation at three Alberta stations. As might be expected, there is more fluctuation in these point precipitation records than there is in the temperature data. The provincial curves in Figures 4 and 5

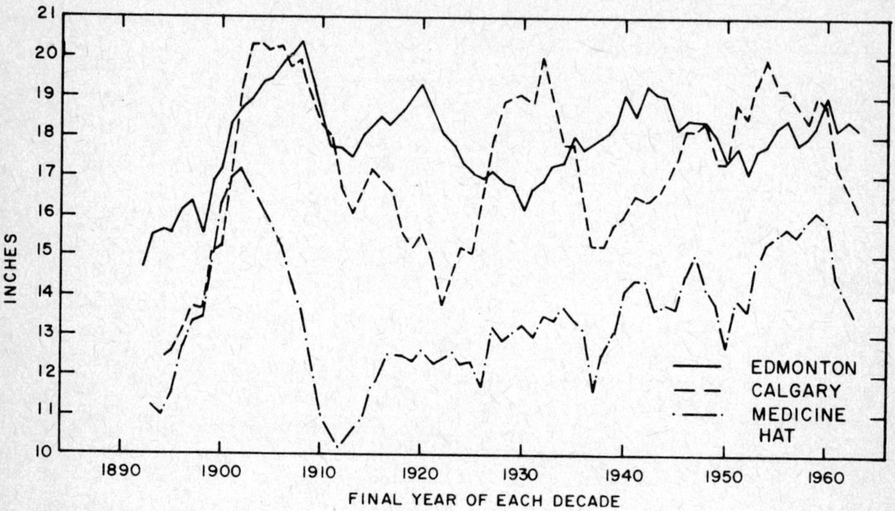

Figure 3. Ten year moving means of annual precipitation at three Alberta stations (inches).

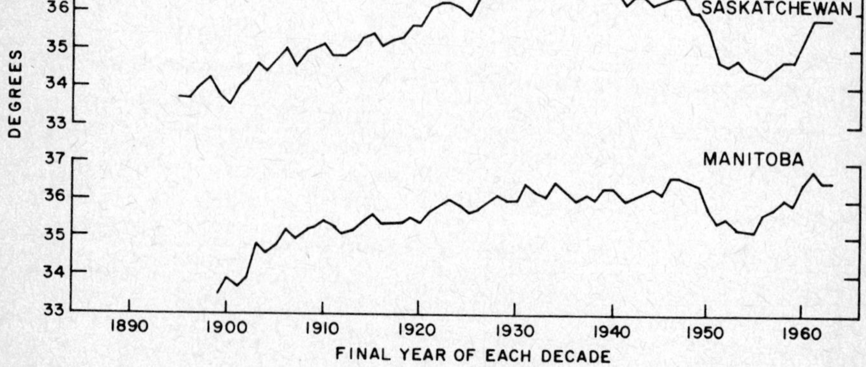

Figure 4. Ten year moving means of annual temperature.

are based on data from Edmonton, Calgary and Medicine Hat in Alberta; Swift Current, North Battleford, Regina and Prince Albert in Saskatchewan; and Winnipeg and Brandon in Manitoba.

Temperature

Considering periods of record that begin in the 1880's, the moving mean trend lines in each province (see Figure 4), show generally increasing temperature trends of 2 to 3 degrees until the early 1930's. In Alberta there followed a slight cooling trend until the late 1940's, after which the temperatures fell more rapidly to the level at which they had been in the 1890's. After 1956 the decadal averages increased, although it is possible that a plateau has been reached during the past three years. In Saskatchewan the slight cooling which began in the early 1930's proceeded more rapidly in the late 1940's, and also reached a minimum in the decade

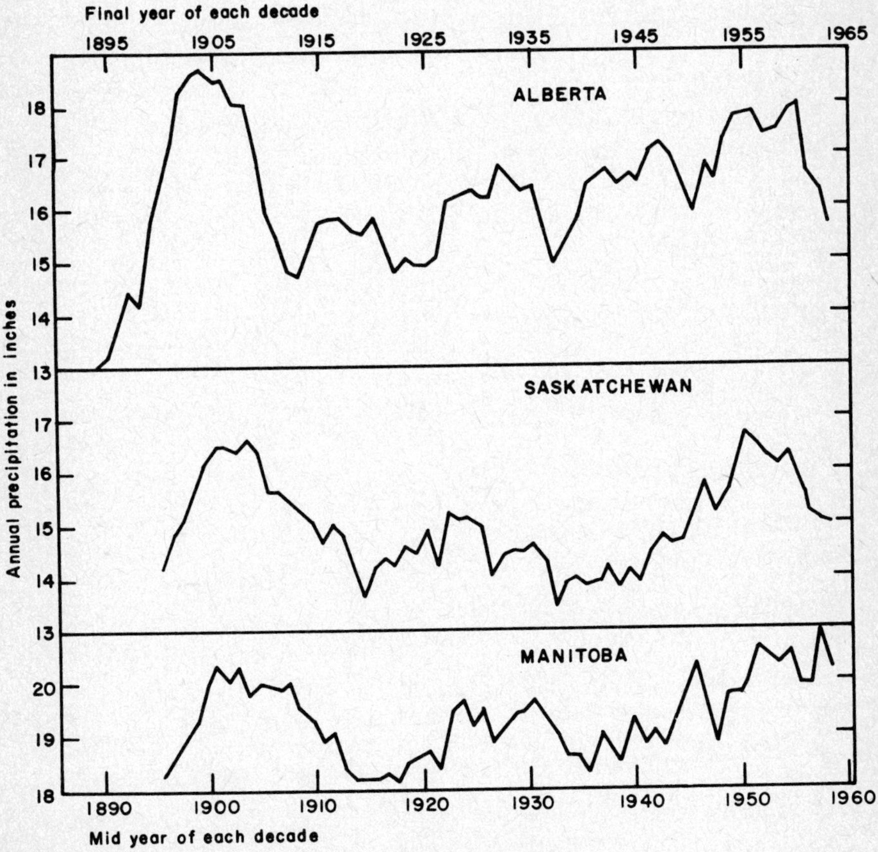

Figure 5. Ten year moving means of annual precipitation.

ending in 1956, but then rose about a degree and a half to a new plateau in the 1960's. In Manitoba, after a marked increase in temperature of 2 degrees over a decade at the turn of the century, there was a slow warming trend which lasted for four decades until the late 1940's. The trend line in Manitoba fell in the 1950's, along with those of Alberta and Saskatchewan, and then started to rise in 1955. In the early 1960's the decadal means are averaging as high as any on record.

Precipitation

The trend lines based on moving decadal means of precipitation (see Figure 5), are perhaps more striking than those of temperature. Precipitation was relatively heavy during the first few years of the century, and the decadal values for periods ending about 1905 were exceptionally high in each of the three provinces. Conditions were then markedly drier, especially in Alberta, where in about ten years the decadal averages dropped about four inches. Except for short periods of low ten-year values ending in the early 1920's, the late 1930's and about 1950, the general trend of precipitation in Alberta has been upward until in 1960 the decadal average was about 18 inches. However, the trend line in Alberta has fallen drastically within the last three years. Conditions were somewhat different in Saskatchewan where, following a particular dry decade ending in 1919, the precipitation trend curve illustrates decreasing precipitation reaching a minimum in the decade ending in 1937. Heavier precipitation was reported in the 1940's and early 1950's, and the trend curve peaks in the decade ending in 1955, with decadal values roughly comparable to the earlier wet periods just after the turn of the century. Over the past eight years the trend in Saskatchewan has been towards smaller amounts of precipitation. In Manitoba there has not been as much variation, and following the dry decades ending around 1915, there has been a gradual increase in precipitation, and in fact, the decade ending in 1962 showed a higher decadal value than any computed since the turn of the century.

Successive Medians

Before accepting moving means as significant trend lines, it might be well to examine further the basic temperature and precipitation data. Table I shows frequency distribution by decades of annual temperature at Winnipeg, and annual precipitation at Edmonton. Inspection of the Winnipeg portion of the table reveals that while there definitely was an increase of temperature from the beginning of record until perhaps 1920, there is no noticeable trend in the frequency table for the succeeding decades. The scatter in the decades is fairly uniform, except for such isolated values as 42 degrees and 32 degrees which stand out as marked extremes in this century. These represent 1931, the warmest year on record when the temperature was above normal every month of the year, and 1950 a generally cool year in which temperatures at Winnipeg were below normal in ten months out of the twelve. The winter months were markedly abnormal in these two years as record high temperatures for January were reported in 1931 on the western Prairies, and January 1950 is considered to be in general the coldest month on record throughout western Canada. It becomes apparent that these two extreme years have influenced the moving means to a marked extent. Similarly, in the Edmonton precipitation frequency table, the effect of the very wet years of 1900 (27.81 inches), and 1901 (27.54 inches), on the rising trend is apparent as are the recent dry years of 1961 and 1963 on the falling trend curve.

TABLE I. DECADAL FREQUENCIES OF ANNUAL TEMPERATURE AT WINNIPEG

Annual Temp. °F.	*Period*									
	1874 - 80	1881 - 90	1891- 1900	1901 - 10	1911 - 20	1921 - 30	1931 - 40	1941 - 50	1951 - 60	1961 - 63
42							1			
41										
40	1									
39						1			2	
38					1	4	2	2	1	2
37				4	2	1	2	3	3	
36	1	1	2	2	5	2	1	3	2	1
35		1	3	2		2	3	1	2	
34			2	1	1		1			
33	1	3	2	1	1					
32	1	2						1		
31	2	2								
30	1	1	1							
	*									*

DECADAL FREQUENCIES OF ANNUAL PRECIPITATION AT EDMONTON

Annual Precip. (In.)	1883 - 90	1891- 1900	1901 - 10	1911 - 20	1921 - 30	1931 - 40	1941 - 50	1951 - 60	1961 - 63
27-28		1	1						
25-26				1				1	
23-24						1	1		
21-22	1	1	2	2	1	1			
19-20	1		1	3	1	4	4	5	
17-18		3	1	2	3	2		2	1
15-16	2	4	2	2	3	2	3	1	
13-14	2		1		1		2	1	1
11-12		1			1				1
9-10	1								
7-8	1								
	*		*						*

*Less Than Ten Values in Decade.

To minimize the effect of these abnormal years, Figures 6 and 7 showing successive medians of eleven-year periods of annual data have been constructed. Each plotted point refers to the median value of an eleven-year period, where five annual values were greater, and five were lower. These temperature and precipitation trend lines are somewhat complementary to those shown in Figures 4 and 5, but the apparent uniformity of the moving means trends disappears. On the temperature graph the same rising trends appear during the first few decades, but the trend lines peak in Alberta early in the 1910's and 1920's. In Saskatchewan the peak does not occur until the 1940's, and the Winnipeg trend has been markedly uniform since 1910. The cold decade of the 1950's shows up at Calgary, Saskatoon and Regina, but not at Winnipeg and Edmonton.

The median trends of precipitation in Figure 7 are more nearly replicas of the moving means in Figure 5 than are the corresponding temperature comparisons. In Alberta the early wet period after the turn of the century is not as pronounced as it appeared on the moving means graph. The variability of Calgary precipitation in comparison to that at Edmonton is evident and the precipitation at Saskatoon is shown to be remarkably uniform when the successive medians graph is inspected. The study of medians gives a different picture of the long-term trend, but one which is just as

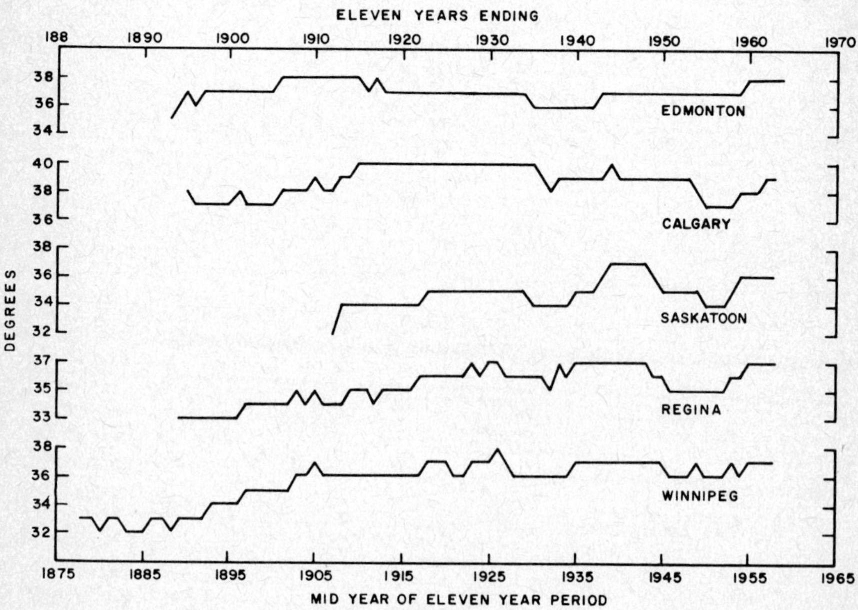

Figure 6. Successive median values of mean annual temperatures for eleven year periods (°F).

valid a representation as the more usual moving mean, with the advantage for the present purpose that the occasion of extremely high or low values is not allowed to exert an undue effect.

Probability Statistics

We may be sure that there will be considerable variations in temperature and precipitation from year to year regardless of the trend. It is also significant that trend studies are limited to history, but what of the future? The meteorologist can look at the record and can report that the precipitation is now averaging higher than it did several decades ago, but when asked whether or not this trend will continue, he must admit that he cannot forecast conditions years, or even months, in advance. He can, however, examine the historical data and by treating it as a stationary process calculate probabilities, and thus provide information useful to the planner and designer. (Stationary process means that the form of the distribution, the mean and the standard deviation do not change with time.)

The frequency distribution of annual values of mean temperature is well represented by the Gaussian normal distribution. This means that by scaling off deviations from the mean in terms of standard deviation, we can make use of the standard normal probability scale. Thus for example, we

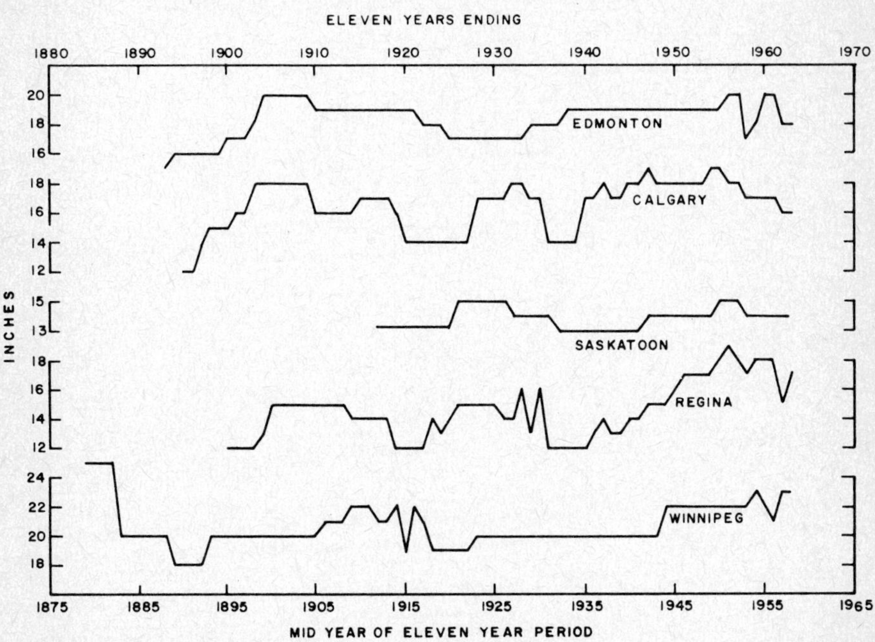

Figure 7. Successive median values of annual precipitation for eleven year periods (inches).

expect that 68% of the time the annual mean temperature will lie within one standard deviation of the mean.

Monthly mean temperatures for any particular month also have a distribution not far from normal, and useful probability estimates can be made on this basis. To make such estimates only the mean and standard deviation are required. The latter has a marked annual variation with a maximum in winter usually in January, and a minimum in summer usually in July. Maps shown as Figure 8 for these two months indicate the range. In January the values exceed 10 degrees over most of Alberta, decreasing to less than 5 degrees in eastern Manitoba, while in July the standard deviations are less than 2 degrees in most of Alberta, and from 2 to slightly over 3 degrees in southern Saskatchewan and Manitoba.

Frequency distributions of precipitation series are in general skewed, but the skewness decreases with the length of period considered, i.e., it is very great for daily values and moderate for monthly values. Annual precipitation is only slightly skewed, and for estimates in the middle part of the probability scale, the normal distribution is a reasonably good fit so that the standard deviation has a direct probability interpretation.

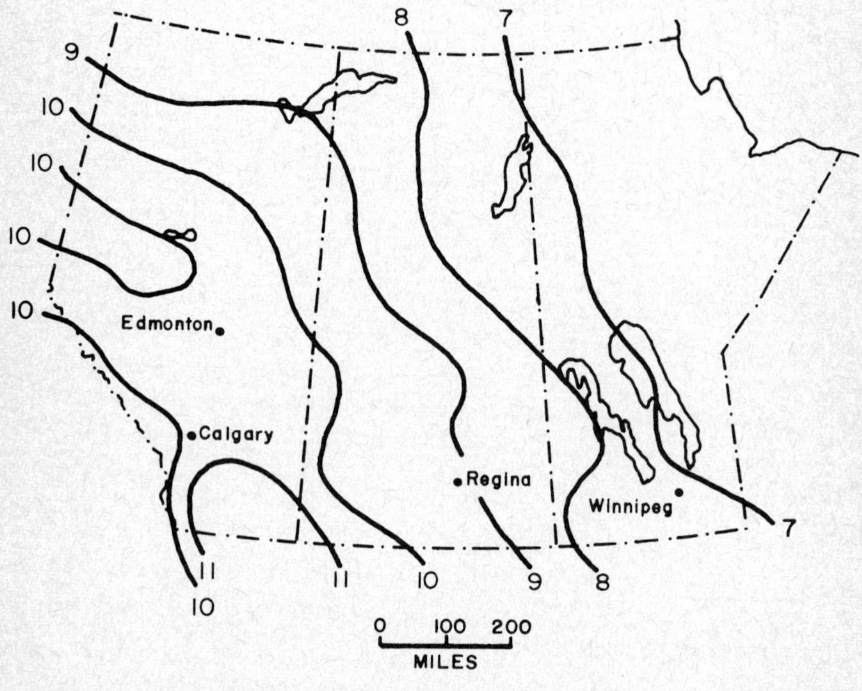

Figure 8(a). Standard deviation of mean January temperature (°F).

A frequency used measure of the variability of precipitation is the coefficient of variation which is the ratio of the standard deviation to the mean expressed as a percentage. Longley (1953) has shown that this is the best and most stable measure of the variability of precipitation. Figure 9 shows the variation of this parameter over the Prairies. The values range from 20% in the Winnipeg and Edmonton areas to more than 28% near Calgary. These values are higher than those found in any other part of Canada.

By plotting annual rainfall data on normal probablity paper, one can obtain a qualitative test of the degree to which the data are fitted by the Gaussian distribution. The better the fit, the more closely the plotted points lie on a straight line. From the plots also we obtain estimates of the percentage of years of total precipitation below any given value. Figure 10 shows the data for three stations thus plotted. As an example, we note that an annual precipitation of 20 inches or less might be expected 50% of the time at Winnipeg, 75% of the time at Edmonton, and 92% at Regina. Graphs based on areal data would be more useful than the point statistics discussed here.

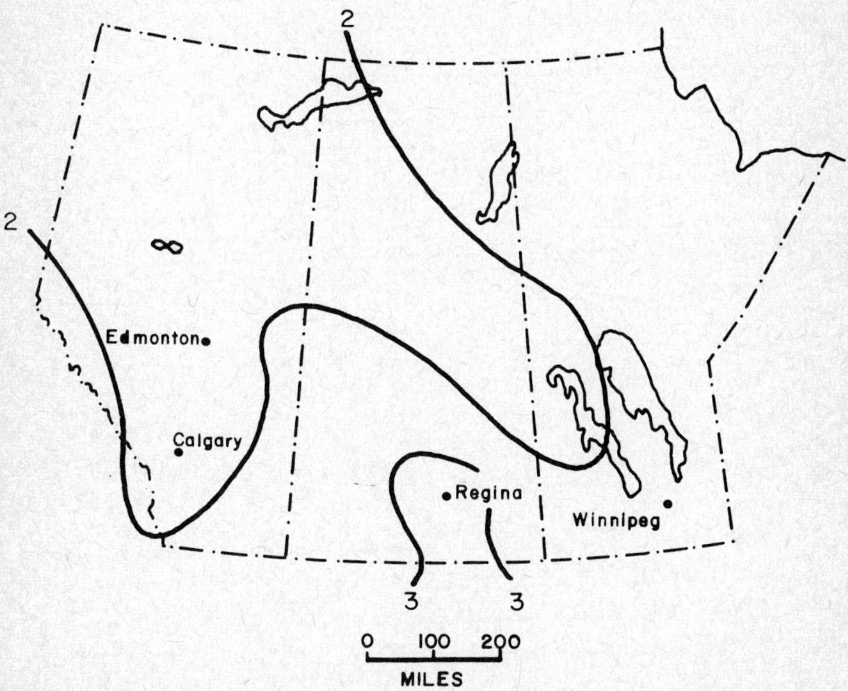

Figure 8(b). Standard deviation of mean July temperature (°F).

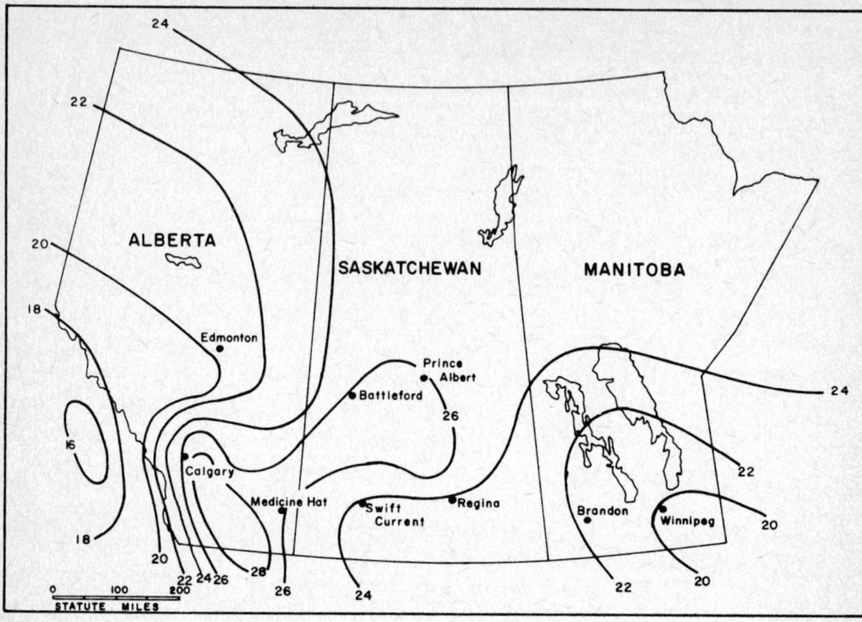

Figure 9. Coefficient of variation of annual precipitation.

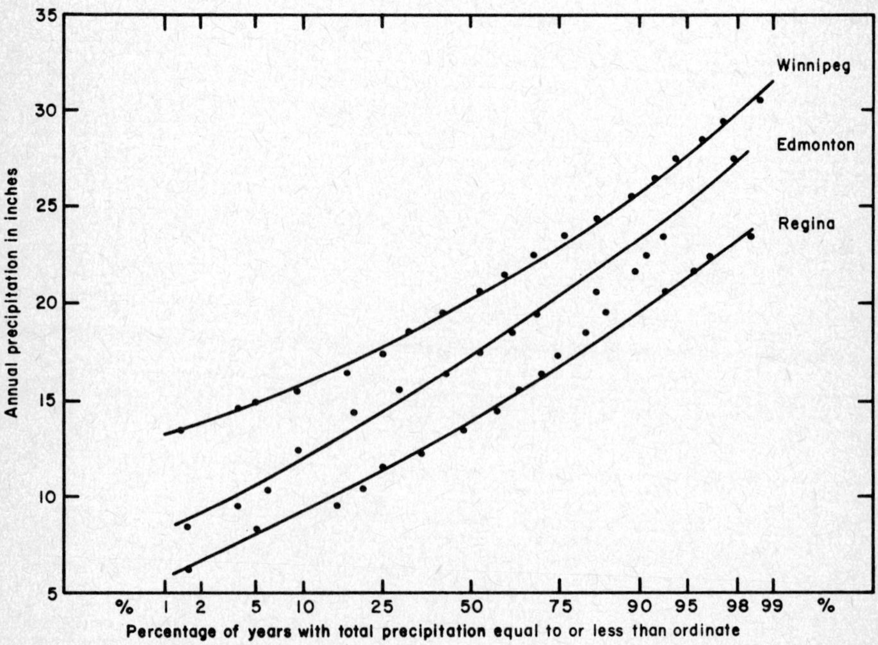

Figure 10. Probabilities of annual precipitation.

TABLE II. TEMPERATURE PROBABILITIES

	JANUARY			JULY		
City	*Mean* °F	*Standard Deviation* °F	*% of Years With Mean Within 1 Standard Deviation Of Period Mean* %	*Mean* °F	*Standard Deviation* °F	*% of Years With Mean Within 1 Standard Deviation Of Period Mean* %
Edmonton	6.6	10.0	60	63.1	1.2	77
Calgary	14.2	10.4	70	62.0	1.6	70
Regina	1.6	9.2	73	66.7	3.0	73
Winnipeg	0.1	7.4	67	68.3	2.1	60

PRECIPITATION PROBABILITIES

City	*Mean* *In.*	*Stanard Deviation* *In.*	*% Of Years With Total Within 1 Standard Deviation Of The Period Mean* %
Edmonton	17.46	3.79	76
Calgary	16.70	4.98	68
Regina	14.37	3.80	64
Winnipeg	20.83	3.91	62

Precipitation Data For Period From Beginning Of Record And Were Taken From Kendall and Thomas Temperature Data For Period From 1931 to 1960.

To illustrate in a slightly different way the adequacy of the normal approximation, we note that two thirds of the time a normal deviate lies within one standard deviation of the mean (more exactly 68.3%). Table 2 shows how well data from the Prairie Provinces fit this criterion. The values vary between 60 and 77%, and this gives a reasonable enough approximation for most practical applications. If more exact results are required, it would be necessary to fit a distribution which takes account of skewness, such as a cube root normal or incomplete gamma distribution.

References

Connor, A. J., "Droughts in Western Canada," The Canada Year Book (1933), pp. 47-59.

Currie, B. W., "Climatic trends on the Canadian Prairies," *Agricultural Institute Review*, Vol. 9 (1954), pp. 21-23.

Hope, E. C., "Weather and crop history in Western Canada," *CSTA Review*, Vol. 16 (March, 1938), pp. 347-58.

Johnson, L. P. V., "Relation of sunspot periodicity to precipitation, temperature and crop yields in Alberta and Saskatchewan," *Canada J. Res.*, Vol. 18, Part C (1940), pp. 79-81.

Kendall, G. R. and Thomas, M. K., "Variability and trends of precipitation in the Prairie Provinces," Canadian Society of Agronomy, Proceedings of the Second Annual Meeting (Toronto, June 1956), pp. C, 17-30.

Longley, R. W., "Variability of annual precipitation in Canada," *Mon. Weath. Rev.*, Vol. 81 (1953), pp. 131-34.

Longley, R. W., "Temperature trends in Canada," *Roy. Meteor. Soc., Proc. Toronto Met. Conference* (1953), pp. 207-311.

Stupart, R. F., "Is the climate changing?" *J. R. Astr. Soc. Can.*, Vol. 11 (1917), pp. 197-207.

Thomson, A. "Sunspots and weather forecasting in Canada," *J. R. Astr. Soc. Can.*, Vol. 30 (1936), pp. 215-32.

Warkentin, J., *The Western Interior of Canada,* The Carleton Library no. 15 (Toronto, McClelland and Stewart Limited, 1964).

Discussion

R. W. Longley. Droughts related to both temperature and precipitation. Thus a correlation between these two variables will be reflected in less or more severe droughts. Have you examined whether there is a correlation between temperature and precipitation?

M. K. Thomas. No. I agree with the statement that there is a correlation but wonder whether or not the gross simple trend statistics used here would identify it.

A. H. Laycock. I agree that drought is frequently local but large areas may be involved in some years. In 1936 and 1937 the southern plains were very dry, well to the east into Manitoba, yet Edmonton and Prince Albert and the Peace River area held below normal deficits. In 1950 the eastern part of Alberta was dry but eastern Saskatchewan was wet. We can't generalize for the Prairies and say that the 1930's were dry – yet some parts were.

M. K. Thomas. In drought studies it would be best to use areal or regional temperature and precipitation data rather than point or station data. A recent drought study in the U.S. (Palmer, 1965) is based on areal data which were already available, to avoid the extreme variability of point weather. Homogeneous areal climatic data have yet to be assembled in Canada.

Reference

Palmer, W. C. "Meteorological drought." United States Weather Bureau Research Paper No. 45, 1965. 58 pp.

22

Human Response to Weather and Climate

W. R. D. Sewell

R. W. Kates

L. E. Phillips

Despite its charm and wit, the oft-quoted aphorism attributed to Mark Twain, "Everybody talks about the weather but nobody does anything about it,"[1] is hardly accurate. Men do, in fact, do something about the weather. They adapt to it or adjust to it, they move toward certain climatic regimes or away from them, and they have long contemplated weather modification. However, although geographers undoubtedly recognize that weather and climate have pervasive effects on human activity, there appears to have been a general decline in their interest in these relationships and in their ability to answer certain fundamental questions. These questions have become more and more important in recent years as man's ability to predict and modify the weather increases.

[1] H. L. Mencken, ed.: A New Dictionary of Quotations on Historical Principles (New York, 1942), p. 1278. "Author unidentified; commonly ascribed to S. L. Clemens (Mark Twain), but not found in his published works."

Reprinted from *Geographical Review,* Vol. LVIII (1968), pp. 262-80. By permission of the American Geographical Society.

The Modern Era of Weather Modification

Traditionally, human response to weather and climate has consisted in adaptation, adjustment, and movement.[2] Men readily *adapt* physiologically to wide ranges of weather and climate. In a broad spectrum of human activity weather and climate have little perceptible effect on the activity pursued or on the rhythm of its pursuit. Other activities men can try to *adjust* to variations of weather and climate. This response functions on several levels. At a minimum level man changes his clothing to adjust to daily changes in the weather. At a more permanent level he insulates his home and installs a furnace or air conditioning to adjust to seasonal fluctuations; he builds stormproof structures to resist high winds; he develops weather-resistant crop varieties to withstand droughts or floods. A more drastic adjustment is *movement* to another place or region. Sometimes man finds that he is unable to withstand, physically or psychologically, severe climatic conditions, such as long periods of cold or heat or persistent rain. Technology helps him to adjust to these conditions, but it may be too costly, or it may result in such an artificial environment that he abandons it.

True, there have been attempts almost since the dawn of civilization to increase or reduce rainfall, to suppress lightning, to disperse hailstorms. But lack of scientific proof of the success of these attempts caused weather modification to be regarded by most people as an interesting idea but impossible to achieve consistently. Most skeptical of all were the research scientists who were studying the atmosphere.[3]

In the past few years, however, views about the possibility of modifying the weather have changed radically. Although the conclusion of the President's Advisory Committee on Weather Control in 1957[4] that precipitation could be substantially increased by artificial means was regarded with much skepticism by professional meteorologists, similar findings of the more recent Special Commission on Weather Modification[5] and the National Academy of Sciences – National Research Council Panel on

---

[2] J. Sonnenfeld (Variable Values in Space and Landscape: An Inquiry into the Nature of Environmental Necessity, *Journ. of Social Issues*, Vol. 22, No. 4, 1966, pp. 71–82) describes two major sets of reactions to the physical environment: adjustment and adaptation. In his definition movement and modification would presumably be a part of adjustment.

[3] See, for example, Horace R. Byers: What Are We Doing about the Weather? *in* Science and Resources: Prospects and Implications of Technological Advance (edited by Henry Jarrett; Baltimore, 1959), pp. 37–53. See also D. S. Halacy, Jr.: The Weather Changers (New York, 1968).

[4] "Final Report of the United States Advisory Committee on Weather Control" (2 vols.; Washington, D.C., 1958).

[5] "Weather and Climate Modification: Report of the Special Commission on Weather Modification," *NSF 66–3*, National Science Foundation [1966].

Weather and Climate Modification[6] have aroused little debate. These later reports conclude that it is now possible to increase precipitation from winter orographic storms, to disperse cold fog, and to suppress lightning. They view with some optimism the possibility of successfully modifying hailstorms in the near future. Alteration of the tracks of major storms such as hurricanes and tornadoes is not ruled out, though it is regarded as not yet on the horizon.

The ability to alter the weather raises a number of questions. What would be the probable effects? Would some activities benefit and others suffer? Would some areas gain at the expense of others? Is modification superior to other possible means of accommodating to weather fluctuations? There are no simple answers to these questions, but it is clear that answers are urgently required, particularly since it seems likely that weather-modification activity in the United States will increase considerably in the next few years. Commercial weather-modification firms feel that the reports of the two scientific committees substantiate claims which they themselves have been making for years. Consequently, commercial operations are almost certain to increase. In addition, several agencies of the federal government have already announced plans for major modification programs. For example, the Bureau of Reclamation has a multimillion-dollar program under way in several western states the aim of which is to find ways of enlarging snowpacks through cloud seeding. The resulting water would be used to expand irrigation. Other federal agencies are busily competing for the main responsibility for managing atmospheric resources.

The ability to modify the weather holds some intriguing possibilities. Most people would prefer to have it rain at night rather than during the day. A reduction in the almost constant drizzle on the Northwest Coast during the winter would be welcomed by most people in Seattle, Portland, and Vancouver. Farmers on the Great Plains would be happy to see their scanty rainfall augmented. Residents in Florida would doubtless be delighted to hear that a way had been found to reduce the incidence of hurricanes. Those in Bismarck, North Dakota, or Winnipeg, Manitoba, would be glad to have much milder temperatures in winter than those they now experience.

It is not certain, however, that alterations in the weather are uniformly beneficial. A rancher might be able to increase his profits as a result of a heavier rainfall on his range at a particular time, but the heavy rain might also fall on the fields of a neighboring farmer and ruin his crop. An electric-power utility might be able to increase its operating efficiency by raising

[6] "Weather and Climate Modification: Problems and Prospects: Final Report of the Panel on Weather and Climate Modification to the Committee on Atmospheric Sciences," *NAS-NRC Publ. No. 1350,* 2 vols., Washington, D.C., 1966.

the level of its reservoirs through cloud seeding, but the precipitation might ruin the holiday of vacationers who happened to be in the reservoir area at the time. The alteration of the track of a hurricane could reduce damage over a large area, but it would also divert precipitation on which city water utilities and electric-power companies depended to maintain the levels of their reservoirs. Weather modification would have differential effects not only within a given area but possibly between areas also. One area might benefit by an increase in precipitation, but this increase might have been achieved only by a reduction somewhere else.[7]

Concomitant with the new possibilities for weather modification have been improvements in the range and quality of forecasts. The use of satellites, the World Weather Watch, and the improved numerical models of the atmosphere linked to larger and faster computers suggest, for example, that eight-day forecasts with positive accuracy on the eighth day are only a decade or so away.

Weather modification and new information systems widen the range within the alternatives from which an accommodation to weather and climate can be chosen. The best choice in any given instance will depend on a careful weighing of the overall gains and loss of the alternatives—adaptation, adjustment, movement, and modification. The following questions give some indication of what it would be desirable to know to enable an objective appraisal. To what extent are various human activities sensitive to changes in the weather? What is the locational impact of weather and climate? How does weather enter into decision making? How do human activities affect the weather? It is pertinent to examine the contribuitons geographers have made so far that might help to answer these questions, and the further research that they could usefully undertake toward this end.

Contributions to the Geographical Literature[8]

From an early stage in the development of their discipline geographers have been interested in the relationship between the distribution of human activity and variations in climate. Indeed, beginning college students in geography might conclude that this is the fundamental factor accounting for the global distribution of man. One reason for this impression could be the emphasis on climate and its effects on human activity found in introductory textbooks. Of twenty-six textbooks commonly used in beginning

---

[7] Evidence from studies in connection with the University of Chicago research project Whitetop seems to raise the possibility that such reductions would occur.

[8] This review is based mainly on North American geographical serials but also includes material in British, New Zealand, and Australian journals. It is incomplete to the extent that geographers may have contributed on the subject to literature outside the field of geography and to the extent that contributions to foreign-language geographical journals were not covered.

geography courses in the United States, fourteen use climatic variations to regionalize the world.[9]

The geographer's view of the relationship between climate and man has changed over time. Simplistic theories of cause and effect, subsumed under the rubric of environmentalism or environmental determinism,[10] have given way to probabilism and possibilism. The shift in emphasis is well recorded, but its implications for geographical thought are still not well understood. In retrospect the retreat from environmentalism appears as a retreat from grand theory in geography. As Rostlund has perceptively put it, "environmentalism was not disproved, only disapproved."[11]

Accompanying the changes in geographical thought about the relationship between weather and climate and human activity has been a general decline in geographical interest in the matter. This is well illustrated by the relative number of articles devoted to the subject in the major American geographical journals in the period 1916–1966 (Figure 1). In the early part of the period more than one-third of the articles in these journals were devoted to weather and climate and to their impact on human affairs. At the end of the period less than 5 percent of the articles dealt with these matters. There are many possible interpretations of this trend, but three distinct phases may be observed, each associated with one or more major shifts in the field of geography. In the first phase, from 1916 to 1925,

---

9 George F. Deasy, Phyllis R. Griess, E. Willard Miller, and Earl C. Case: The World's Nations (Philadelphia, 1958); Loyal Durand, Jr.: World Geography (rev. edit.; New York, 1958); Otis W. Freeman and H. F. Raup: Essentials of Geography (2nd edit.; New York, Toronto, London, 1959); Paul F. Griffin and Ronald L. Chatham: Introductory College Geography (San Francisco, 1965); Oliver H. Heintzelman and Richard M. Highsmith, Jr.: World Regional Geography (3rd edit.; Englewood Cliffs, N.J., 1967); Joseph Bixby Hoyt: Man and the Earth (2nd edit.; Englewood Cliffs, N.J., 1967); George Kish, edit.: An Introduction to World Geography (Englewood Cliffs, N.J., 1956); Alfred H. Meyer and John H. Strietelmeier: Geography in World Society (Philadelphia and New York, 1963); E. Willard Miller, George T. Renner, and associates: Global Geography (2nd edit.; New York, 1957); Richard Joel Russell and Fred Bowerman Kniffen: Culture Worlds (New York, 1951); O. D. Von Engelin and Bruce Carlton Netschert: General Geography for Colleges (New York, 1957); Jesse H. Wheeler, Jr., J. Trenton Kostbade, and Richard S. Thoman: Regional Geography of the World (rev. edit.; New York, 1961); C. Langdon White and George T. Renner: College Geography (New York, 1957); C. Langdon White, George T. Renner, and Robert T. Novak: Essentials of College Geography (New York, 1958).

10 See, for example, Ellsworth Huntington's "Civilization and Climate" (New Haven and London, 1915) and "Mainsprings of Civilization" (New York and London, 1945); and Ellsworth Huntington and Sumner W. Cushing: Principles of Human Geography (2nd edit., rev.; New York and London, 1922). For a recent review see Gordon R. Lewthwaite: Environmentalism and Determinism: A Search for Clarification, *Annals Assn. of Amer. Geogrs.*, Vol. 56, 1966, pp. 1–23.

11 Erhard Rostlund: Twentieth-Century Magic, *in* Readings in Cultural Geography (edited by Philip L. Wagner and Marvin W. Mikesell; Chicago, 1962), pp. 48–53, reference on p. 49.

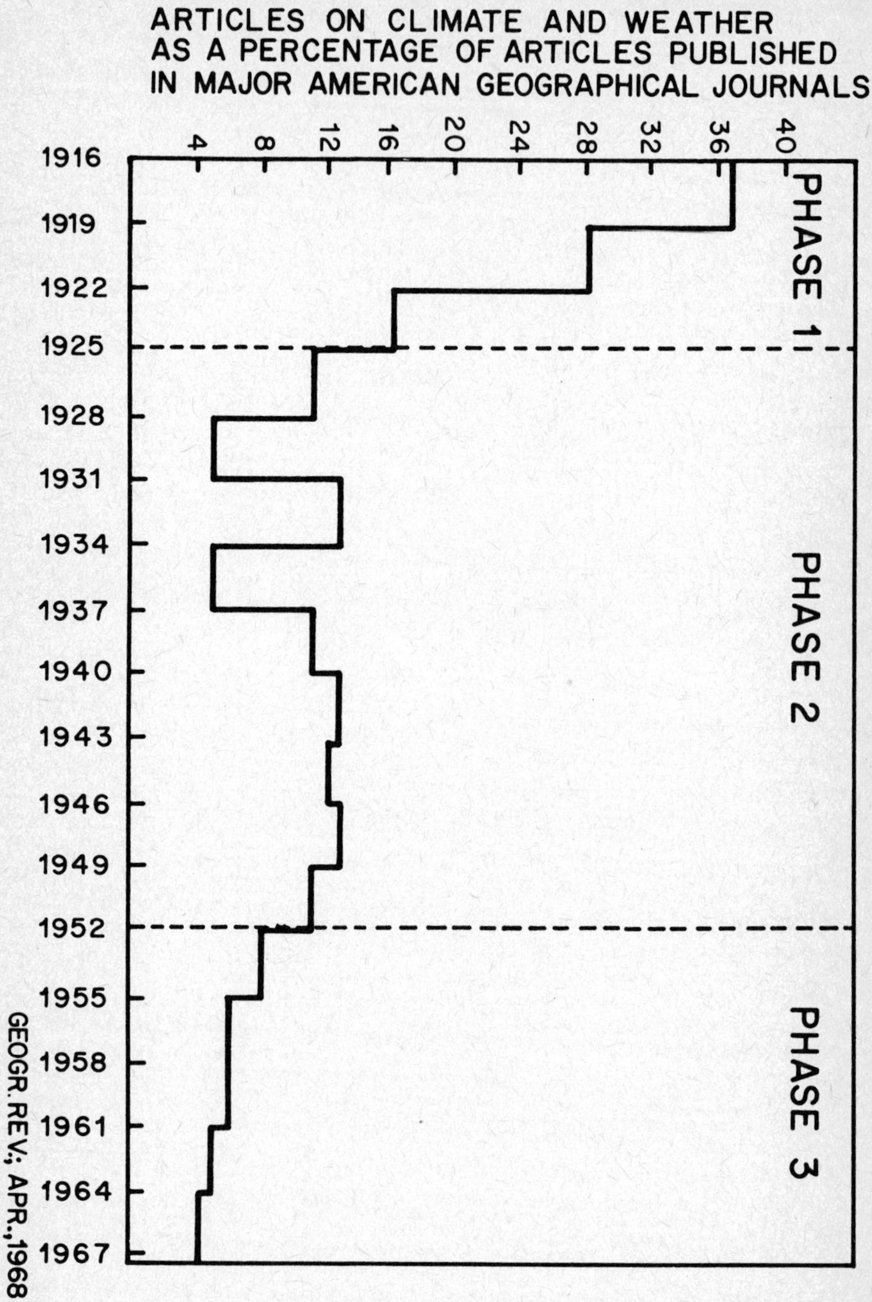

Figure 1. Articles on weather and climate in American geographical journals.

there was a general shift from environmentalism to other interpretations of the distribution of human settlement and activity. The second phase, from 1925 to about 1952, coincided with the concern of geographers with regional description. The third phase, from about 1952 to 1966, coincided with other changes in geographical interest, such as the shift from physical to human geography, from agrarian to urban concern, and, possibly, from the descriptive to the theoretical.

On the whole, geographers appear to have been more interested in the physical dimensions of the relationship between weather and climate and human affairs than in the human dimensions. Of 172 articles on the subject published in the period 1916–1966, only 71 dealt with the human aspects. The rest were concerned with climatological theory and classification, distribution, and description. However, some important contributions were made, notably by Thornthwaite and his associates in their work on the water balance, which has provided the theory and classification prerequisite to an evaluation of human use of the atmosphere.[12] Articles on the human dimensions have been concerned mainly with description and the generalization of the impact of atmospheric variables on various types of economic activities on the one hand, and physiologically on human beings, animals, and plants on the other.

Effects on Agriculture

The fortunes of agriculture are closely related to fluctuations in the weather. Although technological advances have enabled the development of weather-resistant crops, and improvements in forecasting have made possible minor adjustments in planting and harvesting schedules, agricultural production in many areas is still much at the mercy of the weather.

Numerous geographical studies have been made of agriculture in different parts of the world. Most of them note that "weather fluctuations have important economic effects," but few discuss the importance in any detail. Several have analyzed the relationship between rainfall and crop yields,[13] but these have tended to concentrate on a single crop in a single area. There have been some studies of effects of particular weather events on agriculture, such as Calef's article on the winter of 1948–1949 in the

[12] See the following articles by C. Warren Thornthwaite in the *Geographical Review*: "The Climates of North America According to a New Classification" (Vol. 21, 1931, pp. 633–655), "Problems in the Classification of Climates" (Vol. 33, 1943, pp. 233–255), and "An Approach toward a Rational Classification of Climate" (Vol. 38, 1948, pp. 55–94).

[13] For example, Stephen S. Visher: Weather Influences on Crop Yields, *Econ. Geogr.*, Vol. 16, 1940, pp. 437–443; Thomas Frank Barton: Rainfall and Rice in Thailand, *Journ. of Geogr.*, Vol. 62, 1963, pp. 414–418; and P. N. Hore: Rainfall, Rice Yields and Irrigation Needs in West Bengal, *Geography*, Vol. 49, 1964, pp. 114–121.

Great Plains,[14] Subrahmanyam and Subramaniam's on drought in peninsular India,[15] and Hewes's on drought as a cause of wheat failure in the central Great Plains.[16] These studies, like those on the relationship between rainfall and crop yields, are mainly descriptive. Little attempt has been made to assess the economic importance of weather fluctuations either to the agricultural industry or to the economy of a region. Important exceptions are Maunder's study of the implications of climatic variations in New Zealand, in which a model is developed for determining the overall impact of variations in different weather parameters on the economy of a region,[17] and Curry's work in connection with the analysis of seasonal programming of agriculture in the United States and in New Zealand.[18]

Apart from these exceptions, the geographical literature contains little that would enable estimates to be made of the economic and social effects of weather variations on agriculture. And these exceptions, it should be noted, deal with Western society, in which agriculture is undertaken principally for commercial purposes and weather fluctuations mainly affect profits. In some societies the impact of weather changes may filter through the economic life into social relationships.[19] A drought in such societies may, therefore, have more profound effects, and the social impacts may be more critical than the economic ones.

Effects on Transportation

The operations of road, railroad, airline, and ocean transport systems

---

14 Wesley Calef: The Winter of 1948–49 in the Great Plains, *Annals Assn. of Amer. Geogrs.*, Vol. 40, 1950, pp. 267–292.

15 V. P. Subrahmanyam and A. R. Subramaniam: A Climatic Study of Droughts in Peninsular India, *Geogr. Journ.*, Vol. 129, 1963, pp. 122–125.

16 Leslie Hewers: Causes of Wheat Failure in the Dry Farming Region, Central Great Plains, 1939–1957, *Econ. Geogr.*, Vol. 41, 1965, pp. 313–330.

17 W. J. Maunder: Climatic Variations in Agricultural Production in New Zealand, *New Zealand Geographer*, Vol. 22, 1966, pp. 55–69.

18 Leslie Curry: Climate and Economic Life: A New Approach: With Examples from the United States, *Geogr. Rev.*, Vol. 42, 1952, pp. 367–383; *idem.*: The Climatic Resources of Intensive Grassland Farming: The Waikato, New Zealand, *ibid.*, Vol. 52, 1962, pp. 174–194.

19 Vogt has shown that fluctuations in weather have profound importance for the social relationships of certain tribal societies, such as the Navahos and Zuñis of the arid southwestern United States and the Zinacantecos of the humid central highlands of Chiapas, Mexico (Evon Z. Vogt: Some Implications of Weather Modification for the Cultural Patterns of Tribal Societies, *in* Human Dimensions of Weather Modification [edited by W. R. Derrick Sewell], *Univ. of Chicago, Dept. of Geogr., Research Paper No. 105*, 1966, pp. 373–392); and Bennett has investigated social adaptation in arid regions (John W. Bennett: Two Memoranda on Social Organization and Adaptive Selection in a Northern Plains Region, *Plains Anthropologist*, Vol. 8, 1964, pp. 10–12; *idem*: Risk and Rationality: Aspects of Behavioral Adaptation in an Arid Variable Habitat, *ibid.*, Vol. 7, 1963, pp. 8–21; and *idem*: Social Adaptation in a Northern Plains Region: A Saskatchewan Study, *in* Papers Presented at Great Plains Symposium [Institute of Regional Studies, North Dakota State University, Fargo, 1963]).

are highly sensitive to changes in weather, and especially to heavy precipitation, fog, and winds. In the United States these weather phenomena result in losses estimated by the Weather Bureau to exceed four hundred million dollars a year,[20] most of which is attributable to costs of delays in air transportation.[21] Losses are said to be substantial in other countries as well.[22]

Although geographers have made many important contributions to the understanding of transportation development, they have shown comparatively little interest in the relationship of weather to the operation of the various forms of transport. Apart from Van Cleef's pioneering study of streetcar traffic in Duluth, Hay's article on the effects of weather on railroad operation, maintenance, and construction, and Rooney's work on the urban snow hazard,[23] there have been few studies of the effects of various weather phenomena on roads, railroads, airlines, or water transport. Nor have geographers undertaken many investigations of the effects of particular weather events on the transportation of given regions.[24]

Effects on Industry

Fluctuations in temperature, and to some extent in precipitation and wind, lead to fluctuations in the demand for fuel and power,[25] and affect the operation of various energy-supply facilities.[26] Here again, geographers have made useful contributions to the understanding of the development

---

20 Derived from estimates presented in E. Bollay and Associates, Inc.: Economic Impact of Weather Information on Aviation Operations: Final Report to the Federal Aviation Agency (Washington, D.C., 1962); M. A. Senko: Weather Satellite Study (U.S. Weather Bureau, Washington, D.C., 1964); and "Plan for a Marine Weather Service," *Planning Report No. 4*, ESSA, U.S. Weather Bureau, Washington, D.C., 1966.

21 J. C. Thompson: The Potential Economic and Associated Values of World Weather Watch, *World Weather Watch Planning Rept. No. 4*, World Meteorological Organization, Geneva, 1966.

22 *Ibid.*

23 Eugene Van Cleef: The Influence of Weather on Street-Car Traffic in Duluth, Minnesota, *Geogr. Rev.*, Vol. 3, 1917, pp. 126–134; W. W. Hay: Effects of Weather on Railroad Operation, Maintenance, and Construction, *in* Industrial Operations under Extremes of Weather (edited by J. A. Russell), *Meteorol. Monographs*, Vol. 2, No. 9, 1957, pp. 10–36; John F. Rooney, Jr.: The Urban Snow Hazard in the United States: An Appraisal of Disruption, *Geogr. Rev.*, Vol. 57, 1967, pp. 538–559. However, several studies of the influence of weather on transportation have been undertaken by applied meteorologists and others; see, for example, J. C. Tanner: Weather and Road Traffic Flow, *Weather*, Vol. 7, 1952, pp. 270–275, which analyzes the effect of various weather phenomena on daily and holiday road traffic.

24 Here again applied meteorologists have made several interesting contributions. See, for example, Donald L. Champion: Weather and Railway Operation in Britain, *Weather*, Vol. 2, 1947, pp. 373–380.

25 For a discussion see W. J. Sweeney: Importance and Need of Degree-Day Forecasting for the Petroleum Industry, *Weatherwise*, Vol. 10, 1957, pp. 197–201 and 214.

26 See, for example, M. Davies: Grid System Operation and the Weather, *Weather*, Vol. 15, 1960, pp. 18–24; and W. R. Derrick Sewell: Weather Modification and Hydro-electric Power, *Water Power*, September, 1966, pp. 353–357.

of energy industries but have shown comparatively little interest in the economic effects of various weather phenomena on these industries. An exception is Manley's study of climatic fluctuations and fuel requirements,[27] in which he estimates needs for coal where seasonal temperatures fall below certain levels.

A considerable part of geographical research, especially in recent years, has been focused on the location of manufacturing industry. Many valuable studies have been undertaken, including studies of particular industries or of manufacturing activity in particular regions. In addition, geographers have made major contributions to an understanding of the distribution of such activity through the development and application of theoretical models.

In searching for explanations of the location of manufacturing industry, geographers have examined the potential importance of a wide variety of factors, including climate. Most researchers seem to have concluded that climate and weather are seldom major determinants of location, and that they generally do not greatly affect operations. Summarizing the findings of a series of monographs on applied meteorology published in 1957, Russell noted that climate and weather appear to influence only the construction, automotive, and aircraft industries to any large extent.[28] Other industries, he suggested, are either unaffected by variations in precipitation, temperature, pressure, and wind or have been able to protect against them at small cost. Valuable as these studies are for isolating the possible effects of various weather parameters on different industrial processes, no information is given on the economic significance of these effects.

Climate may play an indirect role in industrial location. The migration of population to the American West since World War II is a reflection of the fact that technological advance and increasing affluence have loosened locational ties. The list of footloose industries has grown enormously in the past twenty years. Arizona, California, Texas, and Florida have become magnets for population and for footloose industries.[29] Recognition of these forces led to Ullman's pioneering study of the role of amenities in industrial location.[30] Unfortunately, the amenities factor seems to have received relatively little attention in subsequent geographical analysis.

---

27 Gordon Manley: Climatic Fluctuations and Fuel Requirements, *Scottish Geogr. Mag.*, Vol. 73, 1957, pp. 19–28.

28 J. A. Russell: The Problem, Method, and Conclusions, *in* Industrial Operations under Extremes of Weather [see footnote 23 above], pp. 1–9.

29 There have been a few studies of the growth of urban communities in areas of pleasant climate, such as Margaret T. Parker: Tucson: City of Sunshine, *Econ. Geogr.*, Vol. 24, 1948, pp. 79–113; and Andrew Wilson: The Impact of Climate on Industrial Growth: Tucson, Arizona: A Case Study, *in* Human Dimensions of Weather Modification [see footnote 19 above], pp. 249–260.

30 Edward L. Ullman: Amenities as a Factor in Regional Growth, *Geogr. Rev.*, Vol. 44, 1954, pp. 119–132.

Effects on Tertiary Activities

A principal focus of geographical research in recent years has been urban development, and an important part of this work has been concerned with the growth of tertiary activities, particularly wholesale and retail trade. Here, too, geographers have made major contributions, too numerous to mention, to the understanding of the location of economic activity.

Climate and weather seem to have an influence on various tertiary activities, especially wholesale and retail trade and recreation. Department-store sales, for example, fluctuate with variations in the weather. Sunshine increases receational enjoyment, rainfall reduces it. Snowfall at the right time in ski areas multiplies recreational opportunities. Geographers have undertaken a few studies of the influence of weather on such activities, but none of these have been penetrating. None answer such questions as what difference a wet Monday in spring makes to department-store sales in a particular city; whether similar variations in the weather have similar effects on the retail trade of cities in different parts of the country; what the effect is on baseball of a rainy Sunday in July in Chicago as compared with a rainy Sunday in New York; or whether a period of continuously warm weather affects boating more in the Midwest than in the Pacific Northwest.

Effects on Activities other than Economic

Studies by natural scientists and others have shown that activity of human beings is affected in a wide variety of ways by changes in the weather. Human beings are able to withstand considerable extremes of temperature, but productivity seems to fall rapidly before these extremes are reached. There also seem to be variations with fluctuations in rainfall, wind, and pressure. And weather changes apparently stimulate psychological reactions also.[31]

Geographers have undertaken a variety of studies concerning physiological and psychological reactions to fluctuations in the weather. Maunder has attempted a classification of climate based on indices of human comfort.[32] More recently, Terjung has devised comfort and wind-effect indices to produce a map of annual physioclimatic extremes in the United States.[33]

---

31 For a review of the literature on the effects of climate and weather on human beings see Selco W. Tromp: Medical Biometeorology (Amsterdam, London, New York, 1963); and Marston Bates: The Role of Weather in Human Behavior, *in* Human Dimensions of Weather Modification [see footnote 19 above], pp. 393–407.

32 W. J. Maunder: A Human Classification of Climate, *Weather*, Vol. 17, 1962, pp. 3–12.

33 Werner H. Terjung: Annual Physioclimatic Stresses and Regimes in the United States, *Geogr. Rev.*, Vol. 57, 1967, pp. 225–240. See also his "Physiologic Climates of the Conterminous United States: A Bioclimatic Classification Based on Man," *Annals Assn. of Amer. Geogrs.*, Vol. 56, 1966, pp. 141–179.

Lee and Lemons have discussed man's adaptation to climate as revealed by the adoption of different kinds of clothing.[34] Several geographers have studied geographical aspects of public health. Shoshin, for example, has examined the incidence of various types of disease in the Soviet Union,[35] and Murray has investigated causes of death in England and Wales as they relate to the weather.[36] Lee has also investigated the problems that armed forces face in adapting to foreign climates.[37] Studies by geographers of impacts of weather on human physiology, psychology, and activity complement studies by applied climatologists and others.[38]

Climate, Weather, and the Decision-making Process

Man's accommodation to his atmospheric environment no doubt reflects his perception of the effects of temperature, precipitation, and wind, and of the opportunities for adapting to them. The fact that he rejects certain climatic regions but accepts others indicates that he is unable to pursue some activities where temperatures are very high or very low, where precipitation is scanty or too heavy, or where winds are especially strong. However, accommodation to a given type of climate varies from region to region, according to cultural background, level of technology, degree of economic development, and so on.

The relationship between perception of effects of atmospheric variations, particularly short-term variations, and action to accommodate to

---

34 Douglas H. K. Lee and Hoyt Lemons: Clothing for Global Man, *Geogr. Rev.*, Vol. 39, 1949, pp. 181–213.

35 A. A. Shoshin: Geographical Aspects of Public Health, *Soviet Geography*, Vol. 5, No. 7, 1964, pp. 68–72.

36 Malcolm Murray: The Geography of Death in England and Wales, *Annals Assn. of Amer. Geogrs.*, Vol. 52, 1962, pp. 130–149.

37 Douglas H. K. Lee: The Role of Bioclimatology in the Armed Forces, *Bull. Amer. Meteorol. Soc.*, Vol. 41, 1960, pp. 235–237.

38 Clarence A. Mills: Depression Weather and Health, *Human Biology*, Vol. 10, 1938, pp. 388–399; *idem*: Weather and Health, *Bull. Amer. Meteorol. Soc.*, Vol. 19, 1938, pp. 141–152; *idem*: The Influence of Climate and Geography on Health, *Bull. New York Acad. of Medicine*, Ser. 2, Vol. 17, 1941, pp. 922–933; Alan C. Burton and Otto G. Edholm: Man in a Cold Environment: Physiological and Pathological Effect of Exposure to Low Temperatures (London, 1955); W. S. S. Ladell: The Influence of Environment in Arid Regions on the Biology of Man, *in* Human and Animal Ecology: Reviews of Research (Arid Zone Research No. 8; UNESCO, Paris, 1957), pp. 43–99; R. E. O. Williams: Weather and Health, *New Scientist*, Vol. 4, 1958, pp. 1583–1584; Robert L. Hendrick: An Outdoor Weather-Comfort Index for the Summer Season in Hartford, Connecticut, *Bull. Amer. Meteorol. Soc.*, Vol. 40, 1959, pp. 620–623; "Environmental Physiology and Psychology in Arid Conditions: Reviews of Research" (Arid Zone Research No. 22; UNESCO, Paris, 1963); P. M. Stephenson: An Index of Comfort for Singapore, *Meteorol. Magazine*, Vol. 92, 1963, pp. 338–345; "Environmental Physiology and Psychology in Arid Conditions: Proceedings of the Lucknow Symposium" (Arid Zone Research No. 24; UNESCO, Paris, 1964); and P. Fergusson: Summer Weather at the English Seaside, *Weather*, Vol. 19, 1964, pp. 144–146. The major reference for the effects of weather and climate on physiology and psychology is Tromp, *op. cit.* [see footnote 31 above].

them is only vaguely understood. How much change in temperature, precipitation, or wind does it take to stimulate action? How do people decide whether or not to carry an umbrella? Do factory owners gear operations to changes in the weather, or do they adjust to climate rather than to weather? How do people use weather forecasts? Do they prefer to adapt or adjust to the weather or to try to change it? Answers to questions such as these require an understanding of man's perception of the effects of weather changes and his means of dealing with these changes.

Geographers have undertaken a few investigations that further such understanding. Kates's study of perception of flood hazards[39] and studies by Burton and Kates[40] of other natural hazards have provided a framework for analysis of human adjustment to variable physical phenomena, such as floods and winds. Saarinen's work on adjustment to drought in the Great Plains[41] and Rooney's on adjustment to snow in various parts of the United Staes[42] furnish further insight into the relationship between perception of weather events and adjustment to them.[43] In another vein, studies by Saarinen[44] and by Sewell and Day[45] of the relationships between perception of the effectiveness of weather modification and attitudes toward its adoption suggest new avenues of inquiry into man's accommodation to weather variations. Gould has introduced a relatively new technique of analysis, the theory of games, as a tool of research and as a conceptual framework for human and economic geography.[46] However, as with natural-resources development in general, the role of perception and attitudes in decision making relating to the use of atmospheric resources remains a critically important but still undeveloped field of inquiry.

Effects of Human Activities on Weather and Climate

The physical environment places various limitations on human settlement and activity. But in his attempts to accommodate to this environ-

---

39 Robert William Kates: Hazard and Choice Perception in Flood Plain Management, *Univ. of Chicago, Dept. of Geogr., Research Paper No. 78*, 1962.

40 See, for example, Ian Burton and Robert W. Kates: The Floodplain and the Seashore: A Comparative Analysis of Hazard-Zone Occupance, *Geogr. Rev.*, Vol. 54, 1964, pp. 366–385.

41 Thomas Frederick Saarinen: Perception of the Drought Hazard on the Great Plains, *Univ. of Chicago, Dept. of Geogr. Research Paper No. 106*, 1966.

42 Rooney, *op. cit.* [see footnote 23 above].

43 These four studies have recently been reviewed in Ian Burton, R. W. Kates, and G. F. White: The Human Ecology of Extreme Geophysical Events (forthcoming).

44 T. F. Saarinen: Attitudes towards Weather Modification: A Study of Great Plains Farmers, *in* Human Dimensions of Weather Modification [see footnote 19 above], pp. 323–328.

45 W. R. Derrick Sewell and J. C. Day: Perception of Possibilities of Weather Modification and Attitudes toward Government Involvement, *in* Human Dimensions of Weather Modification [see footnote 19 above], pp. 329–344.

46 Peter R. Gould: Man against His Environment: A Game Theoretic Framework, *Annals Assn. of Amer. Geogrs.*, Vol. 53, 1963, pp. 290–297.

ment, man himself creates other limitations. In many parts of the world concern is now growing about the effects of human activity on the land, the water, and the atmosphere. Geographers have tried to trace the impact of certain activities on the physical environment. Some, for example, have studied the effects of deforestation on the hydrologic cycle.[47] A few have shown particular concern about effects on the atmosphere.[48] Others have investigated the influence of urbanization on climate,[49] particularly the effects of urbanization and industrialization on air pollution and the weather.[50] A recent study of effects of human activities on weather and climate is that of Bryson on the formation of the Rajputana (Thar) Desert in northwestern India.[51]

Purposeful Weather Modification

As was noted earlier, man can adapt to the weather, adjust his activities to weather changes, or move elsewhere. He can also alter the processes that produce these changes. Although geographers have shown considerable interest in other forms of human modification of the physical environment, such as agricultural development, harnessing of water resources, and deforestation, they have undertaken only a few studies of weather modification. Apart from contributions by Ackerman and Löf[52] and by Spilhaus,[53] little is to be found in the geographical literature relating specifically to this subject that would enable an appraisal of accommodation to the atmosphere.

---

47 For example, J. G. Nelson and A. R. Byrne: Man as an Instrument of Landscape Change: Fires, Floods, and National Parks in the Bow Valley, Alberta, *Geogr. Rev.*, Vol. 56, 1966, pp. 226–238.

48 The most comprehensive review of this subject is C. W. Thornthwaite: Modification of Rural Microclimates, *in* Man's Role in Changing the Face of the Earth (edited by William L. Thomas, Jr.; Chicago, 1956), pp. 567–583.

49 For example, H. E. Landsberg: The Climate of Towns, *in* Man's Role in Changing the Face of the Earth, pp. 584–606. Potter has examined the influence of urbanization on snowfall (J. Graham Potter: Changes in Seasonal Snowfall in Cities, *Canadian Geographer*, Vol. 5, No. 1, 1961, pp. 37–42).

50 For example, P. D. Tyson: Some Climatic Factors Affecting Atmospheric Pollution in South Africa, *South African Geogr. Journ.*, Vol. 45, 1963, pp. 44–54; and Philip A. Leighton: Geographical Aspects of Air Pollution, *Geogr. Rev.*, Vol. 56, 1966, pp. 151–174.

51 Reid A. Bryson: Is Man Changing the Climate of the Earth? *Saturday Rev.,* Apr. 1, 1967, pp. 52–55. See also Reid A. Bryson and David A. Baerreis: Possibilities of Major Climatic Modification and Their Implications: Northwest India, A Case for Study, *Bull. Amer. Meteorol. Soc.,* Vol. 48, 1967, pp. 136–142.

52 Edward A. Ackerman and George O. G. Löf: Technology in American Water Development (Baltimore, 1959), especially pp. 356–383 and 630–637; and Edward A. Ackerman: Weather Modification and Public Policy, *in* Science and Resources [see footnote 3 above], pp. 54–62.

53 Athelstan F. Spilhaus: Sea and Air Resources, *Geogr. Rev., Vol. 44,* 1954, pp. 346–351; and *idem*: Control of the World Environment, *ibid.*, Vol. 46, 1956, pp. 451–459.

Only recently have geographers, like other social scientists, begun to examine the possibilities and implications of man's ability to alter the weather. The Symposium on Economic and Social Aspects of Weather Modification held in Boulder, Colorado, in July, 1965, provided a needed stimulus. Several geographers made contributions to this symposium dealing with such matters as the identification and measurement of effects of weather modification, the role of perception and attitudes in acceptance of modification, and the evaluation of weather-modification research.[54] Some of them have undertaken further studies on these topics and participated in a Task Group on the Human Dimensions of the Atmosphere sponsored by the National Science Foundation.[55] More recently a Conference on the Present and Potential Contribution of the Social Sciences to Research and Policy Formulation in the Quality of the Physical Environment[56] brought together representatives of all the social sciences to discuss the status of research on economic, social, and institutional aspects of the management of atmospheric resources.

Priorities for Future Research

Having identified the questions relating to human adjustment to weather and climate that are of interest to geographers, and having reviewed contributions to the geographical literature on these matters, it is pertinent to outline directions that future geographical research should take. Four main directions of inquiry are recommended, corresponding to the four major questions posed earlier in the paper.

Sensitivities to Weather and Climate

Human activities differ considerably in their sensitivity to temperature, precipitation, and wind. Some activities can be undertaken under a wide range of atmospheric conditions; others are profoundly affected by the slightest change in the weather. Our knowledge of these sensitivities, however, is imprecise. We know that a cold spring can result in large losses to agriculture, but we do not know how far these losses extend into the economy as a whole. We can calculate the impact of a ten-inch increase in rainfall in July on corn farming in Iowa, but what effect would the same increase in the same month have on farming elsewhere in the United

---

54 See the contributions by Edward A. Ackerman, Leslie Curry, W. L. Garrison, Robert W. Kates and W. R. Derrick Sewell, T. F. Saarinen, W. R. Derrick Sewell and J. C. Day, Gilbert F. White, and Andrew Wilson in "Human Dimensions of Weather Modification" [see footnote 19 above].

55 Final Report of the Task Group on the Human Dimensions of the Atmosphere (National Science Foundation; forthcoming).

56 Boulder, Colorado, January 31 and February 1 and 2, 1967. (Environmental Science Services Administration). The proceedings are soon to be published, under the title "Social Sciences and the Environment," by the University of Colorado Press.

States? And although information on agricultural impacts is far from complete, it is much more comprehensive than that on most other activities.[57] Data on the effects of weather variations on the manufacturing industry are almost nonexistent. Apart from some studies of effects on the transportation industry,[58] the construction industry,[59] and some tertiary industries,[60] impacts on nonagricultural industries have been investigated to only a small extent. Much more work needs to be done before precise estimates can be made of effects of changes in various weather parameters on given industries or activities, and before the impact of such changes on the economy as a whole can be traced.[61]

Geographers could make valuable contributions to the understanding of the impacts of weather variations by studying the sensitivity of different

---

[57] See, for example, "Weather and Our Food Supply" (Center for Agricultural and Economic Development, Ames, Iowa, 1964); Emery N. Castle and Herbert H. Stoevener: The Economic Evaluation of Weather Modification with Particular Reference to Agriculture, *in* Human Dimensions of Weather Modification [see footnote 19 above], pp. 141–158; and John W. Kirkbride and Harry C. Trelogan: Weather and Crop Production: Some Implications for Weather Modification Programs, *ibid.*, pp. 159–168.

[58] Most of these studies relate to the influence of weather on airline operations. See, for example, "A Preliminary Analysis of the Effects of Weather on Airport Traffic Flows, Airport Capacity and Acceptance Rates" (U. S. Weather Bureau for the Federal Aviation Agency, Washington, D. C., 1961); Elden J. Weigman: Analysis of Wind and Weather Factors on the New York–London Air Route for Peak Traffic Day of August 22, 1962 (Menlo Park, Calif., 1963); Bollay, *op. cit.* [see footnote 20 above]; and W. B. Beckwith: Impacts of Weather on the Airline Industry: The Value of Fog Dispersal Programs, *in* Human Dimensions of Weather Modification [see footnote 19 above], pp. 195-207. A few works have dealt with the influence of weather on railroad operations; for example, Champion, *op. cit.* [see footnote 24 above]; and B. J. Wintle: Railways versus the Weather, *Weather,* Vol. 15, 1960, pp. 137–139.

[59] For example, Palmer W. Roberts: Adverse Weather—Its Effect on Engineering Design and Construction, *Civil Engineering,* Vol. 30, 1960, No. 6, pp. 35–39, and No. 7, pp. 44–47, Vol. 32, No. 2, 1962, pp. 46–49, and Vol. 33, 1963, No. 5, pp. 50–54; and J. A. Russo, Jr., K. Thouern-Trend, and others: The Operational and Economic Impact of Weather on the Construction Industry of the United States (Travelers Research Center, Inc., Hartford, Conn., 1965).

[60] Studies of the effects of weather on consumer sales that have been undertaken by economists, applied meteorologists, and others include Mary T. Petty: Weather and Consumer Sales, *Bull. Amer. Meteorol. Soc.,* Vol. 44, 1963, pp. 68–71; A. T. Steele: Weather's Effect on the Sales of a Department Store, *Journ. of Marketing,* Vol. 15, 1950-1951, pp. 436–443; and R. H. Sutherland: Seasonal Variation in the Sydney Meat Market, *Quart. Rev. of Agric. Economics,* Vol. 17, 1964, pp. 113–119.

[61] For discussions of types of analytical tools required see Edward A. Ackerman: Economic Analysis of Weather: An Ideal Weather Pattern Model, *in* Human Dimensions of Weather Modification [see footnote 19 above], pp. 61–75; and James R. Hibbs: Evaluation of Weather and Climate by Socio-Economic Sensitivity Indices, *ibid.*, pp. 91–109. See also two publications by The Rand Corporation, Santa Monica, California: Richard R. Nelson and Sydney G. Winter, Jr., Weather Information and Economic Decisions: A Preliminary Report (1960); and R. R. Rapp and R. E. Huschke: Weather Information: Its Uses, Actual and Potential (1964).

economic activities to various weather parameters. Two broad types of investigation are needed: studies of impacts on particular industries and studies of impacts on particular regions. Their purpose would be to make it possible to answer such as question as what is the economic (and/or social) impact of an extra inch of rainfall or an extra degree of temperature on a particular activity or what is the economic impact of a hailstorm or a hurricane on a given region.

Models capable of giving preliminary answers to these questions are presently available. Langford[62] has proposed the use of input-output analysis for both intraregional and interregional evaluation of weather-modification and information programs. Such models trace through a regional economic system the alterations in production functions that follow from weather modification or from improvements in informative systems. Potentially, they can take into account the interregional effects as well. The simulation of climatological events has been suggested by McQuigg[63] and by Hufschmidt and Fiering[64] in order to examine the trade-offs between assumed levels of information and modification potential. For example, it should be possible, by use of the methodology developed by the Harvard Water Program,[65] to examine the countervailing effects of hurricane modification that could provide positive benefits in flood prevention and negative benefits to water supply. It should be pointed out, however, that the application of these techniques is not to be taken lightly. The development of the Harvard Water Program Model, and of the Philadelphia Region Input-Output Table developed by Walter Isard and his colleagues at the University of Pennsylvania to illustrate the interconnections between various economic activities, has required many man-years of efforts and hundreds of thousands of dollars of expense.

Locational Impact of Weather and Climate

Although climate has usually been a minor factor in locational decisions, its importance seems to be growing. Great affluence has increased labor mobility, and people are moving to sunnier climates to earn a living as well as for retirement. Some industrialists are competing for skilled labor by offering the fringe benefit of "a pleasant place to live and work." At

---

62 T. Langford: A Proposed Model for the Evaluation of the Economic Aspects of Weather Modification, *in* Final Report of the Task Group on the Human Dimensions of the Atmosphere [see footnote 55 above], Appendix.

63 James D. McQuigg: The Use of Simulation Models as a Tool to Study the Relationships of Weather and Human Activity, *in* Final Report of the Task Group on Human Dimensions of the Atmosphere [see footnote 55 above], Appendix.

64 Maynard M. Hufschmidt and M. B. Fiering: Simulation Techniques for Design of Water-Resource Systems (Cambridge, Mass., 1966).

65 A. A. Maass, M. M. Hufschmidt, and others: Design of Water-Resource Systems (Cambridge, Mass., 1962).

the same time, industry itself has become more mobile. The most rapidly growing industries tend to be those which have great flexibility in locational preference. The two processes of increasing labor mobility and increasing industrial mobility have contributed substantially to the postwar growth of the Far West, the Southwest, and the Southeast.

Although climate has also played a role in the growth of these regions, its relative importance is uncertain. Nor is it obvious whether increased labor mobility or increased industrial mobility has been the more important. Some industries have moved in to take advantage of the growing pools of labor. Others have conducted nationwide campaigns to attract workers to plants located in regions of pleasant climate. Still other industries have been attracted into such regions to serve the needs of retirement populations.

Studies are needed to determine the influence of climate in the postwar expansion of the West, Southwest, and Southeast. They could be of several types, such as investigations of reasons why people move to Tucson or Santa Barbara or of the requirements of industries that have moved into such places. Much could now be added to Ullman's thoughts expressed in 1954 on the role of amenities in regional growth.

The Role of Weather and Climate in Decision Making

How do people decide which of several means of accommodating to different weather parameters is the most appropriate? Much depends on how they perceive these parameters, and on their perception of, and attitude toward, alternative adjustments. Relatively little is known, however, about such perceptions and attitudes.[66]

To illustrate, consider the probabilistic forecasts of precipitation currently issued by the Weather Bureau. Official practice calls for such statements as "There is a 30 percent chance of rain tomorrow." Little is known as to what meanings recipients attach to this forecast—whether, for example, it is perceived as equivalent to "Three chances in ten of rain tomorrow," "There is a small chance of rain tomorrow," or "It is going to rain tomorrow." Research on this question is currently being undertaken by the University of Denver Research Institute.

Even if the meaning attached to the forecast coincides with that perceived by the forecaster, it is still not clear how the recipient will utilize the forecast. This is true even for those who habitually make use of such information in ordering their activities. First, the nature and magnitude of the response or adjustment are significant. Wearing a raincoat requires less effort than canceling a vacation trip. Second, the outcome to which

---

66 See, for example, Douglas H. K. Lee: The Role of Attitude in Response to Environmental Stress, *Journ. of Social Issues,* Vol. 22, No. 4, 1966, pp. 83–91.

the response is directed may vary considerably. Getting wet at the beach is viewed differently from having a brand-new suit of clothes drenched. Third, a weather forecast is only one channel of information, and frequently it is a secondary one. Observation or the evidence of one's own senses may either support or contradict the forecast, depending on whether it is clear or cloudy at the time it is received.

TABLE I. FORECAST PROBABILITIES REQUIRED FOR ADJUSTMENTS FOR DIFFERENT INFORMATION AND OUTCOMES

	ALTERNATIVE INFORMATION			
Outcomes	*Supportive*		*Contradictory*	
	ADJUSTMENTS			
	Minor	Major	Minor	Major
Major	30%	60%	60%	80%
Minor	40%	60%	70%	90%

The hypothetical interaction of these factors is illustrated in Table I. The alternative channel of information may either support or contradict the weather forecast, and the response or adjustment is classified as major or minor. A 30 percent chance of precipitation on a gray, overcast day (supportive information) might induce a rational individual to wear a raincoat (minor adjustment) over his brand-new suit (major outcome). All other conditions would require higher probability forecasts to lead to similar actions.

Despite our inability to specify the use of such simple information as a forecast of rain, some progress has been made in developing normative models to describe how information or weather modification could be used. A number of examples have been prepared dealing with the demand for natural gas or the scheduling of agricultural activities.[67]

[67] See, for example, two publications by The Rand Corporation, Santa Monica, California: L. L. Kolb and R. R. Rapp: The Utility of Weather Forecasts to the Raisin Industry (1961), and R. R. Nelson and S. G. Winter, Jr.: Weather Information and Economic Decision (1963); also J. D. McQuigg: Foreseeing the Future, *Meteorol. Monographs*, Vol. 6, No. 28, 1965, pp. 181–188 (Chap. 12, Forecasts and Decisions); James D. McQuigg, Oscar H. Calvert, and Wayne L. Decker: Using Weather Information to Cut the Cost of Getting a Good Stand of Cotton in Southeast Missouri, *Univ. of Missouri College of Agriculture Bull. 835*, 1965; James D. McQuigg and Russell G. Thompson: Economic Value of Improved Methods of Translating Weather Information into Operational Terms, *Monthly Weather Rev.*, Vol. 94, 1966, pp. 83–87.

The Role of Human Activities

In addition to studies of human response to the atmospheric environment, investigations are needed to increase knowledge of human effects on the atmosphere. There have been, as was noted earlier, a few studies of the influence of urbanization on climate and weather. Cities may provide radically different radiation regimes.[68] The varied man-made topography of the city results in changes in wind patterns and temperature. On a larger scale, man-made lakes may alter regional climates.[69] In a more speculative realm, the earth is seen as heating up from the increased CO_2 content of the air,[70] deserts have been attributed to dust particles,[71] and the contrails of future supersonic jet planes are viewed as a potential source of climatic disturbance.

The foregoing suggestions for geographical studies dealing with human response to weather and climate are intended merely to illustrate the types of problems that merit attention. Many other topics could be proposed. But much more important for geography than a checklist of useful research is a reexamination of its commitment to the study of weather and climate. It seems ironic that today, when tools and instruments of considerable sophistication are available to define with some precision human response to weather and climate, the level of geographical discourse still rests primarily either on Huntingtonian assertion or on introductory textbook generalization.

---

68 For example, Clarence A. Woollum: Notes from a Study of the Microclimatology of the Washington, D. C. Area for the Winter and Spring Seasons, *Weatherwise,* Vol. 17, 1964, pp. 263-271; A. Garnett and W. Bach: An Estimation of the Ratio of Artificial Heat Generation to Natural Radiation Heat in Sheffield, *Monthly Weather Rev.,* Vol. 93, 1965, pp. 383–385; and Richard J. Hutcheon, Richard H. Johnson, and others: Observations of the Urban Heat Island in a Small City, *Bull. Amer. Meteorol. Soc.,* Vol. 48, 1967, pp. 7–9.

69 See, for example, S. L. Vendrov and L. K. Malik: An Attempt to Determine the Influence of Large Reservoirs on Local Climate, *Soviet Geography,* Vol. 6, No. 10, 1965, pp. 25–40.

70 See, for example, by Gilbert N. Plass, "The Carbon Dioxide Theory of Climatic Change," *Tellus,* Vol. 8, 1956, pp. 140–154, and "Carbon Dioxide and Climate," *Scientific American,* Vol. 201, No. 1, 1959, pp. 41–47; and F. Möller: On the Influence of Changes in the CO_2 Concentration in Air on the Radiation Balance of the Earth's Surface and on the Climate, *Journ. of Geophys. Research,* Vol. 68, 1963, pp. 3877–3886.

71 Reid Bryson, for example, has suggested that the Rajputana Desert of northwestern India was once a fertile, inhabited area. Up to 1500 B.C. it was occupied by the Harappan culture. Successive croppings, however, seem to have led to wind erosion on a vast scale, similar to that in the United States in the dust-bowl years. The soil particles were whipped into huge clouds of dust, which reduced radiation and made the climate progressively drier. Attempts to extend the margin of cultivation by plowing up more land merely accelerated the process, and gradually the whole area became a desert. Bryson believes that if grass could somehow be persuaded to grow, the dust cloud that now covers this part of India could be dispersed and the Rajputana Desert become fertile once more. See Bryson and Baerreis, *op. cit.* [see footnote 51 above].

Index

NOTE: **Bold** figures indicate pages containing illustrations.